ELEMENTARY STATISTICS

ELEMENTARY STATISTICS

MARIO F. TRIOLA

Dutchess Community College
Poughkeepsie, New York

THE BENJAMIN/CUMMINGS PUBLISHING COMPANY

Menlo Park, California • Reading, Massachusetts
London • Amsterdam • Don Mills, Ontario • Sydney

To Marc and Scott

Sponsoring editor: *Susan Newman*
Production coordinator: *Ruth Cottrell*
Book and cover design: *John Edeen*
Cover photo: *George B. Fry, III*
Technical art: *Patricia M. Rogondino*
Essay illustrations: *John Edeen*
Photo research: *Ruth Cottrell*

Library of Congress Cataloging in Publication Data

Triola, Mario F
    Elementary Statistics.

    Bibliography: p.
    Includes index.
    1. Statistics.   I.   Title.
QA276.12.T76   519.5   79-24761
ISBN 0-8053-9305-6

    DEFGHIJKL-HA-854321

The Benjamin/Cummings Publishing Company, Inc.
2727 Sand Hill Road
Menlo Park, California 94025

ACKNOWLEDGMENTS

For the photographs used in this text, the
publisher would like to acknowledge the following artists and sources.

Russell Abraham/Jeroboam, p. 229
Suzanne Arms/Jeroboam, pp. 314 and 330
Brown Photo Lab, p. 108
Camera Press/Photo Trends, p. 305
David Glaubinger/Jeroboam, p. 149
Jerry Hirsch/Photo Trends, p. 1
Jean-Claude Le Jeune/The Stockmarket, p. 165
Marshall Licht/The Design Element, p. 8
Dan Madden/Photo Trends, p. 255
Gregg Mancuso/Jeroboam, p. 99
Photo Trends, pp. 11 and 220
Kent Reno/Jeroboam, pp. 4 and 237
Harry Riddle/The Stockmarket, p. 294
Hy Simon/Photo Trends, pp. 62 and 337
Frank Siteman/The Stockmarket, p. 12
James R. Smith/Photo Trends, p. 137
Joseph Stevens/Photo Trends, p. 359
Syndication International/Photo Trends,
    pp. 59 and 189
The Design Element, p. 143
The Times/Photo Trends, p. 371
Mario Triola, pp. 265, 309, and 387
Peeter Vilms/Jeroboam, p. 200
Ben Weaver/Camera Five, p. 75
Lonnie Wilson/Photo Trends, p. 41
Nikolay Zurek/Jeroboam, p. 376

PREFACE

Intended Audience

This book is an introduction to elementary statistics for nonmathematics students. Although a strong mathematics background is not required, students using this book should have at least a high school algebra course. Most of the book can be covered in one semester.

In writing this book, I placed a strong emphasis on clear and readable writing. Real world applications are an important feature of the style. I was careful to include the underlying theory for each topic even though I did not include the mathematical rigor appropriate in a book for math majors.

Special Features

Distinguishing features of this book are:

1. *Student aids.* All chapters (except Chapter 12) begin with an *overview* and a statement of objectives, and they conclude with a *review* of the key concepts and a set of *review exercises* similar to the exercises at the end of each section. Appendix A contains *tables* that will help students, Appendix B contains a *glossary,* Appendix C contains a *bibliography,* and Appendix D contains *answers* to many exercises. There are many helpful *diagrams,* and *flowcharts* are used to clarify standard procedures that are usually considered difficult.

2. *Applications.* Short essays illustrate uses of statistics in very real and practical applications. The following is a sample of some of the areas covered:

   Advertising: Controversy over the Coke-Pepsi taste tests in television commercials.
   Biology: Mendel's experimental data manipulated.
   Business: Money saved by airlines using sampling to determine revenues from split ticket sales.
   Criminology: The number of police officers increased in a New York City experiment.
   Economics: The way unemployment figures are obtained.

Education: Use of aptitude test results criticized by Ralph Nader.

Entertainment: The way the Nielsen T.V. rating system works.

Medicine: The 1954 test of the Salk vaccine as an immunization against polio.

Psychology: Measuring a seemingly qualitative characteristic such as disobedience.

Sociology: Declining standard test scores affected by family size and spacing.

These essays convey ideas that are sometimes missing in a statistics text, and the concepts that are introduced here are important supplements to the traditional core of a statistics course. For example, the essay on measuring disobedience suggests a way of producing numerical data that can be analyzed even though disobedience does not seem to lend itself to measurement. The essay on crime statistics in Washington D.C. points out that figures were intentionally altered and explains how it was done.

3. *Exercises.* The exercises also apply to a variety of real situations, and I have arranged them in order of increasing difficulty. In addition to graduating them in order of difficulty, I have divided them into groups A and B. The B exercises are more difficult because they involve larger data sets or require a stronger mathematical background. In few rare cases, the B exercises introduce a new concept. Here is a sample of the exercise topics:

Energy consumption in homes
Seed germination in farming
Eye color genes
Gasoline rationing
Cholesterol levels in blood
Reliability of instrument readings
I.Q. tests
Car fatality rates for different age groups
Effectiveness of fire detecting devices

To the Instructor

An *Instructor's Resource Guide* contains the answers to almost all of the text exercises. Also included are three sample tests for each chapter and three comprehensive final examinations. These sample tests are especially useful as makeup tests or for testing improvement for students who have done poorly. The solutions are provided for these tests, and the tests can be reproduced directly from the guide. The *Instructor's Resource Guide* also contains 38 different data sets that can be assigned to students for a variety of projects. A central list provides the important statistics for each data set, and transparency masters for key figures and tables provide visual aids.

To the Student

I recommend the use of a hand calculator. You should get one that is capable of addition, subtraction, multiplication, division, and square roots, and it should use algebraic logic instead of chain logic. You can tell the type of logic used by pressing these buttons:

$$2 + 3 \times 4 =$$

If the result is 14, the calculator uses algebraic logic. If the result is 20, the calculator uses chain logic, and you will be likely to make gross errors. Some relatively inexpensive calculators automatically compute the mean and standard deviation and the slope and intercept values of a regression line. Although a calculator may not be required for the course, one will certainly make your life easier.

I also recommend that you read the overview carefully when you begin a chapter. Read the next section quickly to get a general idea of the material, and then return for a careful second reading. Try the exercises. If you encounter difficulty, go back and work some of the examples in the text and compare your solutions with the ones in the text. When you finish a chapter, check the review section to make sure that you didn't miss any topics.

Importance of Statistics

Statistics is used in all aspects of our society, from finding an average for a few test scores to analyzing the experimental results of a new cancer drug. It is used to describe and understand a world dependent on numbers. Modern society demands that the well-educated person possess at least a basic knowledge of statistics. I designed this book to meet that demand.

Acknowledgments

I wish to extend my thanks for the suggestions made by Jim Foster, Weber State College; Joseph Mazanec, Delta College; Loren Radford, West Virginia Northern Community College; Harold Thomas, Pittsburg State University; Jay Devore, California Polytechnic State University; Louis Bush, San Diego City College; Richard Fritz, Moraine Valley Junior College; Vern Crandall, Brigham Young University; Richard Campbell, Butte College of the Butte Community College District; Frederic Fischer, State University of New York at Oswego; Keith Craswell, Western Washington State College; and Gerold Rogers, New Mexico State University. I also thank Ruth Cottrell for her professionalism and competence, which enhanced every phase of production. Finally, I thank the entire Benjamin/Cummings staff.

M.F.T.
LaGrange, New York
November 1979

CONTENTS

ESSAYS

ELEMENTARY STATISTICS

INTRODUCTION TO STATISTICS

1-1 OVERVIEW

To many people, the term statistics is associated with all the numbers that describe the different properties of collections of data. Typical examples can be found in statements such as: "The average family has 2.3 children." and "In 1979, 632 billion cigarettes were smoked in the United States." The subject of statistics is not, however, limited to the collection, tabulation, and summarizing of data. Many uses of statistics involve inferences that go beyond the available evidence. That is, we interpret a relatively small number of observations to make general and meaningful conclusions.

Throughout this book, we present some of the many ways the methods and theories of statistics have been used for the betterment of humanity, as well as some of the ways that statistics has been abused by those who have an ax to grind. The discussion of the ways in which statistics has been used for deceptive purposes does not come close to being a complete compendium of deceptive practices, but it should help you to become critical and analytical when you are presented with statistical claims. As you acquire more knowledge about standard acceptable techniques, you will be better prepared to challenge misleading statistics.

1-2 BACKGROUND

Where did it all begin? In the seventeenth century, a successful store owner named John Graunt (1620–1674) had enough spare time to pursue outside interests. His curiosity led him to study and analyze a weekly church publication, called "Bills of Mortality," which listed births, christenings, and deaths and their causes. Based upon these studies, Graunt published his observations and conclusions in a work with the catchy title of "Natural and Political Observations Made upon the Bills of Mortality." This 1662 publication comprised the first real interpretation of social and biological phenomena based on a mass of raw data, and many people consider this to be the birth of statistics.

Graunt made observations about the differences between the birth and mortality rates of men and women. He noted a surprising consistency among events that would seem to occur by chance. These and other early observations led to conclusions or interpretations that were invaluable in planning, evaluating, controlling, predicting, changing, or simply understanding some facet of the world in which we live.

We say that statistics can be used to predict, but it is very important to understand that we cannot predict with absolute certainty. In fact, statistical conclusions involve an element of uncertainty which can (and often does) lead to incorrect conclusions. It is possible to get ten consecutive heads when an ordinary coin is tossed ten times. Yet a statistical analysis of that experiment would lead to the incorrect conclusion that the coin is biased. That conclusion is not, however, certain. It is only a "likely" conclusion which reflects the very low chance of getting ten heads in ten tosses.

This element of uncertainty sets statistics apart from other areas of applied mathematics which require conclusions that meet the rigid criterion of certainty. For example, in plane geometry we begin with a list of basic rules or axioms and we *prove* without doubt that the sum of the angles of a triangle is 180°. Similarly, in arithmetic we prove that the product of any number and 0 equals 0. In algebra, we prove that if $ac = bc$ while c is not 0, then $a = b$. But in statistics, our ultimate conclusions are not proved at all; they are only shown to be likely or unlikely.

In general, mathematics tends to be **deductive** in nature, meaning that acceptable conclusions are deduced with certainty from previous conclusions or assumptions. Statistics is **inductive** in nature, because inferences are basically generalizations which may or may not be correct.

In statistics, we commonly use the terms population and sample.

DEFINITION	*A **population** is the complete and entire collection of elements (scores, people, measurements, etc.) to be studied.*
DEFINITION	*A **sample** is a subset of a population.*

Statisticians make inferences about an entire population based upon the observed data in a sample. Thus statisticians infer a general conclusion from known particular cases in the sample. In most branches of mathematics the procedure is reversed. That is, we first prove the generalized result and then apply it to the particular case. Geometers first prove that, for the population of all triangles, all possess the property that the sum of their respective angles is 180°. They then apply that established result to specific triangles, and they can be certain that the general property will always hold.

Statisticians, on the other hand, begin with a randomly selected sample of specific triangles and, after extracting the relevant numbers from those triangles, they conclude that "it is likely that all triangles have a sum of 180°." This example is, in one sense, unfair, since it is possible to prove the deductive conclusion, but such proof is not always possible. Flashbulb manufacturers cannot test every one of their products in order to begin with a generalized property; psychologists and educators cannot administer I.Q. tests to all adults; pollsters cannot survey all voters. There are many situations in which sampling must be the first step in an inductive reasoning process.

As we proceed with our study of statistics, you will learn how to extract pertinent data from samples and how to infer conclusions based on the results of those samples. You will also learn how to assess the reliability of conclusions. Yet you should always realize that, while the tools of statistics can enable you to infer the behavior of a population, you can never predict the behavior of any one individual.

AIRLINE COMPANIES SAVE BY SAMPLING

In the past, airline companies used an extensive and expensive accounting system to correctly appropriate revenues from tickets that involved two or more companies. Now, instead of accounting for every ticket involving more than one company, they use a sampling method whereby a small percentage of these split tickets is randomly selected and used as a basis for appropriating all such revenues. The error created by this approach can cause some companies to receive slightly less than their fair share, but these losses are more than offset by the clerical savings accrued by circumventing the 100% accounting method. This concept saves companies millions of dollars each year.

A unique aspect of statistics is its obvious applicability to real and relevant situations. In many branches of mathematics we deal with abstractions that may initially appear to have little or no direct use in the real world, but the elementary concepts of statistics do have direct and practical applications.

1-3 USES AND ABUSES OF STATISTICS

Short essays that use real world examples to illustrate the uses and abuses of statistics appear throughout the book. Among the uses are many applications prevalent today in the fields of business, economics, psychology, biology, computer science, military intelligence, English, physics, chemistry, medicine, sociology, political science, agriculture, and education. Statistical theory applied to these diverse fields often results in changes that benefit humanity. Social reforms are sometimes initiated as a result of statistical analyses of factors such as crime rates and poverty levels. Large-scale population planning can result from projections devised by statisticians. Manufacturers can provide better products at lower costs through the effective use of statistics in quality control. Epidemics and diseases can be controlled and anticipated through application of standard statistical techniques. Endangered species of fish and other wildlife can be protected through regulations and laws that are decided upon in part by statistical conclusions. Educators can discard innovative teaching techniques if statistical analyses show that traditional techniques are more effective.

TRUTH IN ADVERTISING

Two car advertisements were placed only a few pages apart in one issue of a national news magazine. The first ad listed the U.S. Environmental Protection Agency fuel consumption rates for all of the major cars tested by a "simulated average trip under city driving conditions." For that year, the Oldsmobile Cutlass was listed as 10.3 miles per gallon. The second ad had the headline: "Oldsmobile Cutlass—17.6 Miles Per Gallon." In smaller print, the second ad stated that "recent proving-ground tests show 17.6 mpg average at 55 mph." In yet smaller print we are told that "three Cutlass models were driven at a steady 55 mph on a level, straight road at the G.M. Desert Proving Grounds." With broken in and warmed up cars with tuned engines, there was "an average of 17.6 mpg at a constant speed of 55 mph; 12.8 mpg in Suburban-City tests."

Students choose an elementary statistics course for a variety of reasons. Some students plan to major in psychology, economics, sociology, or other fields that require a statistics course. Others can find no alternative mathematics course that fulfills some minimum degree requirement. Still others have been known to elect a statistics course simply because they heard that it was interesting!

Apart from job-motivated or discipline-related reasons, the study of statistics can help you become more critical in your analyses of information so that you are less susceptible to misleading or deceptive claims. You use external data to make decisions, form conclusions, and build your own individual warehouses of knowledge. If you want to build a sound knowledge base, make intelligent decisions, and form worthwhile opinions, you must be careful to filter out the incoming information that is erroneous or deceptive. As educated and responsible members of society, you should sharpen your ability to recognize distorted statistical data, and you should also learn to interpret undistorted data intelligently.

One hundred years ago Benjamin Disraeli said that "there are three kinds of lies: lies, damned lies, and statistics." It has also been said that "figures don't lie; liars figure." Some people have even been accused of using statistics like a drunk uses a lamppost: "more for support than for illumination." All of these statements refer to the abuses of statistics whereby data are presented in a way that is misleading. The typical abuser has personal objectives and is willing to suppress unfavorable data while emphasizing supportive data. Here are a few examples of the many ways that data can be distorted.

The term "average" refers to several different statistical measures. (These will be discussed and defined in Chapter 2.) To most people, the term average means the sum of all values divided by the number of values, but this is only one type of average and it is called the arithmetic mean. There are also the median, the mode, and the midrange. (For example, given the ten annual salaries of $15,000, $15,000, $15,000, $16,000, $18,000, $20,000, $22,000, $24,000, $25,000, and $45,000, we can correctly claim that the average is either $21,500 (arithmetic mean), $19,000 (median), $15,000 (mode), or $30,000 (midrange).) There are no objective criteria that can be used to determine the specific average that is most representative. Consequently, the user of statistics is free to select the average that is most supportive of a favored position.

A union contract negotiator can choose the lowest average in an attempt to emphasize the need for a salary increase, while the management negotiator can choose the highest average to emphasize the well-being of the employees. In actual negotiations both sides are usually adept at exposing such ploys, but the typical citizen often accepts the validity of an average without really knowing which specific average is being presented and without knowing how different the picture would look if another average were used. The educated and thinking citizen is not so susceptible to potentially deceptive information;

the educated and thinking citizen analyzes and criticizes statistical data so that meaningless or illusory contentions are not part of the base upon which his or her decisions are made and opinions formulated. If you read that "the average annual salary of an American is $14,487," you should attempt to find out which of the averages that figure represents. If no additional information is available, you should realize that the given figure may be very misleading. Conversely, as you present statistics on data you have accumulated, you should attempt to provide descriptions identifying the true nature of the data.

Many visual devices such as bar graphs and pie charts can be used to exaggerate or deemphasize the true nature of data. (These are also discussed in Chapter 2.) In Figure 1-1 we show two bar graphs that depict the *same data*, but part (*b*) is designed to exaggerate the decline in values. While both bar graphs represent the same set of scores, they tend to produce very different subjective impressions. Too many of us look at a graph superficially and develop intuitive impressions based upon the pattern we see. Instead, we should scrutinize the graph and search for distortions of the type found in this illustration. We should analyze the numerical information contained in the graph instead of being impressed by its general shape.

FIGURE 1-1
Both bar graphs depict the same data, but part (*b*) exaggerates what is in reality a modest decline.

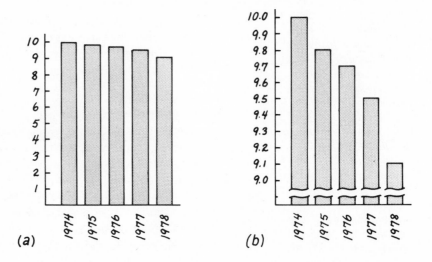

(a)   (b)

Pictures of objects may also be misleading. If the dimensions of a two-dimensional figure (such as a square) are doubled, the area is increased by a multiple of four; if the dimensions of a solid figure are doubled, the volume is increased by a multiple of eight. Solids commonly used to depict data include moneybags, stacks of coins, army tanks, cows, and houses. Let's suppose that school taxes in a small community doubled from one year to the next. By depicting the amounts of taxes as bags of money and by doubling the dimensions of the bag representing the second year, we can easily create the

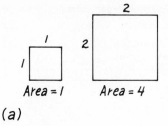

When we double the dimensions of two-dimensional objects, the area increases by a multiple of four.

Area = 1   Area = 4

(a)

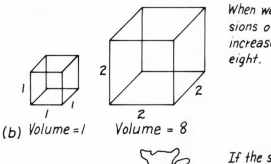

When we double the dimensions of solids, the volume increases by a multiple of eight.

(b) Volume = 1   Volume = 8

If the second moneybag is made twice as tall, twice as wide, and twice as deep as the first, it will appear to be eight times as large.

(c)

FIGURE 1-2

impression that taxes more than doubled (see Figure 1-2). Another variety of statistical "lying" is often inspired by small sample results. The toothpaste preferences of only ten dentists should not be used as a basis for a generalized claim such as "Caressed toothpaste is recommended by seven out of ten dentists." Even if the sample is large, it must be unbiased and representative of the population from which it comes. A popular illustration of a biased sample involves the 1936 presidential election in which an extensive telephone survey indicated that Landon would defeat Roosevelt. Obviously we haven't heard of President Landon so we can conclude that the poll was misleading. Retrospective analysis reveals that a disproportionate number of affluent people had telephones in 1936, and they tended to favor Landon. Consequently the survey did not reach a representative sample of eligible voters. In this case, there was no intent to deceive. The faulty poll is easy to criticize retrospectively, but the basic flaw was not so obvious in 1936.

Sometimes the numbers themselves can be deceptive. A mean annual salary of $14,487.31 sounds precise and tends to instill a high degree of confidence in its accuracy. The figure of $14,400 doesn't convey the same sense of precision and accuracy.

In the late 1970s, gasoline shortages thrust the United States into an energy crisis with long lasting ramifications. Statistical methodology is helpful in determining the best allocation of limited energy resources. Various aspects of our society will change because of the rising cost and decreasing supply of energy, and statistics is used to forecast trends so that we can better plan for tomorrow.

Another source of statistical deception involves numbers that are ultimately guesses, such as the crowd count at a political rally, the dog population of New York City, and the amount of money bet illegally.

"Ninety percent of all our cars sold in this country in the last 10 years are still on the road." The typical consumer hears that commercial message and gets the impression that those cars must be well built in order to persist through those long years of driving. What the auto manufacturer failed to mention was that 90% of the cars they sold in this country were sold within the last 3 years!

A pharmaceutical company advertizes its aspirin as "the one that enters the bloodstream fastest." Hooray! But does it relieve headache pain faster than the others?

Finally, there is the hypothetical story of the race between an American runner and a Russian runner. After the American won, *Pravda* reported that the Russian came in second while the American finished next to last.

The preceding examples comprise a small sampling of the ways in which statistics can be used deceptively. Entire books have been devoted to this subject, including Darrell Huff's *How to Lie with Statistics* and Robert Reichard's *The Figure Finaglers*. Understanding these practices will be extremely helpful in evaluating the statistical data found in everyday situations.

EXERCISES

1-1.   John Graunt studied birth records and observed that the number of male births exceeded the number of female births "by about a thirteenth part." Recently there were 1,703,220 male births and 1,641,736 female births in the United States. How do these figures compare to John Graunt's seventeenth-century claim?

1-2.   If a coin is fair, then heads and tails should have equal chances of occurring.
(a)   Test the fairness of a coin by flipping it 10 times.
(b)   Test the fairness of the same coin by flipping it 50 times.
(c)   If heads and tails never reach 50–50 rate of occurrence, does this mean that the coin is unfair? Explain.

1-3.   How does statistics differ from most other branches of applied mathematics?

1-4.   What is the difference between a population and a sample?

1-5.   How can the concepts of population and sample relate to deductive reasoning and inductive reasoning?

1-6.   Why can't a pollster survey all voters?

1-7.   A newspaper article reports that "this morning's demonstration was attended by 8,437 students." Comment.

1-8.   In 1936, why did voters with telephones tend to favor Landon for the presidency?

1-9.   A district attorney claims that "organized crime paid $89,541 in bribes to local officials last year." Comment.

1-10.   Refer to an almanac to find the number of home accident deaths for the last 5 years. Draw a graph showing the consistency of those deaths.

DESCRIPTIVE STATISTICS

2-1 OVERVIEW

Descriptive statistics involves the methods used to summarize or describe the key characteristics of known data. For example, we compile the 5000 I.Q. scores of all the students attending a certain college. Because a list of 5000 numbers is unmanageable, we compute some average or construct a graph. Such a summary is meaningful and can be used to understand the data more easily. We use descriptive statistics when we simply summarize known raw data.

Inferential statistics goes beyond mere description. We use inferential statistics when we make inferences about a population from sample data.

Suppose that we compute an average of the 5000 I.Q. scores of students at a certain college and obtain a value of 115. If we report that the average I.Q. score of those 5000 students is 115, we have developed a descriptive statistic which simply describes known data. But if we treat those 5000 students as a sample and conclude that the average I.Q. score of all similar college students is 115, we have made an *inference* which goes beyond the known data.

This chapter deals with the basic concepts of descriptive statistics, Chapter 3 includes an introduction to probability theory, and the subsequent chapters deal mostly with inferential statistics. Descriptive statistics and inferential statistics are the two basic divisions of the subject of statistics.

UNITED STATES GOVERNMENT ELIMINATES 500,000 FARMS

When analyzing data from one year to the next, we must be wary of changing definitions which artificially alter the data. In the last century, the Department of Agriculture changed the official definition of a farm at least eight times. A recent change in the definition resulted in decreasing the number of farms by about 500,000. This change involved replacing annual sales of at least $250 (for under 10 acres) or at least $50 (for more than 10 acres) to the new criterion of $1000 in annual sales. If we look for trends in the number of farms and are not aware of these changing definitions, we can easily form invalid conclusions.

2-2 SUMMARIZING DATA

    Congratulations! You have just been appointed acting mayor of a small village. Upon investigation, you determine that the pay is poor, the hours are long, there isn't much chance for advancement, and your job security leaves something to be desired. Nevertheless, your social conscience and civic pride make you want to be an effective and capable leader. You decide to begin your tenure as mayor by obtaining some information about your constituents so that you can better determine their needs. Their earnings are of great concern because you foresee an increase in taxes (higher pay for the mayor, etc.) and you need to know if they can afford a larger tax bill. You commission a survey and find that a random sample of 150 families has the following incomes:

6200	36100	24100	26300	22900	13270
10900	31800	28300	10000	11300	14993
17900	18050	23500	31700	21800	22370
31120	8575	6100	10700	10150	14453
14950	13275	12400	15100	15900	17300
25860	28560	28000	28000	25700	8800
10000	17900	17950	12710	20175	20800
21600	28325	24725	25675	26950	21000
20000	19999	11635	10815	14200	13785
10355	17555	16755	15085	17950	18885
18150	20200	21200	22710	20775	20315
19270	18955	25000	25000	24500	26885
16123	18500	13982	17153	23950	15010
24900	37893	47500	16190	17000	15500
4300	28200	8500	43600	13300	13700
14900	10000	9500	16100	16800	16257
18492	20400	22000	22496	25255	44444
35257	10323	20429	19197	2500	2500
31750	32500	28350	37600	17350	16295
12000	15000	9000	8800	4000	38200
29200	0	0	13445	17275	20500
18450	19275	20875	20345	16780	16100
19900	19995	20800	21800	25555	18760
18930	38700	26600	26010	19000	10000
11575	12000	17500	17500	20500	22222

    A review of the 150 family incomes probably does not lead to any specific conclusions because the human mind usually cannot assimilate and organize that many data. In general, any large collection of raw scores will remain unintelligible until organized and summarized.

    The **frequency table** is an excellent device for making large collections of data much more intelligible. The frequency table is named this way because it lists categories of scores along with their corresponding frequencies. The "frequency" for a category or class is the number of original scores that fall into that class.

    This might seem complicated, but the construction of a frequency table is really a simple process. While construction may be time consuming and monotonous, it is not very difficult. For extremely large collections of scores, the data can be entered in a computer that is programmed to construct the appropriate frequency table automatically. Whether accomplished manually or by computer, the process involves these key steps:

1. Decide on the number of classes your frequency table will contain. A class consists of one grouping of scores such as 0–2999. As a general guideline, the number of different classes in a frequency table should be between 5 and 20. The actual number of classes may be affected by convenience or other subjective factors. We generally begin by examining the highest and lowest scores; they may suggest a convenient number of classes.

For example, if we have 500 I.Q. scores ranging from a low of 90 to a high of 119, there are only 30 possible scores, so it is natural to select either 10 classes of width 3, or 15 classes of width 2.

If we have 347 scores that represent the lives of transistors ranging from 217.2 hours to 638.7 hours, the range of 421.5 hours does not suggest a specific number of classes. There are no objective criteria for determining a number of distinct classes, so we must use our own subjective judgment. (Who says mathematics is not an art?)

2. Find the width of the classes by dividing the number of classes into the range (that is, the difference between the highest and lowest scores). For example, we have I.Q. scores ranging from 90 to 119, and we want 15 class intervals. First we subtract 90 from 119 to get 29 which is the range. Next we divide 29 by 15 (since we want 15 class intervals) and round up the answer to 2 to produce the following class intervals for the frequency table. This rounding *up* is not only convenient, but it guarantees that all of the data will be covered.

90–91
92–93
94–95
etc.

Example   We have 800 scores representing the I.Q.'s of prison inmates. The highest and lowest scores are 142 and 76, respectively. Find the upper and lower limits of each class if we wish to have:

(a) 10 classes

(b) 12 classes

Solution   (a) The difference between the highest and lowest scores is 66. Dividing 66 by 10 we get 6.6 which, for the sake of convenience, we round up to 7. Different starting points such as 74, 75, or 76 may be used. Selecting 74 as the starting point is equivalent to selecting 74 as the lower limit of the first class. With a class width of 7, this means that the next class has 74 + 7 or 81 as its lower limit. (See the definitions of "lower class limits" and "class width" which follow.) The first class must therefore be 74–80 while the second class becomes 81–87. The lower and upper limits of the 10 classes are as follows:

> 74–80
> 81–87
> 88–94
> 95–101
> 102–108
> 109–115
> 116–122
> 123–129
> 130–136
> 137–143

(b) Since the difference between the highest and lowest scores is 66 and we want 12 classes, the class width should be 66 ÷ 12 or 5.5 which we round up to 6. Again the starting point is somewhat arbitrary. Starting with 75 we get the following 12 classes of width 6:

> 75–80
> 81–86
> 87–92
> 93–98
> 99–104
> 105–110
> 111–116
> 117–122
> 123–128
> 129–134
> 135–140
> 141–146

It is important that all classes of a frequency table have the same width, and that all classes are mutually exclusive in the sense that each score will belong to *exactly* one class. The following standard definitions formalize and identify some of the basic terminology generally associated with frequency tables. These definitions may seem difficult, so it might be helpful to examine the example which follows them (see Table 2-1).

DEFINITION   *Lower class limits are the smallest numbers that can actually belong to the different classes.*

DEFINITION   *Upper class limits are the largest numbers that can actually belong to the different classes.*

DEFINITION   *The class boundaries are the numbers that are the upper and lower class limits if there are no gaps between the consecutive classes.*

TABLE 2-1

Score	Frequency
1–5	
6–10	
11–15	
16–20	
21–25	
26–30	

This frequency table has six classes.

The **lower class limits** are 1, 6, 11, 16, 21, 26.

The **upper class limits** are 5, 10, 15, 20, 25, 30.

The **class boundaries** are 0.5, 5.5, 10.5, 15.5, 20.5, 25.5, 30.5.

The **class marks** are 3, 8, 13, 18, 23, 28.

The **class width** is 5.

Let's reconsider the 150 incomes listed earlier in this section. The highest and lowest incomes are 47,500 and 0, respectively. Assume that we want 16 classes in our frequency table; we divide 16 into 47,500 for a class width of 2968.75. If we round up to 3000, we get more convenient classes as follows:

$$0–2999$$
$$3000–5999$$
$$6000–8999$$
$$\text{etc.}$$

We can now proceed to complete the frequency table by recording a tally mark for each score in its proper class and then finding the total number of tally marks in each case (see Table 2-2).

Table 2-2 provides a tremendous advantage by making intelligible the otherwise unintelligible list of 150 incomes. Yet this advantage is not gained without some cost. In constructing frequency tables, we may lose the accuracy of raw data. To see this, consider the last class of 45,000–47,999. Only one family income falls within this class, but the table does not reveal whether that income is 45,000, 47,912, or any other value in that interval.

There is no way to reconstruct the original list of 150 incomes from Table 2-2; the exact values have been compromised for the sake of com-

TABLE 2-2

Family income	Tally marks	Frequency
0–2999	IIII	4
3000–5999	II	2
6000–8999	JHT I	6
9000–11999	JHT JHT JHT	15
12000–14999	JHT JHT JHT I	16
15000–17999	JHT JHT JHT JHT JHT II	27
18000–20999	JHT JHT JHT JHT JHT IIII	29
21000–23999	JHT JHT III	13
24000–26999	JHT JHT JHT I	16
27000–29999	JHT III	8
30000–32999	JHT	5
33000–35999	I	1
36000–38999	JHT	5
39000–41999		0
42000–44999	II	2
45000–47999	I	1

TABLE 2-3

Family income	Frequency
0–50000	150

prehension. To take an extreme and absurd example, the 150 incomes could be put into a frequency table with one class as in Table 2-3. Here the scores have been stripped of any semblance of precision. It is very easy to understand this table, but the data have lost almost all meaning.

Summarizing data generally involves a trade off or compromise between accuracy and simplicity. A frequency table with too few classes is simple but not accurate. A frequency table with too many classes is more accurate but not as easy to understand. The best arrangement is arrived at subjectively, usually in accordance with the common formats used in particular applications.

A variation of the standard frequency table is used when cumulative totals are desired. The cumulative frequency for a class is the sum of the frequencies for that class and all previous classes. Table 2-4 is an example of a **cumulative frequency table,** and it corresponds to the same 150 incomes presented in Table 2-2. A comparison of the frequency column of Table 2-2 and the cumulative frequency column of Table 2-4 reveals that the latter values can be obtained from the former by starting at the top and adding on the successive values. For example, the cumulative frequency of 6 from Table 2-4 represents the sum of 4 and 2 from Table 2-2.

TABLE 2-4

Income by family	Cumulative frequency
Less than 3000	4
Less than 6000	6
Less than 9000	12
Less than 12000	27
Less than 15000	43
Less than 18000	70
Less than 21000	99
Less than 24000	112
Less than 27000	128
Less than 30000	136
Less than 33000	141
Less than 36000	142
Less than 39000	147
Less than 42000	147
Less than 45000	149
Less than 48000	150

**AUTHORS OF THE
FEDERALIST PAPERS
IDENTIFIED**

*In 1787–1788 Alexander
Hamilton, John Jay, and
James Madison
anonymously published the
famous **Federalist** papers in
an attempt to convince New
Yorkers that they should
ratify the Constitution. The
identity of most of the
papers' authors became
known, but the authorship of
12 of the papers was
contested. Through
statistical analysis of the
frequencies of various
words, we can now
conclude that James
Madison is the **likely** author
of these 12 papers. For
many of these disputed
papers, the evidence in
favor of Madison authorship
is overwhelming to the
degree that we can be
almost certain of being
correct.*

In the next section, we will explore various graphic ways to depict data so that they are easily understandable. Frequency tables will be necessary for some of the graphs, and those graphs are often necessary for considering the way the scores are distributed. Frequency tables therefore become important prerequisites for later, more useful concepts.

EXERCISES A:
SUMMARIZING DATA

2-1. Assume that you have 2000 scores representing the weights (rounded up to the nearest pound) of adult males. Also assume that the lowest and highest weights are 110 and 309, respectively.

(a) If you wish to construct a frequency table with 10 classes, identify upper and lower limits of each class. (You may use different starting points.) Also identify the corresponding class boundaries, class marks, and class width.

(b) Do part (a) for 8 classes.

(c) Do part (a) for 16 classes.

2-2. You have been studying the jail sentences of those convicted of public intoxication. You have a list of sentences ranging from 0 days in jail to 95 days in jail, and you intend to publish a frequency table.

    (a)   If you wish to construct a frequency table with 16 classes, identify upper and lower limits of each class. (You may use different starting points.) Also identify the corresponding class boundaries, class marks, and class width.

    (b)   Do part (a) for 12 classes.

    (c)   Do part (a) for 10 classes.

2-3.   A scientist is investigating the time required for a certain chemical reaction to occur. The experiment is repeated 200 times, and the results vary from 17.3 to 42.7 seconds.

    (a)   If you wish to construct a frequency table with 12 classes, identify upper and lower limits of each class. Also identify the corresponding class boundaries, class marks, and class width.

    (b)   Do part (a) for 15 classes.

2-4.   A track star has run the mile in 150 different track meets, and her times ranged from 238.7 seconds to 289.8 seconds. Identify upper and lower limits of each class in a frequency table with a total of 14 classes. Also identify the corresponding class boundaries, class marks, and class width.

2-5.   The following scores represent the time (in hours) between failures of aircraft radios (Stewart model XK–84).

    (a)   Construct a frequency table with 10 classes.

    (b)   Construct a frequency table with 15 classes.

    (c)   Construct a frequency table with 12 classes.

    (d)   Construct a cumulative frequency table with 10 classes.

108	168	152	74	85
136	52	34	62	137
150	175	121	136	126
120	143	174	164	48
42	127	148	81	137
243	74	122	86	128
139	123	86	85	125
103	197	132	49	72
110	123	111	137	77
129	115	215	154	130
115	82	168	107	189
52	124	145	104	55
129	100	163	99	139
75	85	66	81	136
81	155	177	120	164
104	132	85	154	210
92	142	107	81	103
158	100	157	87	125
85	151	126	100	149
115	88	114	75	122

2-6.   A housing development is comprised of homes that are identical in size. They have identical heating systems and lighting arrangements. In a study of energy consumption, the amounts of electricity used in the 150 separate homes for a 1 month period are as follows. (These amounts are in kilowatt hours.) Construct a frequency table with 10 classes.

919	897	968	753	821	978	1021	837	852	774
967	829	1008	1040	924	902	884	836	788	968
950	821	951	815	806	923	926	778	825	851
948	739	909	823	903	1011	1000	894	777	994
945	896	903	932	976	841	806	741	820	1016
981	873	845	811	930	808	827	1040	993	932
861	930	1034	850	897	995	939	948	816	907
858	941	852	883	833	860	855	879	999	884
883	828	897	910	853	948	781	804	793	910
939	877	871	691	957	959	874	876	863	756
873	847	991	861	1087	897	934	900	795	985
809	918	961	854	883	801	852	892	858	909
784	1056	769	906	956	802	969	876	868	808
951	1025	955	764	868	822	889	888	859	849
998	975	840	1030	771	894	766	819	776	869

2-7.   The Bureau of Weights and Measures is investigating consumer complaints that the Curtiss Sugar Company is cheating customers who buy sugar in 5-pound bags. The following scores represent the weights in pounds of 150 bags of sugar produced by the Curtiss Sugar Company. Construct a frequency table with 13 classes.

4.95	4.99	5.04	5.03	4.96	4.99
4.98	4.99	4.93	5.01	5.00	4.92
5.04	5.00	4.97	5.01	4.98	5.02
5.04	4.92	4.98	4.98	5.04	4.96
4.95	4.99	4.92	4.95	4.94	4.98
5.00	4.94	4.98	4.94	5.03	4.92
5.03	4.98	5.04	5.03	4.94	4.97
4.92	5.04	4.97	4.93	5.04	4.97
4.95	4.98	4.94	5.03	4.97	4.92
4.96	4.95	5.02	5.01	5.01	5.03
4.93	5.04	4.92	4.98	5.01	4.96
5.04	4.99	5.01	5.00	4.92	5.02
5.02	5.02	4.93	5.01	5.03	5.04
5.03	5.03	5.02	4.97	4.93	5.03
4.92	4.95	4.96	4.99	5.04	4.92
4.94	5.01	4.96	4.98	4.94	5.01
4.94	5.00	5.00	4.92	5.04	5.00
4.93	5.03	4.95	4.99	5.01	4.94
4.99	4.92	5.00	4.98	5.04	4.98
4.95	4.95	5.02	4.94	4.93	5.04
4.92	4.95	4.96	5.01	5.03	4.93
5.03	4.94	5.04	5.04	5.00	4.98
5.03	4.98	4.99	4.92	4.94	5.00
4.97	5.01	4.95	4.97	5.04	4.92
4.92	4.98	5.01	4.92	5.02	4.96

2-8. The Curtiss Sugar Company also distributes flour. The following scores represent the weights of 150 sacks of flour that bear the Curtiss label. Construct a frequency table with 12 classes.

1.09	0.92	1.00	1.12	0.89	0.98	1.09	1.14
0.75	1.03	1.09	0.95	0.92	1.06	1.20	0.95
1.12	0.95	1.00	1.06	0.95	0.98	0.89	0.98
0.92	1.06	0.84	0.89	0.92	0.78	0.95	1.00
0.92	0.98	0.92	0.84	1.17	1.03	1.00	0.98
0.89	0.75	1.09	1.03	0.98	0.89	0.95	0.92
0.98	1.20	1.00	1.03	0.95	0.95	0.95	1.03
1.14	1.00	0.95	0.95	1.06	1.03	0.95	0.95
1.00	0.92	0.87	0.89	1.00	0.87	0.98	0.92
1.00	1.03	0.98	0.98	1.00	1.00	0.87	0.87
0.89	0.98	1.00	0.87	0.98	0.95	1.00	1.00
0.95	1.00	1.17	0.98	0.87	1.00	1.09	0.92
0.89	0.89	1.00	1.09	1.06	0.87	0.89	0.98
0.92	1.00	0.98	0.89	1.09	0.87	0.89	1.09
1.06	0.92	1.06	1.00	1.03	1.00	1.06	1.06
0.95	0.89	0.92	0.95	0.75	1.09	1.09	0.87
1.03	0.89	1.03	0.87	0.84	0.78	1.09	0.84
0.98	0.95	0.98	0.95	0.98	1.03	1.03	1.00
0.98	1.00	1.00	1.09	1.00	1.03		

EXERCISES B: SUMMARIZING DATA

2-9. Compare the frequency tables of Exercises 2-7 and 2-8. Is there any difference in the *distribution* of the scores? Can such a difference be readily observed by examining raw scores? How can such a difference in distribution occur?

2-10. Suppose the 150 family incomes given in this section were the only responses to a mail survey sent to 500 families. Why might the 150 responses be misleading? That is, what are some likely sources of error?

2-11. (a) Visit your local cemetery and record the life spans of 100 different occupants born before 1875.
(b) Construct a frequency table representing the data from part (a).
(c) Construct a cumulative frequency table representing the data from part (a).

2-3 PICTURES OF DATA

In the preceding section, we saw that frequency tables transform a disorganized collection of raw scores into an organized and understandable summary. In this section we consider ways of representing data in pictorial form. The obvious objective is to promote understanding of the data. We attempt to show that one graphic illustration can be a suitable replacement for hundreds of data.

Pie Charts

It is a simple and obvious characteristic of human physiology that people can comprehend one picture better than a more abstract collection of words or data. Consider the statement: "Of the religious congregations in the United States, 1.5% are Jewish, 7.2% are Roman Catholic, 90.4% are Protestants, and 0.9% of the congregations are of other denominations." What impressions did you develop? How much did you comprehend? Now examine the pie chart in Figure 2-1. While the pie chart depicts the same information as the quotation, it does a better job of dramatizing the data. The abstraction of numbers is overcome by the concrete reality of the "slices" of pie.

The construction of a pie chart from raw data is not too difficult, especially if a calculator, compass, and protractor are available. As an example, let's assume that a family adheres to the following budget:

Food	$2,532
Housing	2,638
Transportation	964
Clothes, medical, and miscellaneous	4,837

We will use the preceding data as a basis for constructing a pie chart with each slice in the proper proportion. To do this, we first find the total budget:

$$
\begin{array}{r}
\$\ 2,532 \\
2,638 \\
964 \\
\underline{4,837} \\
\hline
10,971
\end{array}
$$

FIGURE 2-1
Congregations in the
United States

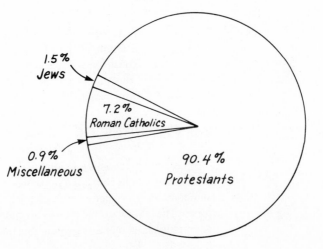

Next each component is divided by this total to get the decimal equivalent of the individual percentages:

$$2532 \div 10971 \overset{\circ}{=} 0.23 \,(\text{or } 23\%)$$
$$2638 \div 10971 \overset{\circ}{=} 0.24 \,(\text{or } 24\%)$$
$$964 \div 10971 = 0.09 \,(\text{or } 9\%)$$
$$4837 \div 10971 \overset{\circ}{=} 0.44 \,(\text{or } 44\%)$$

($\overset{\circ}{=}$ means "approximately equal to")

The third step is designed to reflect the subdivision of every circle (or pie) into 360 equal slices, each of which represents 1°. Multiply each decimal obtained from the previous step by 360:

$$0.23 \times 360 = 82.8$$
$$0.24 \times 360 = 86.4$$
$$0.09 \times 360 = 32.4$$
$$0.44 \times 360 = 158.4$$

The theory underlying this whole process is simple. We divide the individual components by the total to find the percentage that each component represents. We then multiply the decimal equivalent of those percentages by 360 to find the number of degrees (or the size of the slice) that each component comprises. If the calculations are correct, these last products should total 360. There may be some error due to rounding, but a large discrepancy indicates a mistake and the calculations should be redone.

We now know that of the 360 standard unit slices (degrees) in the pie, 82.8 represent food; 86.4 represent housing; 32.4 represent transportation; 158.4 represent clothing, medical, and miscellaneous expenses. Use the compass to draw a circle. The markings on the protractor can be used to determine slices of 82.8 degrees, 86.4 degrees, 32.4 degrees, and 158.4 degrees. The result should resemble the pie chart of Figure 2-2.

FIGURE 2-2
Family budget

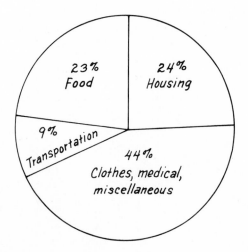

Histograms

The histogram is another common graphic way of presenting data. Histograms are especially important in the study of statistics because they display the distribution of the data, and this distribution is often unattainable without histograms.

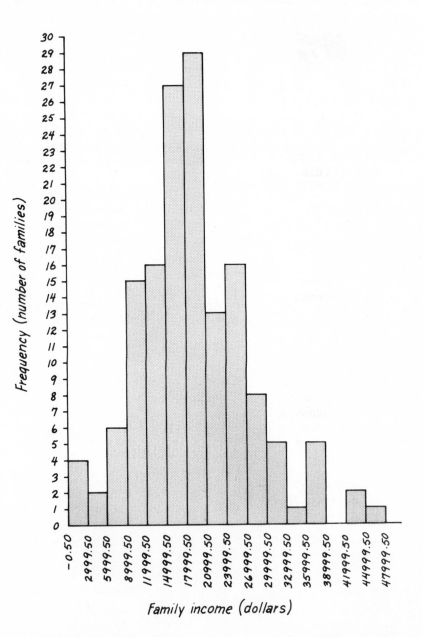

FIGURE 2-3
Histogram

DECLINE IN TEST SCORES ATTRIBUTED TO FAMILY SIZE AND SPACING

Dr. Robert Zajonc believes that the size of the family and the spacing of children are key factors which affect declining test scores more than television watching or increased permissiveness. Dr. Zajonc says that "while it is difficult to determine how big a factor family configuration plays in the drop of test scores, we estimate that 30% to 50% of the drop can be attributed to it." His conclusions are based on analyses of changing family configurations (number of children and spacing) along with the declining SAT scores over a 12-year period.

We generally construct a histogram to represent a set of scores after we have completed a frequency table. The standard format for a histogram usually involves a vertical scale which delineates frequencies and a horizontal scale which identifies class boundaries (or class marks). Shaded bars represent the individual classes of the frequency table by associating the class boundaries with the corresponding class frequencies. As an example, Figure 2-3 is the histogram that corresponds directly to Table 2-2 in the previous section.

Before proceeding from a completed frequency table to the actual construction of a histogram, we must give some consideration to the scales used on the vertical and horizontal axes. We can begin by determining the maximum frequency, and the maximum frequency should suggest a suitable scale for the vertical axis. In Table 2-2 the maximum frequency of 29 will correspond to the tallest bar on the histogram, so 29 (or the next highest convenient number) should be located at the top of the vertical scale with 0 at the bottom. In Figure 2-3, we designed the vertical scale to run from 0 to 30. The horizontal scale should be designed to accommodate all of the classes of the frequency table. Consider the desired length of the horizontal axis along with the number of classes that must be incorporated. Both axes should be clearly labeled.

The histogram of Figure 2-4 relates the same data as Table 2-2 and Figure 2-3, but the vertical axis is scaled down to create an impression of smaller differences. The flexibility in manipulating the scales for the vertical and horizontal axes can serve as a great device for deception. A clever person can exaggerate small differences or deemphasize large differences to suit his or her own purpose. However, conscientious and critical students of statistics become aware of such misleading presentations and carefully examine histograms by analyzing the objective numerical data, instead of superficially glancing at the overall picture.

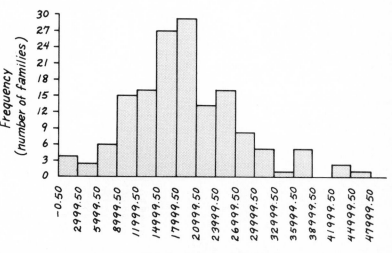

FIGURE 2-4
Histogram

Frequency Polygons

The **frequency polygon** is a variation of the histogram in which the vertical bars are replaced by dots that are subsequently connected to form a line graph. When delineating the horizontal scale of a frequency polygon, we usually use class marks so that the dots can be located directly above them. In Table 2-2 the class marks are 1499.50, 4499.50, 7499.50, . . . , and

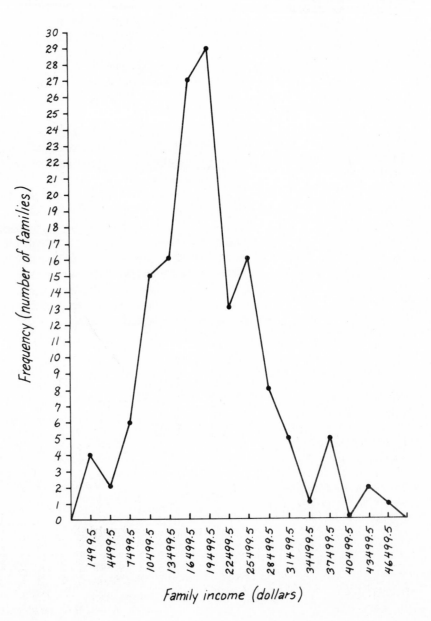

FIGURE 2-5
Frequency polygon

46499.50. In Figure 2-5 we identify those values along the horizontal axis, plot the points directly above them, connect those points with straight line segments, and then extend the resulting line graph at the beginning and end so that it starts and ends with a frequency of 0.

Ogives

Another common pictorial display is the **ogive** (pronounced "oh-jive") or **cumulative frequency polygon**. This differs from an ordinary frequency polygon in that the frequencies are cumulative. The vertical scale is again used to delineate frequencies, but we must now adjust that scale so that it accommodates the total of all individual frequencies. The horizontal scale should depict the class boundaries (see Figure 2-6).

FIGURE 2-6
Ogive

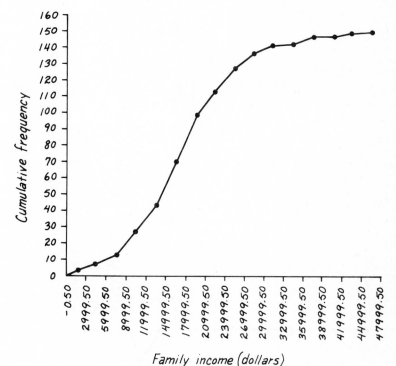

Miscellaneous Graphics

Numerous pictorial displays other than the ones just described can be used to represent data dramatically and effectively. Some examples are soldiers, tanks, airplanes, stacks of coins, and moneybags. This list comprises a very meager sampling of the diverse graphic methods used to convey the nature of statistical data. In fact, there is almost no limit to the variety of different ways that data can be illustrated. However, pie charts, histograms, frequency polygons, and ogives are among the most common devices used (see Table 2-5 and Figures 2-7, 2-8, and 2-9). These graphic representations often convey the *distribution* of the data, and an understanding of the distribution is often critically important to the statistician. In the following section we consider ways of measuring other characteristics of data.

TABLE 2-5

Score	Frequency
1–5	1
6–10	0
11–15	3
16–20	5
21–25	2

FIGURE 2-7
Histogram representing the data summarized in Table 2-5. Note that Figures 2-8 and 2-9 are different ways of illustrating the same data.

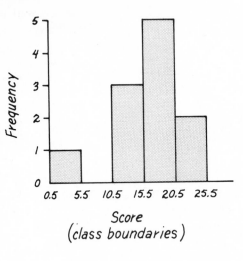

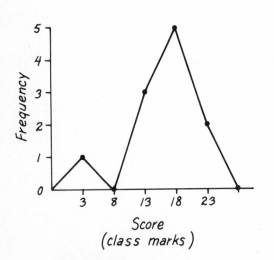

FIGURE 2-8
Frequency polygon representing the data summarized in Table 2-5.

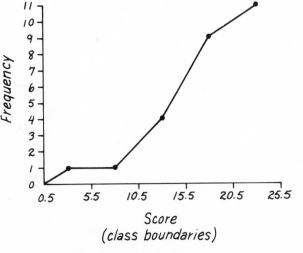

FIGURE 2-9
Ogive representing the data summarized in Table 2-5.

EXERCISES A: PICTURES OF DATA

Exercises 2-12 through 2-23 refer to data found on pages 19–21.

2-12. Refer to Exercise 2-5 for the times between failures of 100 aircraft radios. Construct a histogram that corresponds to a frequency table with 10 classes.

2-13. Refer to Exercise 2-6 for the energy consumption amounts in 150 homes. Construct a histogram that corresponds to a frequency table with 10 classes.

2-14. Refer to Exercise 2-7 for the actual weights of 150 bags of sugar. Construct a histogram that corresponds to the frequency table with 13 classes.

2-15. Refer to Exercise 2-8 for the actual weights of 150 sacks of flour. Construct a histogram that corresponds to the frequency table with 12 classes.

2-16. Refer to Exercise 2-5 for the times between failures of 100 aircraft radios. Construct a frequency polygon that corresponds to a frequency table with 10 classes.

2-17. Refer to Exercise 2-6 for the energy consumption amounts in 150 homes. Construct a frequency polygon that corresponds to a frequency table with 10 classes.

2-18. Refer to Exercise 2-7 for the actual weights of 150 bags of sugar. Construct a frequency polygon that corresponds to a frequency table with 13 classes.

2-19. Refer to Exercise 2-8 for the actual weights of 150 sacks of flour. Construct a frequency polygon that corresponds to a frequency table with 12 classes.

2-20. Refer to Exercise 2-5 for the times between failures of 100 aircraft radios. Construct a cumulative frequency table with 10 classes and construct the corresponding ogive.

2-21. Refer to Exercise 2-6 for the energy consumption amounts in 150 homes. Construct a cumulative frequency table with 10 classes and construct the corresponding ogive.

2-22. Refer to Exercise 2-7 for the actual weights of 150 bags of sugar. Construct a cumulative frequency table with 13 classes and construct the corresponding ogive.

2-23. Refer to Exercise 2-8 for the actual weights of 150 sacks of flour. Construct a cumulative frequency table with 12 classes and construct the corresponding ogive.

In Exercises 2-24 to 2-27, develop a pie chart for the given data by:

(a) *Finding the total.*
(b) *Finding the percentage of the total for each item.*
(c) *Finding the number of degrees corresponding to each item.*
(d) *Constructing the completed pie chart.*

2-24. New York State: "Where your tax dollars go."
62¢ local assistance (health, housing, schools, etc.)
33¢ state functions (hospitals, parks, State Universities)
5¢ capital construction and debt services

2-25. Income of federal government (in billions of dollars):

Individual income tax	112
Corporate income taxes	41
Social Security contributions	67
Excise taxes	18
Miscellaneous	12

2-26. Federal expenditures for education (in millions of dollars):

Colleges	6,000
Elementary and secondary education	4,500
Vocational education	2,500
Research	1,500
Lunches and milk	1,500
Miscellaneous	2,000

2-27. Family budget:

Food	$2,532
Housing	2,638
Transportation	964
Taxes	1,366
Clothes	1,196
Medical	612
Other	1,663

2-28. Given the histogram in Figure 2-10:
   (a) Construct the corresponding frequency table. Use the 11 classes 15, 16, 17, . . . , 25.
   (b) Construct the corresponding ogive.
   (c) Refer to the frequency table of part (a) to determine how many police trainees scored below 21.
   (d) If 17 is a passing score, what percentage of trainees failed this test?

2-29. Given the frequency polygon in Figure 2-11:
   (a) Construct the corresponding frequency table.
   (b) Construct the corresponding ogive.
   (c) How many subjects had a pulse rate between 75 and 84?
   (d) What percentage of subjects had a pulse rate below 90?

FIGURE 2-10
Histogram

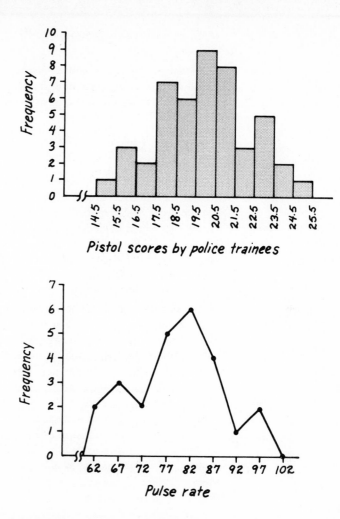

Pistol scores by police trainees

FIGURE 2-11
Frequency polygon

Pulse rate

EXERCISES B: PICTURES OF DATA

Exercises 2-30 through 2-33 refer to data found on pages 19–21.

2-30. (a) Reconstruct the histogram of Exercise 2-12 so that the differences in frequencies are exaggerated.
      (b) Reconstruct the histogram of Exercise 2-12 so that the differences in frequencies are deemphasized.

2-31. (a) Reconstruct the frequency polygon of Exercise 2-19 so that the differences in frequencies are exaggerated.
      (b) Reconstruct the frequency polygon of Exercise 2-19 so that the differences in frequencies are deemphasized.

2-32. Attempt to answer Exercise 2-9 after examining the appropriate histograms.

2-33. If the data for Exercise 2-11 were compiled from a cemetery, construct:
   (a) The corresponding histogram.
   (b) The corresponding frequency polygon.
   (c) The corresponding ogive.

2-4 AVERAGES

In this chapter we describe some of the fundamental methods and tools generally categorized as descriptive statistics. We began with frequency tables that are used to summarize raw data (Section 2-2) and then considered a variety of graphic devices that represent the data in pictorial form (Section 2-3). However, summary and graphic methods cannot be used easily if we are limited to verbal exchanges. In addition they are very difficult to use when we are making statistical inferences.

Suppose, for example, that we wish to compare the reaction times of two different groups of people. We could collect the two sets of raw data, organize them into two frequency tables, and then construct the appropriate histograms. Next we could compare the relative shapes and positions of the two histograms, but we would be evaluating the strengths of their similarities subjectively. Our conclusions and inferences would be much stronger and more reliable if we could develop more objective procedures. Fortunately, we can.

The more objective procedures do require numerical measurements that describe certain characteristics of the data. In this and the following section we are concerned with some of these quantitative measurements. This section deals mainly with **averages** which are also called **measures of central tendency** since they are measurements that are intended to capture the value at the center of the scores.

An important and basic point to remember is that there are different ways to compute an average. We have already stated that, given a list of scores and instructions to find the average, most people will obligingly proceed to first total the scores and then divide that total by the number of scores. However, this particular computation is only one of several procedures that are classified under the umbrella description of an average. The following are the most commonly used averages: mean, median, mode, and midrange.

Mean

Technically called the **arithmetic mean,** this computation is the most important of all numerical descriptive measurements, and it corresponds to what most people call an average. The arithmetic mean of a list of scores is obtained by adding the scores and dividing the total by the number of scores. This particular average will be employed frequently in the remainder of this text, and it will be referred to simply as the mean (see Figure 2-12).

FIGURE 2-12

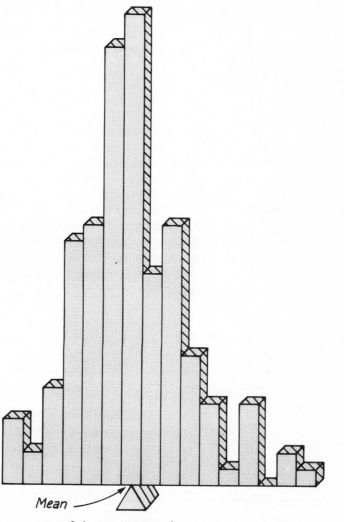

Mean —

If a fulcrum is placed at the position of the mean, it will balance the histogram.

Example Find the mean of the following quiz scores:

2, 3, 6, 7, 7, 8, 9, 9, 9, 10

Solution First add the scores

$$2 + 3 + 6 + 7 + 7 + 8 + 9 + 9 + 9 + 10 = 70$$

and then divide the total by the number of scores present.

$$\frac{70}{10} = 7$$

The mean score is therefore 7.

In many cases, we can provide a formula for computing some measurement. In statistics, these formulas often involve the Greek letter Σ (capital sigma) which is intended to denote the summation of a set of values. Σx means "sum the values that x can assume." If we are working with a specific set of scores, the letter x is used as a variable which can assume the value of any one of those scores. Thus, Σx means the sum of all of the scores. In the preceding example, as the variable x assumes the values of 2, 3, 6, 7, 7, 8, 9, 9, 9, and 10, Σx becomes $2 + 3 + 6 + 7 + 7 + 8 + 9 + 9 + 9 + 10$ or 70. In future formulas, we may encounter expressions such as Σx^2. Standard usage of the symbol Σ requires that the values to be added must be in the form of the variable following Σ. Thus, Σx^2 indicates that the individual scores must be squared *before* they are summed. Referring to the preceding scores, Σx^2 becomes:

$$2^2 + 3^2 + 6^2 + 7^2 + 7^2 + 8^2 + 9^2 + 9^2 + 9^2 + 10^2$$
$$= 4 + 9 + 36 + 49 + 49 + 64 + 81 + 81 + 81 + 100$$
$$= 554$$

NOTATION

Σ	denotes **summation** of a set of values.
x	is the **variable** usually used to represent the individual raw scores.
n	represents the **number** of scores being considered.
$\bar{x}$	denotes the **mean** of a set of **sample** scores.
μ	denotes the **mean** of all scores in some **population**.

Since the mean is the sum of all scores (Σx) divided by the number of scores (n), we get:

Formula 2-1 $\text{mean} = \dfrac{\Sigma x}{n}$

The result can be denoted by $\bar{x}$ if the available scores are samples from a larger population; if all scores of the population are available, then we can denote the computed mean by μ (Greek letter mu). Applying the formula to the preceding quiz scores, we get:

$$\text{mean} = \frac{\Sigma x}{n} = \frac{70}{10} = 7$$

According to the definition of mean, 7 is the central value of the ten given quiz scores. Other definitions of averages involve different perceptions as to how the center is arrived at. The median reflects one different approach.

Median

The **median** of a set of scores is the middle value when the scores are arranged in order of magnitude.

After first arranging the original scores in increasing (or decreasing) order, the median will be either of the following:

1. The number which is exactly in the middle of the list (if the number of scores is odd).

2. The mean of the two middle numbers (if the number of scores is even).

Example Find the median of the scores 7, 2, 3, 7, 6, 9, 10, 8, 9, 9, 10.

Solution Begin by arranging the scores in increasing order.

2, 3, 6, 7, 7, 8, 9, 9, 9, 10, 10

With these eleven scores, the number 8 is located in the exact middle so that 8 is the median.

Example Find the median of the scores 7, 2, 3, 7, 6, 9, 10, 8, 9, 9.

Solution Again, begin by arranging the scores in increasing order.

2, 3, 6, 7, 7, 8, 9, 9, 9, 10

With these ten scores, there is no single score that is at the exact middle. Instead, the two scores of 7 and 8 share the middle. We therefore proceed to find the mean of those two scores.

$$\frac{7 + 8}{2} = \frac{15}{2} = 7.5$$

The median is therefore 7.5.

Mode

The **mode** is obtained from a collection of scores by selecting the one score that occurs most frequently. In those cases where no score is repeated, or more than one score is repeated "most often," we stipulate that there is no mode.

Example   The scores 2, 3, 6, 7, 8, 9, 10 have no mode since no score is repeated.

Example   The scores 1, 2, 2, 2, 3, 4, 5, 6, 6, 6, 7, 9 have no mode since 2 and 6 both occur with the same highest frequency.

Example   The scores 1, 2, 2, 2, 3, 4, 5, 6, 7, 9 have a mode of 2.

Midrange

The **midrange** is that average obtained by adding the highest score to the lowest score and then dividing the result by 2.

$$\text{midrange} = \frac{\text{highest score} + \text{lowest score}}{2}$$

Example   Find the midrange of the scores 2, 3, 6, 7, 7, 8, 9, 9, 9, 10.

Solution   Midrange $= \dfrac{10 + 2}{2} = 6$.

A simple rule for rounding off answers is to carry one more decimal place than was present in the original data. We should round off only the final answer and not intermediate values. For example, the mean of 2, 3, 5 is expressed as 3.3. Since the original data were whole numbers, we rounded off the answer to the nearest tenth. The mean of 2.1, 3.4, 5.7 is rounded off to 3.73.

You should know and understand the preceding four averages (mean, median, mode, and midrange). Other averages (geometric mean, harmonic mean, and quadratic mean) are not used as often, and they will be included only in Exercises B at the end of this section (see Exercises 2-61, 2-62, and 2-63).

Weighted Mean

The **weighted mean** is useful in many situations where the scores vary in their degree of importance. An obvious example occurs frequently in the determination of a final average for a course that includes four tests plus a final examination. If the respective grades are 70, 80, 75, 85, and 90, the mean of 80 does not reflect the greater importance placed on the final exam. Let's suppose that the instructor counts the respective tests as 15%, 15%, 15%, 15%, and 40%. The weighted mean then becomes:

$$\frac{(70 \times 15) + (80 \times 15) + (75 \times 15) + (85 \times 15) + (90 \times 40)}{100}$$

$$= \frac{1050 + 1200 + 1125 + 1275 + 3600}{100}$$

$$= \frac{8250}{100} = 82.5$$

This computation suggests a general procedure for determining a weighted mean. Given a list of scores $x_1, x_2, x_3 \ldots , x_n$, and a corresponding list of weights $w_1, w_2, w_3, \ldots , w_n$, the weighted mean is obtained by computing

Formula 2-2 $\text{weighted mean} = \dfrac{\Sigma w \cdot x}{\Sigma w}$

That is, first multiply each score by its corresponding weight; then find the total of the resulting products thereby evaluating Σwx. Now add up the values of the weights to find Σw and divide the latter value into the former.

Formula 2-2 can be modified so that we can compute the mean from a frequency table when we don't know the original raw scores. Given the data in Table 2-6, on page 38, we can use the class marks as representative scores and the frequencies as weights. Then the formula for the weighted mean leads directly to Formula 2-3, which can be used to compute the mean of a set of scores in a frequency table.

Formula 2-3 $\bar{x} = \dfrac{\Sigma f \cdot x}{\Sigma f}$

where

    x = class mark
    f = class frequency

Recall that the class mark of a class interval is simply the midpoint value. For example, the class interval of 55–59 has 57 as its class mark. Thus Table 2-6 has class marks of 57, 62, 67, 72, 77, 82, 87, 92, 97, and 102. We can now compute the weighted mean.

TABLE 2-6

Pulse rate	Frequency	Class mark
55–59	7	57
60–64	9	62
65–69	14	67
70–74	16	72
75–79	24	77
80–84	14	82
85–89	8	87
90–94	2	92
95–99	2	97
100–104	1	102

$$\bar{x} = \frac{\Sigma f \cdot x}{\Sigma f} = \frac{(7 \cdot 57) + (9 \cdot 62) + (14 \cdot 67) + (16 \cdot 72) + (24 \cdot 77) + (14 \cdot 82) + (8 \cdot 87) + (2 \cdot 92) + (2 \cdot 97) + (1 \cdot 102)}{7 + 9 + 14 + 16 + 24 + 14 + 8 + 2 + 2 + 1}$$

$$= \frac{399 + 558 + 938 + 1152 + 1848 + 1148 + 696 + 184 + 194 + 102}{97}$$

$$= \frac{7219}{97}$$

$$\stackrel{\circ}{=} 74.4$$

This procedure is justified by the fact that a class like 55–59 can be represented by its class mark of 57, and the frequency number indicates that the representative score of 57 occurs seven times. In essence, we are treating Table 2-6 as if it contained seven 57's, nine 62's, fourteen 67's, etc. The total of all the scores is therefore $(7 \cdot 57) + (9 \cdot 62) + (14 \cdot 67) + \cdots + (1 \cdot 102)$ while the total number of scores present is $7 + 9 + 14 + \cdots + 1$; the mean is the former sum divided by the latter and it corresponds exactly to Formula 2-3.

Quartiles, Deciles, and Percentiles

Just as the weighted mean is often a useful variation of the mean, there are useful variations of the median. Quartiles, deciles, and percentiles are measurements similar to the median in that they all partition the data according to rankings and fixed percentages. Such measurements are often called **measures of position**.

Just as the median divides the data into two equal parts, the three **quartiles**, denoted Q_1, Q_2, and Q_3, divide the ranked scores into four equal parts. Roughly speaking, Q_1 separates the bottom 25% of the ranked scores from the top 75%, Q_2 is the median, and Q_3 separates the top 25% from the bottom 75%. To be more precise, at least 25% of the data will be less than or equal to Q_1 and at least 75% will be greater than or equal to Q_1. At least 75% of the data will be less than or equal to Q_3, while at least 25% will be equal to or greater than Q_3. Q_2 is actually the median.

Example

Find the three quartiles for these pulse rates:

$$67, \ 79, \ 77, \ 57, \ 79, \ 77, \ 76, \ 63, \ 64, \ 78, \ 92,$$
$$59, \ 80, \ 79, \ 86, \ 97, \ 91, \ 76, \ 82, \ 76$$

Solution

First arrange the scores in increasing order:

$$57, \ 59, \ 63, \ 64, \ 67, \ 76, \ 76, \ 76, \ 77, \ 77, \ 78, \ 79, \ 79, \ 79, \ 80, \ 82, \ 86, \ 91, \ 92, \ 97$$

Since there are 20 scores, the first quartile Q_1 must separate the bottom 25% (or 5 scores) from the top 75%. Examining the list, we see that Q_1 is between 67 and 76, so we select the midvalue of 71.5. (The midvalue of any two numbers is one-half of their sum.) Q_2 is the median or middle score and must therefore be between the tenth and eleventh scores of 77 and 78. Thus, $Q_2 = 77.5$. Finally, Q_3 separates the top 25% from the lower 75% and must be between the fifteenth and sixteenth scores of 80 and 82. Hence, $Q_3 = 81$.

Given a list of 100 scores $x_1, x_2, x_3, \ldots, x_{100}$ that are already arranged in increasing order, we can compute the three quartiles as follows:

$$Q_1 = \frac{x_{25} + x_{26}}{2}$$

$$Q_2 = \frac{x_{50} + x_{51}}{2}$$

$$Q_3 = \frac{x_{75} + x_{76}}{2}$$

In those cases where the number of scores present isn't a convenient number (divisible by 4) like 100, the arithmetic may be more complicated.

There are nine **deciles**, denoted $D_1, D_2, D_3, \ldots, D_9$, that partition the ranked data into ten groups with 10% of the data in each group. For example, given the 100 scores $x_1, x_2, x_3, \ldots, x_{100}$ that have already been arranged in increasing order, we can compute the deciles as follows:

$$D_1 = \frac{x_{10} + x_{11}}{2}$$

$$D_2 = \frac{x_{20} + x_{21}}{2}$$

$$D_3 = \frac{x_{30} + x_{31}}{2}$$

$$\cdot$$
$$\cdot$$
$$\cdot$$

$$D_9 = \frac{x_{90} + x_{91}}{2}$$

Given a large collection of data, we may wish to obtain the 99 **percentiles** P_1, P_2, P_3 . . . , P_{99} that partition the ranked data into 100 groups with 1% of the scores in each group. A student taking a competitive college entrance examination might learn that he scored in the ''ninety-second percentile.'' This does not mean that he received a grade of 92% on the test; it indicates roughly that whatever score he did achieve was higher than 92% of his peers who took a similar test (and also lower than 8% of his colleagues). Percentiles are useful for converting meaningless raw scores into meaningful comparative scores. For this reason, percentiles are used extensively in educational testing. A raw score of 750 on a college entrance exam means nothing to most people, but the corresponding percentile of 93% provides useful comparative information.

Having deviated from the basic objective of this section to discuss the variations of quartiles, deciles, and percentiles, let us now return to the four basic averages of mean, median, mode, and midrange. We have stressed that these statistics can all be properly called averages and that the freedom to select a particular average can be a source of deception. Consider the extreme example of a class of students from which the following statistics have been computed.

mean I.Q. = 110
median I.Q. = 120
mode I.Q. = 95
midrange I.Q. = 100

The teacher of this class can technically use any of the preceding figures as the average I.Q. At the end of the academic year, the teacher could refer to the median value of 120 so that his achievements and performance are enhanced. Even when we are not trying to enhance our achievements, the selection of the best average is not always easy. The different averages are characterized by different advantages and disadvantages. Consequently there are no objective criteria which determine the most representative average for a given set of data.

CALL A COP

The Law Enforcement Assistance Administration (L.E.A.A.) made a study of 1000 crime victims and found that it took a median time of 6 minutes and 17 seconds for the police to be called. In an article on that study, Time *made the following statement: "If crimes were reported in 1 minute instead of 5 minutes, the L.E.A.A. study added, in one of those statistical computations that always inspire wonder, the probability of an arrest would increase by up to 15%."*

The mean has the distinct advantage of being the most familiar of the four basic averages. Also, for any finite set of scores, the mean always exists as a single unique number which can be used in further statistical analyses. As another advantage, the mean takes every score into account. One serious disadvantage of the mean is that it is affected by exceptional extremes (high or low scores).

For any finite set of scores, the median always exists as a single unique number which is relatively easy to find. Unlike the mean, the median exhibits the advantage of not being affected by some exceptional extreme values. However, a disadvantage is that the median is not sensitive to the value of every score.

When it exists, the mode is the easiest average to find, but there are many collections of data for which the mode does not exist. A most serious disadvantage of the mode is its total insensitivity to all other scores. The mode is a useful measure in certain cases, but it may be a poor choice in many cases. A unique advantage of the mode is its applicability to qualitative data. For example, take a survey of 70 artists to determine their favorite colors. How can you average "18 reds, 23 blues, and 29 greens?" The modal choice is green, but the other averages are useless in this example.

For any finite set of scores, the midrange always exists as a unique number that is easy to compute, but it is too strongly affected by extreme values, and it does not take every score into account.

The preceding list of advantages and disadvantages emphasizes the lack of objectivity in the selection of a particular average. An understanding of this situation is a prerequisite for evading intended and accidental statistical deceptions that arise from the use of averages.

In the next section, we explore **measures of dispersion** that give information about variations within sets of data.

EXERCISES A: AVERAGES

In Exercises 2-34 to 2-45, find the (a) mean, (b) median, (c) mode, and (d) midrange for the given sample data.

2-34. Ten cars were tested for fuel consumption and the following mpg ratings were obtained:

$$24, \; 21, \; 21, \; 21, \; 26, \; 22, \; 25, \; 23, \; 23, \; 25$$

2-35. Twenty subjects were tested on a device used in a biofeedback experiment and the following readings (in microvolts) were obtained:

11.9, 7.8, 6.4, 8.2, 6.4, 9.0, 7.2, 6.0, 3.6, 6.4, 12.7, 7.9, 5.6, 5.6, 9.5, 3.8, 4.1, 7.3, 8.2, 6.5

2-36. Twenty snow tires were tested and their braking distances (in feet) on ice were:

160, 148, 148, 175, 145, 158, 151, 135,
141, 144, 162, 154, 156, 162, 154, 159,
157, 111, 150, 161,

2-37. A survey was made of 15 housewives to determine the number of hours worked each week and the following results were obtained:

52, 50, 46, 41, 45, 46, 47, 50, 42, 56, 61, 39, 57, 44, 45

2-38. In Section 1 of a statistics class, ten test scores were randomly selected and the following results were obtained:

5329   5929   4624   5929   4356
74, 73, 77, 77, 71, 68, 65, 77, 67, 66
5476   5929   5041   4225   4489

2-39. In Section 2 of a statistics class, ten test scores were randomly selected and the following results were obtained:

42, 100, 77, 54, 93, 85, 67, 77, 62, 58

2-40. The reaction times (in seconds) of a group of adult men were found to be:

0.74, 0.71, 0.41, 0.82, 0.74, 0.85, 0.99, 0.71, 0.57, 0.85, 0.57, 0.55

2-41. The college entrance examination scores of 15 randomly selected high school seniors are:

448, 403, 278, 448, 322, 403, 535, 210, 322, 425, 308, 370, 425, 320, 632

2-42. A farmer measures the heights (in centimeters) of 20 seedlings that have been allowed to grow for 2 years and the following results were obtained:

62, 76, 89, 56, 79, 67, 101, 63, 49, 76, 74,
61, 72, 70, 57, 82, 62, 67, 60, 65

2-43. Job applicants are given an aptitude test and their results are:

186, 159, 173, 176, 216, 197, 206, 168, 204, 213,
240, 190, 180, 233, 202, 213, 182, 203, 223, 227

2-44. A study was made of a regional center of the Internal Revenue Service to determine the number of income tax returns opened and sorted in 1 hour. The results for 11 employees are:

248, 260, 259, 240, 248, 226, 250, 266, 255, 256, 234

2-45. A study was made to determine the number of defective items produced by 15 different employees in 1 day and the results are:

1, 1, 1, 2, 3, 7, 7, 8, 8, 9, 11, 12, 18, 19, 500

In Exercises 2-46 through 2-49, use the given frequency table and:

(a) Identify the class mark for each class interval.

(b) Find the mean using the class marks and frequencies.

2-46.

Time		Frequency
0–59	*70* *29.5*	1
60–119	*90* *89.5*	2
120–179	*150* *149.5*	5
180–239	*210* *209.5*	14
240–299	*270* *269.5*	7
300–359	*330* *329.5*	12
360–419	*390* *389.5*	9

50

Frequency table of the time (in seconds) it takes the victim of a crime to call the police.

2-47.

Number of trials	Frequency
1–5	2
6–10	1
11–15	12
16–20	10
21–25	5

Frequency table of the number of trials required by 30 different monkeys to learn a certain task.

2-48.

Number	Frequency
0–9	5
10–19	19
20–29	32
30–39	14
40–49	1

Frequency table of the number of violent deaths seen on television by children in one week.

2-49.

Index	Frequency
41–50	12
51–60	10
61–70	8
71–80	18
81–90	27
91–100	3

Frequency table of the pollution indices for a certain city.

EXERCISES B: AVERAGES

Exercises 2-50 through 2-52 refer to data found on pages 19–20.

2-50. Using the data of Exercise 2-5, find:
    (a) The three quartiles.
    (b) The nine deciles.

2-51. Using the data of Exercise 2-6, find:
    (a) The three quartiles.
    (b) The nine deciles.

2-52. Using the data of Exercise 2-7, find:
    (a) The three quartiles.
    (b) The nine deciles.

2-53. Using the 150 family incomes given in Section 2-2 find:
    (a) The three quartiles.
    (b) The nine deciles.

2-54. Compare the averages that comprise the solutions to Exercises 2-38 and 2-39. Do these averages discriminate or differentiate between the two lists of scores? Is there any apparent difference between the two scores that is not reflected in the averages?

2-55. In Exercise 2-45 which averages are misleading? Which averages are representative of the data?

In Exercises 2-56 through 2-58, use a calculator or computer to find the mean of the raw data in the exercise specified. The data appear on pages 19–20.

2-56. Exercise 2-5.

2-57. Exercise 2-6.

2-58. Exercise 2-7.

2-59. If the data for Exercise 2-11 were obtained, find the mean, median, mode, and midrange.

2-60. For each of the Exercises 2-34, 2-35, and 2-36, find:
(a) Σx
(b) Σx^2
(c) $(\Sigma x)^2$

2-61. To obtain the **harmonic mean** of a set of scores, divide the number of scores n by the sum of the reciprocals of all scores.

$$\text{harmonic mean} = \frac{n}{\Sigma \frac{1}{x}}$$

For example, the harmonic mean of 2, 3, 6, 7, 7, 8 is

$$\frac{6}{\frac{1}{2} + \frac{1}{3} + \frac{1}{6} + \frac{1}{7} + \frac{1}{7} + \frac{1}{8}} \doteq \frac{6}{1.4} \doteq 4.3$$

(Note that 0 cannot be included in the scores.)
(a) Find the harmonic mean of 2, 3, 6, 7, 7, 8, 9, 9, 9, 10.
(b) A group of students drives from New York to Florida (1200 miles) at a speed of 40 miles per hour and returns at a speed of 60 miles per hour. What is their average speed for the round trip? (The harmonic mean is used in averaging speeds.)

2-62. Given a collection of n scores (all of which are positive), the **geometric mean** is the nth root of their product. For example, to find the geometric mean of 2, 3, 6, 7, 7, 8, 9, 9, 9, 10, first multiply the scores

$$2 \cdot 3 \cdot 6 \cdot 7 \cdot 7 \cdot 8 \cdot 9 \cdot 9 \cdot 9 \cdot 10 = 102876480$$

and then take the tenth root of the product.

$$\sqrt[10]{102876480} \doteq 6.3$$

The geometric mean is often used in business and economics for finding average rates of change, average rates of growth, or average ratios. Find the geometric mean of 5, 5, 6, 8, 9, 10, 15, 20, 25, 50.

2-63. The **quadratic mean** (or root mean square or R.M.S.) of a set of scores is obtained by squaring each score, adding the results, dividing by the number of scores n, and then taking the square root of that result.

$$\text{quadratic mean} = \sqrt{\frac{\Sigma x^2}{n}}$$

For example, the quadratic mean of 2, 3, 6, 7, 7, 8, 9, 9, 9, 10 is

$$\sqrt{\frac{4 + 9 + 36 + 49 + 49 + 64 + 81 + 81 + 81 + 100}{10}} = \sqrt{\frac{554}{10}} = \sqrt{55.4} \triangleq 7.4$$

The quadratic mean is usually used in physical applications. Find the quadratic mean of 1, 2, 2, 3, 5.

2-5 DISPERSION STATISTICS

The preceding section was concerned with averages or measures of central tendency. Those statistics were designed to reflect a certain characteristic of the data from which they came. That is, the averages are supposed to be *central* scores. However, other features of the data may not be reflected at all by the averages. Suppose, for example, that two different groups of 10 students are given identical quizzes with the following results:

Group A	Group B
65	42
66	54
67	58
68	62
71	67
73	77
74	77
77	85
77	93
77	100

Proceeding to compute the averages, we get:

Group A	Group B
mean = 71.5	mean = 71.5
median = 72	median = 72
mode = 77	mode = 77
midrange = 71	midrange = 71

From looking at these averages, we can see no difference between the two groups. Yet an intuitive perusal of both groups shows an obvious difference: the scores of Group B are much more widely scattered than those of Group A. This variability among data is one characteristic to which averages are not sensitive. Consequently, statisticians have tried to design statistics that measure this variability or dispersion. The three basic measures of dispersion are range, variance, and standard deviation.

Range

The **range** is simply the difference between the highest value and the lowest value. For Group A, the range of 12 is the difference between 77 and 65. The range of Group B is 58 (100 − 42). This much larger range suggests greater dispersion. Be sure you avoid confusion between the midrange (an average) and the range (a measure of dispersion). The range is extremely easy to compute, but it's often inferior to other measures of dispersion. The rather extreme example of Groups C and D should illustrate this point.

The larger range for Group C suggests more dispersion than in Group D. But the scores of Group C are very close together while those of Group D are much more scattered. The range may be misleading in this case (and in many other circumstances) because it depends only on the maximum and minimum scores. Better measures of dispersion have been developed, but the improvement is not achieved without some loss. The loss occurs in the ease of computation. The variance and standard deviation are measures which generally have greater statistical significance, but they require computations that are much more difficult.

Group C	Group D
1	2
9	2
10	2
10	2
10	2
10	18
10	18
10	18
10	18
20	18
Range = 19	Range = 16

Variance

The **variance** is computed by applying the formula*

Formula 2-4
$$\text{variance} = \frac{\Sigma(x - \bar{x})^2}{n - 1}$$

to the given set of values.

*Many authors define the variance with a denominator of n instead of $n - 1$. However, the use of $n - 1$ provides a better estimate of the variance for a population when the formula is applied to a random sample from that population. Consequently, if the data are an entire population of scores, divide by n, but divide by $n - 1$ when dealing with sample data. For large values of n (such as those greater than 30) it really doesn't make too much difference which choice is made. For example, computations involving 100 scores will result in division by 100 or by 99 so that we might get values like 22/100 or 22/99. These values in decimal form to three places are 0.220 and 0.222, respectively, and they are essentially the same.

σ^2 (where σ is the lowercase Greek sigma) denotes the variance of all scores in a *population*. (If all scores in a population are known, $\bar{x}$ in Formula 2-4 actually becomes μ.)

s^2 denotes the variance of a set of *sample* scores.

To use Formula 2-4 you should follow this procedure:

1. Find the mean of the scores ($\bar{x}$).

2. Subtract the mean from each individual score ($x - \bar{x}$).

3. Square each of the differences obtained from Step 2. That is, multiply each value by itself. (This produces numbers of the form $(x - \bar{x})^2$.)

4. Add all of the squares obtained from Step 3 to get $\Sigma(x - \bar{x})^2$.

5. Divide the preceding total by the number ($n - 1$); that is, one less than the total number of scores present. The result is the variance.

Upon students' initial exposure to the variance, there is sometimes a feeling of awe, mystery, or even an urge to acquire a withdrawal slip. These feelings can, of course, be overcome. Computation of the variance involves only the simple arithmetic operations of addition, subtraction, multiplication, and division. The mystery associated with this particular procedure must remain for some time, since a complete explanation of it is not possible at this stage.

We can point out that after the mean $\bar{x}$ has been obtained, any individual score x will differ or deviate from that mean by an amount of $(x - \bar{x})$. The sum of all such deviations, denoted by $\Sigma(x - \bar{x})$, might initially seem like a reasonable measure of dispersion of variance, but that sum will *always* equal 0. Scores greater than the mean will cause $(x - \bar{x})$ to be positive, while scores below the mean will cause $(x - \bar{x})$ to be negative. The positive values will cancel out exactly with the negative values since the mean $\bar{x}$ is at the center of the scores.

TABLE 2-7. Sample of five scores: 2, 3, 5, 6, 9. The variance is $30 \div 4$ or 7.5.

x	$(x - \bar{x})$	$(x - \bar{x})^2$
2	-3	9
3	-2	4
5	0	0
6	1	1
9	4	16
	$\Sigma(x - \bar{x}) = 0$	$\Sigma(x - \bar{x})^2 = 30$

How do we prevent this undesirable canceling out of positive and negative values so that the measure of dispersion is not always 0? One alternative is to use absolute values as in $\Sigma |x - \bar{x}|$. This leads to the **mean deviation** which will be mentioned later, but this approach tends to be unsuitable for the important methods of statistical inference. Instead, we make all of the terms $(x - \bar{x})$ nonnegative by squaring them (see Table 2-7).

Perhaps some feeling for the concept of variance can be cultivated through a comparative study of the five examples of Figure 2-13.

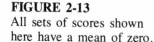

FIGURE 2-13
All sets of scores shown here have a mean of zero.

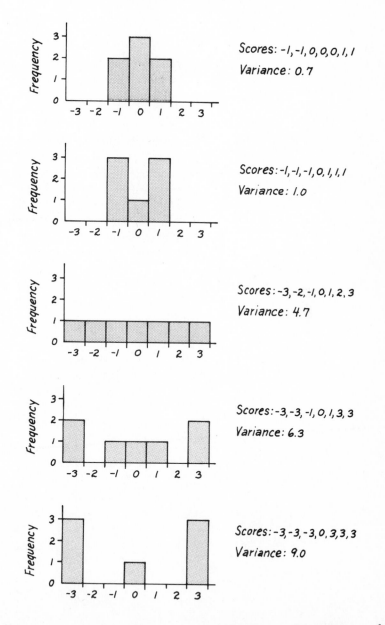

Scores: -1, -1, 0, 0, 0, 1, 1
Variance: 0.7

Scores: -1, -1, -1, 0, 1, 1, 1
Variance: 1.0

Scores: -3, -2, -1, 0, 1, 2, 3
Variance: 4.7

Scores: -3, -3, -1, 0, 1, 3, 3
Variance: 6.3

Scores: -3, -3, -3, 0, 3, 3, 3
Variance: 9.0

We now proceed to apply the variance formula to a set of data.

Example Find the variance for these quiz scores from Group D: 2, 2, 2, 2, 2, 18, 18, 18, 18, 18.

Solution Step 1. The mean is obtained by adding the scores (100) and then dividing the total by the number of scores present (10). The mean is 10.

Step 2. Subtracting the mean of 10 from each score, we get −8, −8, −8, −8, −8, 8, 8, 8, 8, 8.

Step 3. Squaring each value obtained from Step 2, we get 64, 64, 64, 64, 64, 64, 64, 64, 64, 64.

Step 4. Adding all of the preceding squares we get a total of 640.

Step 5. Ten scores are present so we divide 640 by one less than 10. 640/9 $\overset{\circ}{=}$ 71.1. The variance for the scores of Group D is therefore 71.1.

The same procedure applied to the data of Group C should produce a variance of 20.2. (Verification of this is left to the students.) The misleading ranges implied that Group D had a dispersion smaller than that of Group C, but the more reliable variance suggests the correct conclusion that Group C has less dispersion.

Returning to the quiz scores for Group A and Group B, we compute their respective variances as 22.7 and 331.8. While the averages did not distinguish between these two groups, the variances emphatically reflect the different degrees of dispersion. Table 2-8 depicts an organized format in which the scores of Group A are applied to the variance formula.

TABLE 2-8

x	$(x - \bar{x})$	$(x - \bar{x})^2$
65	−6.5	42.25
66	−5.5	30.25
67	−4.5	20.25
68	−3.5	12.25
71	−0.5	0.25
73	1.5	2.25
74	2.5	6.25
77	5.5	30.25
77	5.5	30.25
77	5.5	30.25
715		204.5

$$\bar{x} = \frac{715}{10} = 71.5$$

$$s^2 = \frac{\Sigma (x - \bar{x})^2}{n - 1} = \frac{204.5}{9} = 22.7$$

Formula 2-4 can be expressed in an equivalent form as follows:

Formula 2-5 $$s^2 = \frac{n(\Sigma x^2) - (\Sigma x)^2}{n(n - 1)}$$

Formulas 2-4 and 2-5 are equivalent in the sense that they will always produce the same results. This means that we can select the more convenient of the two formulas when we must compute a variance. Formula 2-5 will probably be more convenient if the mean is either unknown or if it is known to be something other than a whole number. If the mean is a known whole number, Formula 2-4 will probably be more convenient.

Several brands of inexpensive pocket calculators are designed to compute variances automatically. A calculator which computes variances according to Formula 2-5 will require only three memory registers in which n, Σx, and Σx^2 can be stored as the data are entered. (These are the required components of Formula 2-5). If a calculator were to use Formula 2-4, it would need a separate memory register for each score (since all scores must be known before the mean can be obtained), and that is obviously impractical.

In the previous section we developed a procedure for computing the mean when the scores are incorporated in a frequency table. The fundamental importance of measures of dispersion demands that we develop a procedure for finding the variance of scores summarized in a frequency table. In some cases, we may not have access to the original list of raw scores and the frequency table may have the only data available. In other cases, the number of raw scores may be large enough to render any individual treatment of the scores impractical. We again use the class marks and class frequencies and we obtain Formula 2-6, the variance computed from a frequency table.

Formula 2-6 $$s^2 = \frac{n(\Sigma f \cdot x^2) - (\Sigma f \cdot x)^2}{n(n - 1)}$$

where
   x = class mark
   f = class frequency
   n = sample size

To illustrate the use of Formula 2-6, we present Table 2-9 which summarizes the various steps.

TABLE 2-9. Computation of variance of 42 job aptitude test scores.

Score	Frequency f	Class mark x	$f \cdot x$	$f \cdot x^2$ or $f \cdot x \cdot x$
60–64	2	62	124	7,688
65–69	1	67	67	4,489
70–74	4	72	288	20,736
75–79	8	77	616	47,432
80–84	12	82	984	80,688
85–89	9	87	783	68,121
90–94	5	92	460	42,320
95–99	1	97	97	9,409
	$n = \Sigma f$ $= 42$		$\Sigma f \cdot x$ $= 3419$	$\Sigma f \cdot x^2$ $= 280,883$

$$s^2 = \frac{n(\Sigma f \cdot x^2) - (\Sigma f \cdot x)^2}{n(n-1)}$$

$$= \frac{42(280883) - (3419)^2}{42(42-1)}$$

$$= \frac{107525}{1722}$$

$$\stackrel{\circ}{=} 62.4$$

Standard Deviation

DEFINITION

> *The **standard deviation** of a set of scores is the square root of the variance of those scores. That is, the standard deviation s is found by*
>
> $$s = \sqrt{\frac{\Sigma(x-\bar{x})^2}{n-1}} \quad \text{or} \quad \sqrt{\frac{n(\Sigma x^2) - (\Sigma x)^2}{n(n-1)}}$$

If the standard deviation is to be found for a set of scores summarized in a frequency table, we can simply compute the variance using Formula 2-6 and then take the square root of that result to obtain the corresponding standard deviation. For the job aptitude test scores of Table 2-9, the variance was found to be 62.4. Consequently the standard deviation is $\sqrt{62.4} = 7.9$.

Recall that the square root of a number is another number having the property that, when multiplied by itself, the result is the original number. Three is the square root of 9 because multiplying 3 by itself produces 9.

How do we interpret the values we compute for ranges, variances, and standard deviations? The following chart summarizes the measures of dispersion for the quiz scores of Group A and Group B on page 46. What do these numbers really mean?

	Group A	Group B
Range	12	58
Variance	22.7	331.8
Standard deviation	4.8	18.2

It isn't difficult to develop a strong intuitive sense for relating to the range. A **range** of 12 indicates data that span a much smaller gap than comparable data with a range of 58. The **variances** of 22.7 and 331.8 provide good comparative information indicating that Group A has less dispersion, but those numbers alone probably mean very little at this point. Similarly, the two **standard deviations** serve as a good comparison, but they too are without meaning when considered separately. Subsequent discussions will involve these values and make them more meaningful. Exercise 2-93 uses Chebyshev's theorem to impose some meaning on individual standard deviations.

So far in this section we have talked about the three most common measures of dispersion, but there are others. The **mean deviation** is obtained by evaluating $\Sigma|x - \bar{x}| \div n$ where the absolute value signs (vertical lines) require that each value of $(x - \bar{x})$ be recorded as the *positive* difference (or zero) between the score and the mean. The **semi-interquartile range** is obtained by evaluating $(Q_3 - Q_1)/2$ where Q_3 and Q_1 are the quartiles discussed in the previous section. The **10–90 percentile range** is simply $P_{90} - P_{10}$ where P_{90} and P_{10} are the percentile scores we have already discussed. Note that, except for variance, the measures of dispersion are expressed in the same units as the original data. The standard deviation for a list of heights might be 3 inches or 0.25 foot. The variance is in square units; for a set of heights, the variance might be 9 square inches or 0.0625 square foot.

Measures of dispersion are extremely important in many practical circumstances. Manufacturers interested in producing items of consistent quality are very concerned with statistics such as standard deviations. A producer of car batteries might be pleased to learn that a product has a mean life of 4 years, but that pleasure would become distress if the standard deviation indicated a very large dispersion that would correspond to many battery failures long before the mean of 4 years. Quality control requires consistency, and consistency requires a relatively small standard deviation.

There is often a need to compare scores taken from two separate populations with different means and standard deviations. The standard score (or z score) can be used to help make such comparisons.

S M T W T F S   S M T W T F S
JAN   **JULY**
1 2 3 4 5   1 2 3 4 5
6 7 8 9 10 11 12   6 7 8 9 10 11 12
13 14 15 16 17 18 19   13 14 15 16 17 18 19
20 21 22 23 24 25 26   20 21 22 23 24 25 26
27 28 29 30 31   27 28 29 30 31

FEB   **AUG**
1 2   1 2
3 4 5 6 7 8 9   3 4 5 6 7 8 9
10 11 12 13 14 15 16   10 11 12 13 14 15 16
17 18 19 20 21 22 23   17 18 19 20 21 22 23
24 25 26 27 28 29   24 25 26 27 28 29 30
31

MAR   **SEPT**
1   1 2 3 4 5 6
2 3 4 5 6 7 8   7 8 9 10 11 12 13
9 10 11 12 13 14 15   14 15 16 17 18 19 20
16 17 18 19 20 21 22   21 22 23 24 25 26 27
23 24 25 26 27 28 29   28 29 30
30 31

APR   **OCT**
1 2 3 4 5   1 2 3 4
6 7 8 9 10 11 12   5 6 7 8 9 10 11
13 14 15 16 17 18 19   12 13 14 15 16 17 18
20 21 22 23 24 25 26   19 20 21 22 23 24 25
27 28 29 30   26 27 28 29 30 31

MAY   **NOV**
1 2 3   1
4 5 6 7 8 9 10   2 3 4 5 6 7 8
11 12 13 14 15 16 17   9 10 11 12 13 14 15
18 19 20 21 22 23 24   16 17 18 19 20 21 22
25 26 27 28 29 30 31   23 24 25 26 27 28 29
30

JUNE   **DEC**
1 2 3 4 5 6 7   1 2 3 4 5 6
8 9 10 11 12 13 14   7 8 9 10 11 12 13
15 16 17 18 19 20 21   14 15 16 17 18 19 20
22 23 24 25 26 27 28   21 22 23 24 25 26 27
29 30   28 29 30 31

EMPLOYEES WORK FOR THE GOVERNMENT UNTIL MAY 1

The typical taxpayer must work until May 1 to pay federal, state, and local taxes. It takes all of the money earned in the first 4 months to pay the annual tax bills.

DEFINITION

> *The **standard score** or **z score** is the number of standard deviations that a given value is above or below the mean, and it is found by*
>
> $$z = \frac{x - \bar{x}}{s}$$

For example, which is better: a score of 65 on Test A or a score of 29 on Test B? The class statistics for the two tests are as follows:

Test A	Test B
$\bar{x} = 50$	$\bar{x} = 20$
$s = 10$	$s = 5$

For the score of 65 on Test A we get a z score of 1.5 since

$$z = \frac{x - \bar{x}}{s} = \frac{65 - 50}{10} = \frac{15}{10} = 1.5$$

For the score of 29 on Test B we get a z score of 1.8 since

$$z = \frac{x - \bar{x}}{s} = \frac{29 - 20}{5} = \frac{9}{5} = 1.8$$

That is, a score of 65 on Test A is 1.5 standard deviations above the mean while a score of 29 on Test B is 1.8 standard deviations above the mean. This implies that the 29 on Test B is the better score. While 29 is below 65, it has a better *relative* position when considered in the context of the other test results. Later, we will make extensive use of these standard or z scores.

EXERCISES A: DISPERSION STATISTICS

In Exercises 2-64 through 2-75, find the range, variance, and standard deviation for the data in the exercises named. The data are found on pages 42–43.

2-64.	Exercise 2-34.	2-65.	Exercise 2-35.
2-66.	Exercise 2-36.	2-67.	Exercise 2-37.
2-68.	Exercise 2-38.	2-69.	Exercise 2-39.
2-70.	Exercise 2-40.	2-71.	Exercise 2-41.
2-72.	Exercise 2-42.	2-73.	Exercise 2-43.
2-74.	Exercise 2-44.	2-75.	Exercise 2-45.

2-76. The following scores represent the number of speeding tickets given by ten different police officers in one week. Find the range, variance, and standard deviation.

$$3, 3, 4, 5, 6, 6, 7, 8, 9, 9$$

2-77. Add 14 to each score given in Exercise 2-76 and then find the range, variance, and standard deviation. Compare the results to those of Exercise 2-76.

2-78. Subtract 3 from each score given in Exercise 2-76 and then find the range, variance, and standard deviation. Compare the results to those of Exercise 2-76.

2-79. Multiply each score given in Exercise 2-76 by 10 and then find the range, variance, and standard deviation. Compare the results to those of Exercise 2-76.

2-80. Double each score in Exercise 2-76 and then find the range, variance, and standard deviation. Compare the results to those of Exercise 2-76.

2-81. Which would you expect to have a higher variance: the I.Q. scores of a class of 25 statistics students or the I.Q. scores of 25 randomly selected adults? Explain.

2-82. Which would you expect to have a higher standard deviation: the reaction times of 30 sober adults or the reaction times of 30 adults who have recently consumed three martinis each?

2-83. Is it possible for a set of scores to have a variance of zero? If so, how? Is it possible for a set of scores to have a negative variance?

2-84. The scores on a test have a mean of 72 and a standard deviation of 8. For each of the given test scores, find the corresponding z score.

(a) 80 1
(b) 84 1.5
(c) 60 -1.5
(d) 52 -2.5

2-85. Assume that I.Q. scores have a mean of 100 and a standard deviation of 15. For each of the given I.Q. scores, find the corresponding z score.

(a) 120
(b) 100
(c)   70
(d) 127
(e)   82

EXERCISES B:
DISPERSION STATISTICS

In Exercises 2-86 through 2-88, use a calculator or computer to find the range, variance, and standard deviation for the data in the exercises named. The data are found on pages 19–20.

2-86. Exercise 2-5.

2-87. Exercise 2-6.

2-88. Exercise 2-7.

2-89. Find the mean deviation, the semi-interquartile range, and the 10–90 percentile range for the data in Exercise 2-34.

2-90. Find the mean deviation, the semi-interquartile range, and the 10–90 percentile range for the data in Exercise 2-35.

2-91. Find the variance for the data found in Table 2-6. (Hint: use class marks.)

2-92. Find the mean and standard deviation for the scores 18, 19, 20, . . . , 180, 181, 182.

2-93. Chebyshev's theorem states that the *proportion* of any set of data lying within K standard deviations of the mean is always *at least* $1 - 1/K^2$ where K is any positive number greater than 1.
   (a) Given a mean I.Q. of 100 and a standard deviation of 15, what does Chebyshev's theorem say about the number of scores within two standard deviations of the mean (i.e., 70–130)?
   (b) Given a mean of 100 and a standard deviation of 15, what does Chebyshev's theorem say about the number of scores between 55 and 145?
   (c) The mean score on the College Entrance Examination Board Scholastic Aptitude Test is 500 and the standard deviation is 100. What does Chebyshev's theorem say about the number of scores between 300 and 700?
   (d) Using the data of part (c), what does Chebyshev's theorem say about the number of scores between 200 and 800?

REVIEW

Chapter 2 deals mainly with the methods and techniques of descriptive statistics. The main objective of this chapter is to develop the ability to organize, summarize, and illustrate data, and to extract from data some meaningful measurements. In Section 2-2 we considered the **frequency table** as an excellent device for summarizing data, while Section 2-3 dealt with graphic illustrations, including **pie charts, histograms, frequency polygons,** and **ogives**. These visual illustrations help us to determine the position and distribution of a set of scores. In Section 2-4 we defined the common **averages** or measures of central tendency. The **mean, median, mode,** and **midrange** represent different notions as to what constitutes the central value of a collection of data. The **weighted mean** is used to find the average of a set of scores that may vary in importance. Section 2-4 included the common measures of position: **quartiles, deciles,** and **percentiles.** In Section 2-5 we presented the usual **measures of dispersion,** including the **range, variance,** and **standard deviation;** these descriptive statistics are designed to measure the variability among a set of scores. The **standard score** or *z* **score** was introduced as a way of measuring the number of standard deviations by which a given score differs from the mean.

By now you should be able to organize, present, and describe collections of data composed of single scores. You should be able to compute the key descriptive statistics that will be used in later applications.

The following are the most important of the formulas presented in this chapter:

$$\bar{x} = \frac{\Sigma x}{n}$$
Mean

$$\bar{x} = \frac{\Sigma f \cdot x}{\Sigma f}$$
Computing the mean when the data is in a frequency table

$$s^2 = \frac{\Sigma (x - \bar{x})^2}{n - 1}$$
Variance

or

$$s^2 = \frac{n(\Sigma x^2) - (\Sigma x)^2}{n(n - 1)}$$
Short-cut formula for variance

$$s^2 = \frac{n(\Sigma f \cdot x^2) - (\Sigma f \cdot x)^2}{n(n - 1)}$$
Computing the variance when the data is in a frequency table.

The standard deviation is the square root of the variance.

$$z = \frac{x - \bar{x}}{s}$$
Standard score or *z* score

REVIEW EXERCISES

The given scores represent the number of cars rejected in 1 day at an automobile assembly plant; the 50 scores correspond to 50 different days.

29	58	80	35	30	23	88	49	35	97
12	73	54	91	45	28	61	61	45	81
83	23	71	63	47	87	36	8	94	26
95	63	86	42	22	44	8	27	20	33
28	91	87	15	67	10	45	67	26	19

2-94. Given the preceding scores, construct a frequency table with 10 classes. Identify the class limits, class boundaries, class marks, and class width.

2-95. Construct a histogram that corresponds to the frequency table from Exercise 2-94.

2-96. For the preceding scores, find the:
   (a) Mean
   (b) Median
   (c) Mode
   (d) Midrange

2-97. For the preceding scores, find:
   (a) Q_1
   (b) Q_3
   (c) D_3
   (d) P_{45}

2-98. Using only the first two columns of 10 scores, find the:
   (a) Range
   (b) Variance
   (c) Standard deviation

2-99. Use the frequency table from Exercise 2-94 to find the mean and standard deviation for the number of rejects.

PROBABILITY

3-1 OVERVIEW

Suppose you discover a way to mark each molecule in an 8-oz glass of water so that each one is recognizable. You then proceed to the nearest ocean beach and pour the water into the first wave. After waiting 20 years for the waters of the world to mix, you board a plane for Venice, Italy. Upon your arrival, you scoop a glass of water from the first canal you come to and examine the water in search of one of the molecules that you dumped in 20 years ago. Can you really expect to find one of the molecules in Venice? As absurd as this problem seems, the answer is actually yes. (If you find this incredible, you are not alone.) In fact, you can expect about 1000 of the original molecules to reappear in Venice.

In a class of 25 students, each is asked to identify the month and day of his or her birth. What are the chances that at least two students will share the same birthday? Again our intuition is misleading; it happens that at least two students will have the same birthday in more than half of the classes with 25 students.

The preceding conclusions are based upon simple principles of probability which play a critical role in the theory of statistics. These introductory examples are not at all intuitively obvious, and they do require nontrivial calculations.

All of us now form simple probability conclusions in our daily lives. Sometimes these determinations are based on fact, while other probability determinations are subjective. In addition to being an important aspect of the study of statistics, probability theory is playing an increasingly important role in a society that must attempt to measure uncertainties.

In Chapter 1 we stated that inferential statistics involves the use of sample evidence in formulating inferences or conclusions about a population. These inferential decisions are based on probabilities or likelihoods of events. As an example, let's suppose that a statistician plans to test a coin for fairness. She tosses this coin 100 times, and the result is heads each time. Either the coin is fair and an *extremely* rare event has occurred, or the coin is not fair and it favors heads. What should we infer? The statistician, along with most reasonable people, would conclude that the coin is not fair. This decision is based on the very low probability of getting 100 consecutive heads with a normal coin.

In subsequent chapters we develop methods of statistical inference that rely on this type of thinking. It is therefore important to acquire a basic understanding of probability theory. We want to cultivate some intuitive feeling for what probabilities are, and we want to develop some very basic skills in calculating the probabilities of certain events (see Figure 3-1).

While probability theory is important for statistical inferences, we should note a fundamental difference between probability theory and the theory of statistical inferences. **In a typical probability problem, the population is known and we want to determine the likelihood of observing a**

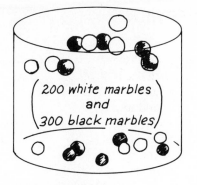

(a)   Probability

(b)   Statistics

With <u>known contents</u>, we can see that if one marble is randomly selected by a blindfolded person, there are 200 chances out of 500 of picking a white marble. This corresponds to a probability of 200/500 or 0.4.) The population of all marbles in the jar is known, and we are concerned with the likelihood of obtaining a particular sample (that is, a white marble). We are making a conclusion about a sample based upon our knowledge of the population.

With unknown contents, we can obtain a fairly good idea of what is inside by randomly selecting one hundred marbles and assuming that this sample is representative of the whole. After making the random selections, our sample is known and we can use it to make inferences about the population of all marbles in the jar. We are making a conclusion about the population based upon our knowledge of the sample.

FIGURE 3-1

sample; in a typical statistical inference problem, a sample is known and we want to form a conclusion about the population. In this sense, probability and statistics are opposites.

3-2 FUNDAMENTALS

In considering probability problems, we deal with experiments and events.

DEFINITION

> *An **experiment** is a process that allows us to obtain observations.*

YOU BET?

Only 40% of the money bet in the New York State lottery is returned in prizes, so the "house" has a 60% advantage. The house advantage at racetracks is usually about 15%. In casinos, the house advantage is 5.26% for roulette, 5.9% for blackjack, 1.4% for craps, and 3% to 22% for slot machines. Some professional gamblers can systematically beat the house at blackjack if they use sophisticated card-counting techniques. The basic idea is to keep a record of cards that have been played so that the player will know when the deck has a disproportionately large number of high cards. When this occurs, the casino is more likely to lose and the player begins to place large bets. Many casinos eject card counters as soon as they detect them.

DEFINITION

> *An **event** is a result or outcome of an experiment.*

For example, the rolling of a single die is an experiment, and the occurrence of an odd number is an event. In this case, the event of an odd number will occur if the die shows a 1, 3, or 5. There is often a need to consider events which cannot be decomposed into simpler outcomes.

DEFINITION

> *A **simple event** is an outcome or an event which cannot be broken down any further.*

When conducting the experiment of rolling a single die, the occurrence of a 5 is a simple event, but the occurrence of an odd number is not a simple event since it can be broken down into the three simple events of 1, 3, or 5. When flipping two coins, there are four simple events:

heads-heads
heads-tails
tails-heads
tails-tails

DEFINITION

> *The **sample space** for an experiment consists of all possible simple events. That is, the sample space consists of all outcomes which cannot be broken down any further.*

There is no universal agreement for the actual definition of the probability of an event, but among the various theories and schools of thought, two basic approaches emerge most often. The approaches will be embodied in two rules for finding probabilities. The notation employed will relate P to probability, while capital letters such as A, B, and C will denote specific events. For example, A might represent the event of winning a million-dollar state lottery; $P(A)$ denotes the probability of event A occurring.

RULE 1

Empirical approximation of probability. *Conduct (or observe) an experiment a large number of times and count the number of times that event A actually occurs. Then*

$$P(A) = \frac{\text{number of times } A \text{ occurred}}{\text{number of times experiment was repeated}}$$

RULE 2

Classical approach to probability. *Assume that a given experiment has n different simple events, each of which has an **equal chance** of occurring. If event A can occur in s of these n ways, then*

$$P(A) = \frac{s}{n}$$

The first rule is often referred to as the **empirical** or **relative frequency** interpretation of probability. The second rule, often referred to as the **classical** approach, is useful in the analysis of simple experiments that involve equally likely outcomes, such as flipping coins, rolling dice, etc. (see Figure 3-2).

Complicated experiments, such as those involving events that are not equally likely, all require the relative frequency approximation. For example,

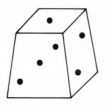

When trying to determine P(2) on a "shaved" die, we must repeat the experiment of rolling it many times and then form the ratio of the number of times 2 occurred to the total number of rolls.

(a) *Empirical approach (Rule 1)*

FIGURE 3-2

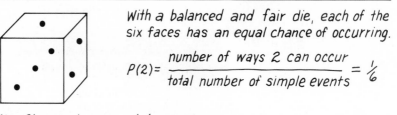

With a balanced and fair die, each of the six faces has an equal chance of occurring.

$$P(2) = \frac{\text{number of ways 2 can occur}}{\text{total number of simple events}} = \frac{1}{6}$$

(b) *Classical approach (Rule 2)*

to determine the probability of an 18-year-old male living to age 65, we must study past records in order to arrive at a reasonable approximation. The events of living to age 65 and dying before age 65 are not equally likely, so we cannot use Rule 2.

Many experiments that do involve equally likely outcomes are so complicated that Rule 1 is used as a practical alternative to extremely complex computations. Consider, for example, the probability of winning at solitaire. In theory, Rule 2 applies, but in reality the better approach requires that we settle for an approximation obtained through Rule 1. That is, we should play solitaire many times and record the number of wins along with the number of attempts. The ratio of wins to trials is the approximation we seek.

The following examples are intended to illustrate the use of Rules 1 and 2. In some of these examples we use the term random. Throughout this chapter, a random selection means that each element that can be chosen has an equal chance of being chosen.

Example   Twenty smokers and 30 nonsmokers are given routine medical examinations. If one of the subjects is randomly selected for more detailed tests, what is the probability that the selected subject smokes?

Solution   When one subject is selected, there are 50 possible outcomes (simple events) of which 20 correspond to the event in question. Assuming that each subject has the same chance of being chosen, Rule 2 applies and we get

$$P(\text{smoker is selected}) = \frac{20}{50} \quad \text{or} \quad 0.4$$

Example

On a college entrance examination, each question has 4 possible answers. If an examinee makes a random guess on the last question, what is the probability that the response is incorrect?

Solution

There are 4 possible outcomes or answers, and there are 3 ways to answer incorrectly. Random guessing implies that the outcomes are equally likely, so we apply Rule 2 to get

$$P(\text{wrong answer}) = \frac{3}{4} \quad \text{or} \quad 0.75$$

Example

Find the probability of a 20-year-old student living to be 21 years of age.

Solution

Here the two outcomes of living and dying are not equally likely, so the relative frequency approximation must be used. This requires that we observe a large number of 20-year-old students and then count those who live to be 21. Suppose that we survey 10,000 20-year-old students and find that 9961 lived to be 21 (these are realistic figures). We then conclude that

$$P(\text{20-year-old student living to 21}) \doteq \frac{9,961}{10,000} \quad \text{or} \quad 0.9961$$

This is the basic approach used by insurance companies in the development of mortality tables.

Example

A magazine is developing a telephone solicitation department. What is the probability of getting an order on any given call?

Solution

Again the outcomes of getting or not getting an order are not equally likely, so the relative frequency approximation applies. In essence, past experience is projected, so we need to accumulate some experience before making any conclusions. Let's suppose that on the first 100 calls, there are 17 orders. We therefore state that

$$P(\text{order}) = \frac{17}{100} \quad \text{or} \quad 0.17$$

Example

Find the probability that a couple having 3 children will have exactly 2 boys.

Solution

If the couple has exactly 2 boys in 3 births, there must be 1 girl. The possible outcomes are listed and each is assumed to be equally likely. In reality,

$$P(\text{boy}) \doteq \frac{105}{205} \doteq 0.51 \text{ instead of } 0.5$$

boy-boy-boy   girl-boy-boy
boy-boy-girl   girl-boy-girl
boy-girl-boy   girl-girl-boy
boy-girl-girl   girl-girl-girl

Of the 8 different possible outcomes, 3 correspond to exactly 2 boys.

$$P(2 \text{ boys in 3 births}) = \frac{3}{8} \quad \text{or} \quad 0.375$$

Example   Find the probability of rolling one die and getting an even number.

Solution   A die is a cube with dots on each side that represent the numbers 1 through 6. When the die is cast, the number that faces up can be any one of 6 equally likely outcomes: 1, 2, 3, 4, 5, 6. Applying Rule 2, we see that 3 of the possible outcomes are even so that

$$P(\text{even number}) = \frac{3}{6} = \frac{1}{2} \quad \text{or} \quad 0.5$$

Example   Find the probability of rolling one die and getting 8.

Solution   Of the 6 possible outcomes, there are 0 ways of getting 8, so

$$P(8) = \frac{0}{6} = 0$$

The probability of any impossible event is 0.

Example   Find the probability of rolling one die and getting a number between 0 and 8.

Solution   Of the 6 possible outcomes, there are 6 ways of getting a number between 0 and 8, so

$$P(\text{number between 0 and 8}) = \frac{6}{6} = 1$$

The probability of any event that is certain to occur is 1.

Since any event imaginable is either impossible, certain, or somewhere in between, it follows that the mathematical probability of any event is either 0, 1, or a number between 0 and 1 (see Figure 3-3).

This property can be expressed as follows.

$$0 \leq P(A) \leq 1 \quad \text{for any event } A$$

FIGURE 3-3
Possible values for
probabilities.

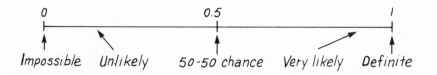

Example

Your 6-cylinder car is running roughly because of one faulty spark plug, and you have enough time to check and replace only 1 spark plug. What is the probability that you will select the defective one?

Solution

There are 6 possible outcomes corresponding to the 6 different spark plugs. Since 1 plug is defective, we conclude that

$$P(\text{defective plug}) = \frac{1}{6} = 0.167$$

Example

An experiment in ESP involves blindfolding the subject and allowing him to make 1 selection from a standard deck of cards. The subject must then identify the correct suit (i.e., clubs, diamonds, hearts, or spades). What is the probability of making a correct random guess?

Solution

The standard deck contains 52 cards with 13 of each suit. There are therefore 52 possible outcomes, with 13 outcomes corresponding to whatever suit the subject selects. Therefore

$$P(\text{correct guess}) = \frac{13}{52} = \frac{1}{4} = 0.25$$

The 1/4 probability of the previous example implies that, on the average, for each four guesses, there will be one correct selection. If our ESP subject does *significantly* better than one correct response in four, then we can conclude that his selections are not made solely on the basis of chance. But what is significant? Thirty correct guesses in 100 trials? Fifty correct guesses in 100 trials? Ninety-nine correct guesses in 100 trials? This decision relates to a key concept of statistical inference and will be explored later in the book.

GAMBLING PAYS

In 1978, the State of Nevada collected $106 million from gambling taxes. In order to produce that amount of tax dollars, the 10 million gamblers had to lose 1671 million dollars in the 108 casinos. Nevada residents pay no income, sales, or inheritance taxes.

EXERCISES A: FUNDAMENTALS

3-1.   In a recent national election, there were 11,022,000 citizens in the 18-to-20-year age bracket. Of these, 5,318,000 actually voted. Find the probability that a person randomly selected from this group of 18-to-20-year-old citizens did vote in that national election.

3-2.   In the 1972 presidential elections, 47 million votes were in favor of the Republican candidate (Nixon), while 27 million votes were in favor of the Democratic candidate (McGovern). Find the probability that a voter selected at random cast their ballot in favor of Nixon.

3-3.   A couple plans to have 2 children.
(a)   List the different possible outcomes according to the sex of each child. Assume that these outcomes are equally likely.

(b) Find the probability of getting 2 boys.

(c) Find the probability of getting exactly 1 child of each sex.

3-4. A couple plans to have 4 children.

(a) List the 16 different possible outcomes according to the sex of each child. Assume that these outcomes are equally likely.

(b) Find the probability of getting all girls.

(c) Find the probability of getting *at least* 1 child of each sex.

(d) Find the probability of getting *exactly* 2 children of each sex.

3-5. Which of the following could *not* be a probability?

3/7, 2, −1/2, 3/4, 99/101, 0, 1, 5, 1.11, 1.0001, 0.001, 0.9999

3-6. A vaccine is administered to 87 subjects and found to be effective in 56 cases. What is the approximate probability that the vaccine will be effective?

3-7. A traffic survey indicates that of 3,756 cars approaching a shopping plaza, 857 turned into the parking lot. Find the empirical probability of a car *not* entering the parking lot.

3-8. A quick quiz consists of 3 true-false questions and an unprepared student must guess at each one. The guesses will be random.

(a) List the different possible solutions.

(b) What is the probability of answering the three questions correctly?

(c) What is the probability of guessing incorrectly for all three questions?

(d) What is the probability of passing by guessing correctly for at least two questions?

3-9. A farmer plants 1000 seeds of which 451 are productive. What is the empirical probability of a similar seed being nonproductive?

3-10. Both parents have the brown-blue pair of eye-color genes, and each parent contributes one gene to a child. Assume that if the child has at least one brown gene, that color will dominate and the eyes will be brown. (Actually, the determination of eye color is somewhat more complex.)

(a) List the different possible outcomes. Assume that these outcomes are equally likely.

(b) What is the probability that a child of these parents will have the blue-blue pair of genes?

(c) What is the probability that a child will have brown eyes?

3-11. Find the probability of selecting a slip indicating a date in May if 1 slip is randomly drawn from a bowl containing 366 slips denoting the different possible dates in a leap year.

3-12. A roulette wheel has 38 possible outcomes consisting of 0, 00, and the numbers 1 through 36. What is the probability of *randomly* getting an odd number?

3-13. What is the probability of getting 0 or 00 on the roulette wheel of Exercise 3-12?

3-14. What is the probability of getting a number *greater than* 20 on the roulette wheel of Exercise 3-12?

3-15. For each of the following events, determine whether the situation involves equally likely outcomes:
   (a) Passing a statistics course.
   (b) Rolling a die and getting 3.
   (c) Living from 18 years of age to 39 years of age.
   (d) Winning the Master's Golf Tournament.
   (e) Bowling a 300 game.
   (f) Winning the New York State Lottery.
   (g) Picking the winning number on a roulette wheel.
   (h) Rolling 7 on a pair of dice.
   (i) Having an auto accident in the next 3 years.
   (j) Selecting the winning horse at a race track.

EXERCISES B: FUNDAMENTALS

3-16. On a distant planet, one parent from each of three different sexes is necessary for reproduction. Each parent contributes one gene from a triplet of genes that determine the color of a child's antenna. Three parents have the following genes:

   red-white-blue
   red-white-white
   red-red-white

   (a) List the 27 different possible outcomes. Assume that these outcomes are equally likely.
   (b) What is the probability that a child receives the blue-red-red combination of genes?
   (c) The color of a child's antenna is determined by majority. For example, a child receiving the blue-white-white combination would have a white antenna. If one of each color is present, the antenna will be blue. What is the probability that a child has a white antenna?
   (d) What is the probability that a child has a blue antenna (See part (c).)
   (e) What is the probability that a child has an antenna that is not white? (See part (c).)

3-17. Determine empirically the approximate probability that a tossed thumbtack will land with the point up.

3-18. A red die and a white die are both tossed:
   (a) List the 36 different possible outcomes. Assume that these outcomes are equally likely.

(b)   Find the probability of rolling a total of 7.

(c)   Find the probability of rolling a total of 3.

(d)   Find the probability of rolling a total of 1.

(e)   Find the probability of rolling a total that exceeds 1.

(f)   Find the probability of rolling a total of 7 if the white die is known to be a 4.

(g)   Find the probability of rolling a total that exceeds 6 if the white die is known to be a 3.

3-3 ADDITION RULE

The first section, designed to introduce the basic concept of probability, dealt largely with simple experiments and simple events. However, reality dictates that we also consider compound events and some of their attendant rules.

DEFINITION

> *A **compound event** is an amalgam or combination of simple events.*

For example, rolling an odd number on a single die can be considered a compound event since it corresponds to the simple events of 1, 3, or 5; 1, 3, and 5 combine to form the category of "odd number." Given any two events *A* and *B*, the two most basic ways they can be combined into a compound event are:

A or *B*

A and *B*

For example, suppose that *A* is the event of selecting a woman while *B* is the event of selecting a Republican. Then (*A* or *B*) is the event of selecting someone who is a woman or a Republican or both, while (*A* and *B*) is the event of selecting a woman Republican. This section treats compound events of the form (*A* **or** *B*), while the following section deals with compound events of the form (*A* **and** *B*). (Throughout this chapter, we use the **inclusive or**, which means "either one, or the other, or both." We will *not* consider the exclusive "or," which means "either one, or the other, but not both.")

Our initial concern is the development of a rule for finding a probability of the form *P*(*A* or *B*). Let's consider an example that will help us to understand the underlying rationale that justifies the rule we will present. We define our experiment to be the rolling of one fair die, and we will let *A* represent the event of rolling an odd number while *B* represents the event of rolling a number less than 4. Thus, the occurrence of (*A* or *B*) is the occurrence of "an odd number or a number less than 4." Similarly, *P*(*A* or *B*) is the probability of rolling an odd number or a number less than 4. Let us now determine that probability through a direct and simple analysis. We know that the single die has 6 possible outcomes, so we can see which outcomes are either odd or less than 4 (or both).

> 1 is an odd number or a number less than 4.
> 2 is an odd number or a number less than 4.
> 3 is an odd number or a number less than 4.
> 4 is neither odd nor less than 4.
> 5 is an odd number or a number less than 4.
> 6 is neither odd nor less than 4.

Examining the preceding list we see that of the 6 possible outcomes, exactly 4 of them are odd or less than 4. This means that of the 6 simple events that comprise the sample space, there are exactly 4 ways that (A or B) can occur. Since the 6 simple events are all equally likely, Rule 2 of Section 3-2 can be applied and we get $P(A$ or $B) = 4/6$.

If all problems were as simple and direct as this, we would have no real need for more abstract rules and formulas. But important problems of greater complexity do require formal generalized rules.

Let's use this last example as a basis for constructing one such rule. We know that $P(A) = P(\text{odd number}) = 3/6$ and $P(B) = P(\text{number less than 4}) = 3/6$ from Section 3-2. We also know that $P(A$ or $B) = 4/6$. It would be nice if we could express $P(A$ or $B)$ in terms of $P(A)$ and $P(B)$, but how can we do that?

$P(A$ or $B) = 4/6$ does not seem to suggest any combination of $P(A) = 3/6$ and $P(B) = 3/6$. The problem really reduces to that of counting the number of ways A can occur (3), the number of ways B can occur (3), and getting a total of 4.

The obstacle here is the overlapping of A and B. There are two ways that A and B will both occur simultaneously. If either 1 or 3 is the outcome of the roll of the die, then we have an odd number that is also less than 4. **When combining the number of ways A can occur with the number of ways B can occur, we must avoid double counting of those outcomes wherein A and B happen simultaneously.** If we were to state that $P(A$ or $B) = P(A) + P(B)$ in this example, we would be wrong because $4/6 \neq 3/6 + 3/6$. The error arises from the two outcomes of 1 and 3 which were each counted twice. The last equation is corrected by a subtraction which compensates for that double counting:

$$\frac{4}{6} = \frac{3}{6} + \frac{3}{6} - \frac{2}{6}$$

The right-hand side of this last equation reflects the three ways of getting an odd number, the three ways of getting a number 4, and the compensating subtraction of the two ways of getting both. This corrected equation can be generalized as follows.

Addition rule $P(A$ or $B) = P(A) + P(B) - P(A$ and $B)$

FIGURE 3-4
The probability of *A* or *B* equals the probability of *A* (left circle) plus the probability of *B* (right circle) minus the probability of *A* and *B* (football-shaped middle region). This figure should show that the addition of the areas of the two circles will cause a double counting of the football-shaped middle region. To compensate for that double counting, we subtract the area of that region. This is the basic concept that underlies the addition rule.

See Figure 3-4. Remember that $P(A$ and $B)$ is the probability that A and B both occur as an outcome in an experiment.

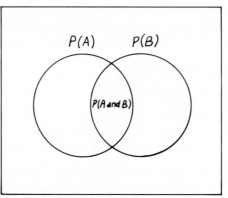

Total area = 1

$P(A)$ $P(B)$

$P(A$ and $B)$

Example

A pollster surveys 100 subjects consisting of 40 Democrats (of which half are female) and 60 Republicans (of which half are female). Find the probability of randomly selecting one of these subjects and getting a Democrat or a female (see Figure 3-5).

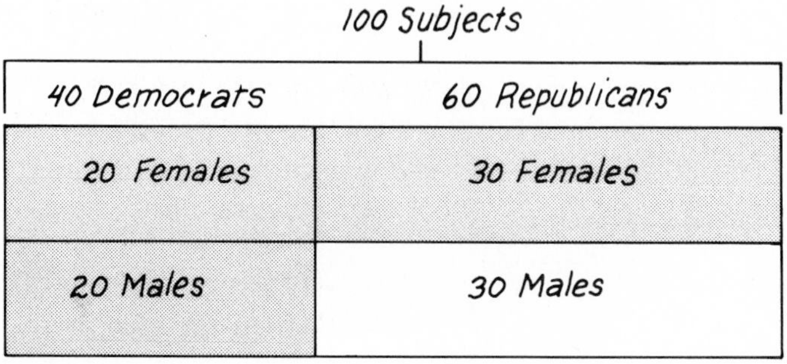

100 Subjects

40 Democrats	60 Republicans
20 Females	30 Females
20 Males	30 Males

FIGURE 3-5
Shaded region represents "Democrats or females."

Solution

By the addition rule,

P(Democrat or female)

$$= P(\text{Democrat}) + P(\text{female}) - P(\text{Democrat and female})$$

We now proceed to evaluate the terms on the right side of this equation. Of the 100 subjects there are 40 Democrats, so $P(\text{Democrat}) = 40/100$. The total number of females is 50, so $P(\text{female}) = 50/100$. Also, there are 20 Democratic females (half of the 40) so that $P(\text{Democrat and female}) = 20/100$.

Therefore,

P(Democrat or female)

$$= P(\text{Democrat}) + P(\text{female}) - P(\text{Democrat and female})$$

$$= \frac{40}{100} + \frac{50}{100} - \frac{20}{100}$$

$$= \frac{70}{100}$$

See Figure 3-5.

The addition rule is simplified whenever A and B cannot occur simultaneously so that $P(A \text{ and } B)$ becomes zero.

DEFINITION

> *Events A and B are **mutually exclusive** if they cannot occur simultaneously.*

In the experiment of rolling one die, the event of getting a 2 and the event of getting a 5 are mutually exclusive events, since no outcome can be both a 2 and 5 simultaneously. The following pairs of events are other examples of mutually exclusive pairs in a single experiment:

$\begin{cases} \text{Selecting a voter who is a registered Democrat.} \\ \text{Selecting a voter who is a registered Republican.} \end{cases}$

$\begin{cases} \text{Testing a subject with an I.Q. above 100.} \\ \text{Testing a subject with an I.Q. below 95.} \end{cases}$

$\begin{cases} \text{Drawing a single club from a deck of cards.} \\ \text{Drawing a single heart from a deck of cards.} \end{cases}$

The following pairs of events are examples that are *not* mutually exclusive in a single trial:

$\begin{cases} \text{Selecting a voter who is a registered Democrat.} \\ \text{Selecting a voter who is under 30 years of age.} \end{cases}$

$\begin{cases} \text{Testing a subject with an I.Q. above 100.} \\ \text{Testing a subject with an I.Q. above 110.} \end{cases}$

$\begin{cases} \text{Drawing a single club from a deck of cards.} \\ \text{Drawing a single ace from a deck of cards.} \end{cases}$

Example   One die is rolled. Find the probability that the outcome is a 2 or 5.

Solution   We have already noted that 2 and 5 are mutually exclusive events. This means that it is impossible for both 2 and 5 to occur together, so $P(2 \text{ and } 5) = 0$ and the addition law is applied as follows:

$$P(2 \text{ or } 5) = P(2) + P(5) - P(2 \text{ and } 5)$$

$$= \frac{1}{6} + \frac{1}{6} - 0$$

$$= \frac{2}{6}$$

In Figure 3-6 we see that the addition rule can be simplified when the events in question are known to be mutually exclusive. The following examples involve application of the addition rule.

Total Area = 1

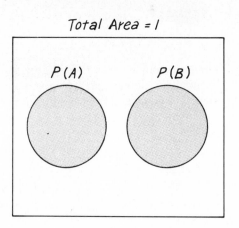

FIGURE 3-6
If events A and B are known to be mutually exclusive. They are completely disjointed and involve no overlapping. When this occurs, the addition rule can be simplified to $P(A \text{ or } B) = P(A) + P(B)$.

Example   In a convention of 100 guests, there are 37 Democrats and 41 Republicans. Find the probability that a conventioneer selected at random is a Democrat or a Republican.

Solution   If D is the event of selecting a Democrat and R is the event of selecting a Republican, then D and R are mutually exclusive, so

$$P(D \text{ or } R) = P(D) + P(R) = \frac{37}{100} + \frac{41}{100} = \frac{78}{100}$$

We can ignore the subtraction of $P(D \text{ and } R)$. That probability is zero since D and R are mutually exclusive events.

NUCLEAR POWER PLANT HAS AN UNPLANNED "EVENT"

In 1974 a team of scientists, led by an M.I.T. professor of nuclear engineering, estimated the likelihood of a serious nuclear accident to be about one in a million. At 4:00 A.M. on Wednesday, April 4, 1979, a siren signaled the beginning of an "event" at a nuclear power plant on Three Mile Island in Pennsylvania. Residents of the area prepared for a full-scale evacuation since there was a danger of a major catastrophe which, fortunately, never materialized. This accident proved that the backup and fail-safe systems were not as infallible as nuclear energy proponents had claimed. There were at least five separate equipment failures. The ensuing reassessment of the likelihood of a nuclear catastrophe is certain to have a strong impact on future energy policies. Scientists will undoubtedly conduct a careful review of the design of nuclear power plants, including the use of redundant or backup components. Compare the two systems below and determine which is likely to be more reliable. Assume that p, q, r, and s are valves and that each system supplies water necessary for cooling purposes.

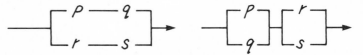

Example In the same convention of 100 guests with 37 Democrats and 41 Republicans, there are 35 conventioneers under 30 years of age, including 17 of the Democrats and 13 of the Republicans. Find the probability that a conventioneer selected at random is a Democrat or is under 30 years of age.

Solution Let Y denote the event of selecting someone under 30 years of age. We seek $P(D \text{ or } Y)$, but D and Y are not mutually exclusive because they do occur simultaneously whenever a Democrat under 30 is chosen. Consequently,

$$P(D \text{ or } Y) = P(D) + P(Y) - P(D \text{ and } Y)$$

$$= \frac{37}{100} + \frac{35}{100} - \frac{17}{100} = \frac{55}{100}$$

This example illustrates the method of compensating for the double counting that inevitably results when overlapping events are involved.

Example The 366 different possible birthdays are written on separate slips of paper and mixed in a bowl. Find the probability of making one selection that is a birthday in May or November.

Solution Let M denote the event of drawing a May date, while N denotes the event of drawing a November date. Clearly, M and N are mutually exclusive because no date is in both May and November. Thus

$$P(M \text{ or } N) = P(M) + P(N) = \frac{31}{366} + \frac{30}{366} = \frac{61}{366}$$

Again the subtraction of $P(M \text{ and } N)$ can be ignored since it is zero.

Example Using the same population of 366 different birthdays, find the probability of making one selection that is the first day of a month or a November date.

Solution Let F denote the event of selecting a date that is the first of the month. Now F and N are not mutually exclusive because they can occur simultaneously (as on November 1). Thus

$$P(F \text{ or } N) = P(F) + P(N) - P(F \text{ and } N)$$

$$= \frac{12}{366} + \frac{30}{366} - \frac{1}{366}$$

$$= \frac{41}{366}$$

If there is any doubt as to whether the subtraction of $P(A \text{ and } B)$ should be included, then include it. If it turns out that A and B are mutually exclusive, then $P(A \text{ and } B) = 0$, so the formula automatically reduces to $P(A \text{ or } B) =$

$P(A) + P(B)$. Mistakes are made when the subtraction of $P(A$ and $B)$ is ignored in situations that require it. That is, when events that are not mutually exclusive are treated as being mutually exclusive! One positive indication of such an error is a total probability that exceeds 1, but errors involving the addition rule do not necessarily cause the probability to exceed 1.

EXERCISES A: ADDITION RULE

3-19. For each pair of events given, determine whether the two events are mutually exclusive for a single experiment:
   (a) Selecting a dominant personality type.
       Selecting a submissive personality type.
   (b) Selecting a voter who favors legalization of marijuana.
       Selecting a Conservative.
   (c) Selecting a blonde.
       Selecting a person with brown eyes.
   (d) Selecting a woman.
       Selecting a surgeon.
   (e) Selecting an honest politician.
       Selecting an incumbent politician.
   (f) Selecting an ace from a deck of cards.
       Selecting a red card from a deck of cards.
   (g) Selecting a registered voter.
       Selecting someone over 65 years of age.
   (h) Selecting a required course.
       Selecting an elective course.
   (i) Selecting a novel.
       Selecting a biography.
   (j) Selecting a summer day.
       Selecting a legal holiday.

3-20. A neighborhood is comprised of 82 Italians, 75 Irish, 40 Germans, and 2 Austrians. Find the probability of randomly selecting one resident and getting an Italian or an Austrian.

3-21. An empty pond is filled with water and stocked with 500 trout, 400 bass, and 300 sunfish. Find the probability of randomly catching the first fish and getting a bass or a sunfish.

3-22. A statistics class is attended by 12 psychology majors, 8 business majors, 4 biology majors, and 6 education majors. Find the probability of randomly selecting one student and getting a biology or business major.

3-23. Using the statistics class of Exercise 3-22, find the probability of randomly selecting one student and getting a psychology or biology or education major.

3-24. A statistics class contains 15 females and 12 males. Five of the females are business majors while the remaining females are psychology majors. All of the males are psychology majors. Find the probability of randomly selecting a student who is a psychology major or a female.

3-25. Using the statistics class of Exercise 3-24, find the probability of randomly selecting a student who is a business major or a male.

3-26. Using the statistics class of Exercise 3-24, find the probability of randomly selecting a student who is a psychology major or a male.

3-27. An insurance firm serves 5673 clients, of which 4138 are males. There are 2565 clients under 30 years of age, and 1875 of these are males. Find the probability of making one random selection and getting a female or a client under 30 years of age.

3-28. Using the clients from Exercise 3-27, find the probability of making one random selection and getting a male or a client under 30 years of age.

3-29. A psychology study related to personality characteristics involves 1000 subjects, of which 200 are neurotic, 800 are submissive, and 155 are both neurotic and submissive. What is the probability that a person selected at random from this group is either neurotic or submissive?

3-30. Find the probability of rolling one die and getting an even number or a number less than 5.

3-31. A pollster surveys 100 subjects consisting of 40 Democrats (of which half are female) and 60 Republicans (of which half are female). Find the probability of randomly selecting one of these subjects and getting a Republican or a male.

3-32. Using the subjects from Exercise 3-31, find the probability of randomly selecting one subject and getting a Democrat or a male.

3-33. Using the subjects from Exercise 3-31, find the probability of randomly selecting one subject and getting a Republican or a female.

3-34. A convention of 100 guests consists of 37 Democrats, 41 Republicans, and 22 Independents. Find the probability of randomly selecting one voter and getting a Democrat or an Independent.

3-35. In a convention of 100 guests with 37 Democrats and 41 Republicans, there are 35 conventioneers under 30 years of age, including 17 of the Democrats and 13 of the Republicans. Find the probability that a conventioneer selected at random is a Republican or is under 30 years of age.

3-36. In one local survey, 100 subjects indicated their opinions on the legalization of marijuana. Of the 62 favorable responses, there were

40 males. Of the 38 unfavorable responses, there were 15 males. Find the probability of randomly selecting one of these subjects and getting a male or a favorable response.

3-37. Find the probability of getting a black card or a seven if one card is randomly selected from a standard deck.

3-38. Find the probability of getting a seven or an ace if one card is randomly selected from a standard deck.

3-39. The 12 residents of one house include exactly 5 registered voters of which 4 are males; there are 3 other males in the house. If a resident is randomly selected, find the probability of getting a female who is a registered voter.

EXERCISES B: ADDITION RULE

3-40. How is the addition rule changed if the exclusive "or" is used instead of the inclusive "or"? Recall that the exclusive "or" means "either one, or the other, but not both."

3-41. (a) If $P(A) = 0.4$ and $P(B) = 0.5$, what is known about $P(A \text{ or } B)$ if A and B are mutually exclusive events?

(b) If $P(A) = 0.4$ and $P(B) = 0.5$, what is known about $P(A \text{ or } B)$ if A and B are not mutually exclusive events?

3-42. $P(A \text{ or } B) = 0.8$ while $P(A) = 0.4$. What is known about events A and B?

3-43. Find $P(B)$ if $P(A \text{ or } B) = 0.6$, $P(A) = 0.6$, and A and B are mutually exclusive events.

3-4 MULTIPLICATION RULE

In Section 3-3 we developed a rule for finding the probability that A or B will occur in a given experiment. In the addition rule, we used the term $P(A \text{ and } B)$ which denoted the probability that A **and** B both occurred. Determination of a value for $P(A \text{ and } B)$ is often easy when we are dealing with a single simple experiment, but now we want to develop a more general rule for finding $P(A \text{ and } B)$ so that A can occur in one experiment while B occurs in another. Again we begin with simple examples that will suggest more general rules.

Suppose we have a two-part experiment that consists of flipping a fair coin and then rolling a fair die, and we wish to determine the probability of getting (heads and 5) which we will denote by $P(\text{heads and } 5)$.

heads-1
heads-2
heads-3
heads-4
heads-5
heads-6
tails-1
tails-2
tails-3
tails-4
tails-5
tails-6

The preceding list itemizes the 12 equally likely outcomes, and we can see that the result of (heads and 5) occurs only once so that

$$P(\text{heads and } 5) = \frac{1}{12}$$

Considering the component events of ''heads'' and ''5'' separately, we see that $P(\text{heads}) = 1/2$ while $P(5) = 1/6$. Recognizing that $1/12$ is the product of $1/2$ and $1/6$, we observe that $P(\text{heads and } 5) = P(\text{heads}) \cdot P(5)$ and use this observation as a basis for formulating the more general multiplication rule.

Multiplication rule $P(A \text{ and } B) = P(A) \cdot P(B)$

CONVICTED BY PROBABILITY

In 1964, Mrs. Juanita Brooks was robbed in Los Angeles. According to witnesses, the robber was a Caucasian woman with blond hair in a ponytail who escaped in a yellow car driven by a black male with a mustache and beard. Janet and Malcolm Collins were arrested and convicted after a college mathematics instructor testified that there is only about one chance in 12 million that any couple would have the characteristics described by the witnesses. The following estimated probabilities were presented in court.

> P(yellow car) = 1/10
> P(man with mustache) = 1/4
> P(girl with hair in a ponytail) = 1/10
> P(girl with blond hair) = 1/3
> P(black man with beard) = 1/10
> P(interracial couple in car) = 1/1000

The convictions were later reversed after the Supreme Court of California noted that no evidence was presented to support the stated probabilities and the independence of the characteristics was not established.

Example

Three separate slips of paper are marked a, b, and c. One slip is randomly selected and replaced, and then a second slip is drawn. Find the probability of getting (a and b) in that order.

Solution

By the multiplication rule, $P(a$ and $b) = P(a) \cdot P(b)$. $P(a)$ is the probability of getting an a on the first selection, so $P(a) = 1/3$. Similarly, $P(b) = 1/3$ so that

$$P(a) \cdot P(b) = \frac{1}{3} \cdot \frac{1}{3} = \frac{1}{9}$$

Therefore $P(a$ and $b) = 1/9$.

Unfortunately, we cannot be sure of the reliability of any generalization based upon a specific case. We would be wise to test the multiplication rule under a variety of cases to see if any errors arise. The next example does show that the above multiplication rule is sometimes inadequate.

Example

Let's again assume that we have three separate slips of paper marked a, b, and c and we wish to find the probability of randomly selecting a and b (in that order). Two slips are to be drawn, and the first slip is *not* replaced before the second selection is made.

Solution

Without replacement of the first slip, the following 6 outcomes are equally likely, and they constitute the entire sample space:

a–b
a–c
b–a
b–c
c–a
c–b

By examining the preceding sample space, we see that only one outcome corresponds to (a and b) in that order, so $P(a$ and $b) = 1/6$. But on any single selection we have $P(a) = P(b) = 1/3$, so the multiplication rule would again give us $P(a$ and $b) = P(a) \cdot P(b) = 1/3 \cdot 1/3 = 1/9$, which does not agree with our sample space analysis. One-ninth is incorrect, so the multiplication rule does not fit this case.

In this example, the multiplication rule does not take into account the fact that the first slip is not replaced. $P(a)$ is again equal to $1/3$, but after drawing an a on the first slip, there would be two slips (b and c) remaining, so $P(b)$ on the second selection would be $1/2$.

Here is the key concept of this example: **Without replacement of the first slip, the second probability is affected by the first result.** Since this dependence of the second event on the first result is so important, we formulate a special definition.

DEFINITION

> *Two events* A *and* B *are **independent** if the occurrence of one does not affect the probability of the occurrence of the other. (Several events are similarly independent if the occurrence of any one does not affect the probabilities of the occurrence of the others.) If* A *and* B *are not independent, they are said to be **dependent***.

The multiplication rule holds if the events A and B are independent; in that case $P(B)$ is not affected by the occurrence of A. But in our last example, where the first slip was not replaced before the second selection, $P(b)$ was affected by the result of the first selection. If the first slip is marked b, then $P(b)$ on the second selection becomes zero. If the first slip is either a or c, then $P(b)$ on the second selection is 1/2. Note that $P(b)$ definitely does depend upon the first result. To conclude, we note that $P(a)$ on the first selection is 1/3 while $P(b)$ on the second selection becomes 1/2 if we assume that the first result was a. This is a necessary and reasonable assumption, since we are attempting to determine $P(a$ and $b)$. Multiplying 1/3 and 1/2 we get the correct probability of $P(a$ and $b) = 1/6$.

Apart from the concept of independence, this last example suggests an improved multiplication rule.

NOTATION

> Let $P(B|A)$ represent the probability of B occurring after assuming that A has already occurred:
>
> **Multiplication rule for dependent events**
>
> $$P(A \text{ and } B) = P(A) \cdot P(B|A)$$

Example A box contains 10 fuses of which 3 are defective. Two fuses are randomly selected, and the first one is not replaced before the second selection. Find the probability that both fuses are defective.

Solution Let A denote the event of getting a defective fuse on the first selection, while B denotes the event of selecting a defective fuse on the second selection. We therefore seek $P(A$ and $B)$ where A and B are dependent since the probability of a defective fuse on the second selection does depend upon the result of the first selection. $P(A) = 3/10$ since we begin with 10 fuses of which 3 are defective. $P(B|A)$ is the probability of getting a defective fuse on the second drawing, assuming that the first selected fuse was defective. With that assumption there would remain 9 fuses of which 2 are defective so that $P(B|A)$ = 2/9. Therefore

$$P(A \text{ and } B) = P(A) \cdot P(B|A)$$

$$= \frac{3}{10} \cdot \frac{2}{9} = \frac{6}{90}$$

This improved multiplication rule does seem to accommodate dependent events, but how do we treat independent events? If A and B are independent, $P(B|A)$ must equal $P(B)$, since the definition of independence states that the occurrence of A should not affect $P(B)$. That is, if A and B are independent, we can use the first multiplication rule.

Multiplication rule

$$P(A \text{ and } B) = P(A) \cdot P(B) \text{ if } A \text{ and } B \text{ are } \textbf{independent}$$

$$P(A \text{ and } B) = P(A) \cdot P(B|A) \text{ if } A \text{ and } B \text{ are } \textbf{dependent}$$

In the following examples we attempt to illustrate both of these cases.

Example

Find the probability of getting two consecutive tails when a fair coin is flipped twice.

Solution

The fair coin has no memory, so the second outcome is unaffected by the first result. The two events

A — tails on the first flip
B — tails on the second flip

are therefore **independent** and we get

$$P(A \text{ and } B) = P(A) \cdot P(B)$$

$$= \frac{1}{2} \cdot \frac{1}{2}$$

$$= \frac{1}{4}$$

The same principle used in the preceding example is easily extended to several events. For example, the probability of getting five consecutive tails is $1/2 \cdot 1/2 \cdot 1/2 \cdot 1/2 \cdot 1/2 = 1/32$. In general, **the probability of any sequence of independent events is simply the product of their corresponding probabilities.**

The next three examples illustrate this extension of the multiplication rule.

Example

Find P(heads and 2 and November date) for the compound experiment that consists of flipping a coin, rolling a fair die, and randomly selecting one birthdate from the 366 possibilities.

Solution

The components of the experiment are obviously independent, so

$$P(\text{heads and 2 and November date}) = P(\text{heads}) \cdot P(2) \cdot P(\text{November date})$$

$$= \frac{1}{2} \cdot \frac{1}{6} \cdot \frac{30}{366}$$

$$= \frac{30}{4392}$$

Example

If license plate numbers are issued on a random basis, what is the probability that the next seven plates you get will all end with even numbers?

Solution

The probability of getting an even-numbered plate is 1/2 and the different plates are independent of each other so that

$$P(7 \text{ even-numbered plates}) = \frac{1}{2} \cdot \frac{1}{2} \cdot \frac{1}{2} \cdot \frac{1}{2} \cdot \frac{1}{2} \cdot \frac{1}{2} \cdot \frac{1}{2}$$

$$= \frac{1}{128}$$

Example

A pill designed to present certain physiological reactions is advertized as 95% effective. Find the probability that the pill will work in all of 15 separate and independent applications.

Solution

Since the 15 applications are independent, we get

$$P(15 \text{ successes}) = 0.95 \cdot 0.95 \cdot 0.95 \cdot \ldots \cdot 0.95 \, (15 \text{ times})$$

$$= 0.95^{15}$$

$$\overset{\circ}{=} 0.463$$

While the preceding three examples illustrate an extension of the multiplication rule for independent events, the next example illustrates a similar extension of the multiplication rule for dependent events.

Example

A meeting is attended by 4 men and 6 women. If a reporter randomly selects 3 different attendees for interviews, find the probability that they are all women.

Solution Letting A, B, and C represent the respective events of randomly selecting a woman on the first, second, and third attempts, we see that those events are dependent since successive probabilities are affected by previous results. The probability of getting two different women on the first two selections can be found through direct application of the multiplication rule for dependent events.

$$P(A \text{ and } B) = P(A) \cdot P(B|A)$$

$$= \frac{6}{10} \cdot \frac{5}{9}$$

$$= \frac{30}{90}$$

If the probability of the first woman is $6/10$ and the probability of the second woman is $5/9$, then the probability of the third woman must be $4/8$. This seems reasonable since, after the first two women are selected, there would be 8 remaining attendees of which 4 are women. Therefore

$$P(A \text{ and } B \text{ and } C) = P(A) \cdot P(B|A) \cdot P(C|A \text{ and } B)$$

$$= \frac{6}{10} \cdot \frac{5}{9} \cdot \frac{4}{8}$$

$$= \frac{120}{720} = \frac{1}{6}$$

In this section, we have seen that **and** in a probability problem suggests multiplication in the same way that **or** suggested addition in the previous section. The independence of events must be considered in the determination of probabilities like $P(A \text{ and } B)$, since the component probabilities may be affected. Replacement of randomly selected items usually involves independent events, while failure to replace selected items usually causes the events to be dependent.

Example Three defective transistors and two good transistors are mixed in a box. Two transistors are randomly selected. Find the probability that they are both defective if the selections are made:

(a) With replacement.

(b) Without replacement.

Solution (a) With replacement, the result of the first selection has no affect on the second, so the two events

A: First selection is defective.

B: Second selection is defective.

are independent. We apply the multiplication rule since we need $P(A$ **and** $B)$. The multiplication rule for independent events shows that

$$P(A \text{ and } B) = P(A) \cdot P(B)$$

$$= \frac{3}{5} \cdot \frac{3}{5}$$

$$= \frac{9}{25}$$

(b) Without replacement, events A and B are dependent since the result of the first selection does affect $P(B)$. Applying the multiplication rule for dependent events we get

$$P(A \text{ and } B) = P(A) \cdot P(B \,|\, A)$$

$$= \frac{3}{5} \cdot \frac{2}{4}$$

$$= \frac{6}{20}$$

In both cases of this last example, $P(A)$ is 3/5 since we begin with 5 items, 3 of which are defective. To compute $P(A \text{ and } B)$ we must assume that A does occur, so we determine the probability of getting the second defective transistor. With replacement, there are again 5 items of which 3 are defective, but without replacement we would be left with 4 items of which 2 are defective. Considerations such as these help to make the multiplication rule intuitively obvious instead of an abstract mystery.

When using the multiplication rule for finding probabilities in compound events, a **tree diagram** is sometimes helpful in determining the number of possible outcomes. In a tree diagram, we depict schematically the possible outcomes of an experiment as line segments emanating from one starting point. In Figure 3-7 we show the tree diagram that summarizes the possibilities for parents who plan to have 3 children. Note that the tree diagram in Figure 3-7 presents the 8 different possible outcomes as the 8 different possible paths begin at the left and end at the right. Assuming that boys and girls are equally likely, we see that the 8 paths are equally likely, so each of the 8 possible outcomes has a probability of 1/8. Such diagrams are often helpful in counting the number of possible outcomes provided that the number of possibilities is not too large. In cases involving large numbers of choices, the use of tree diagrams is impractical. However, they are useful as visual aids that help to provide insight into the multiplication rule. Suppose, for example, that an experiment consists of flipping a coin and then rolling a die, and we wish to determine the probability of getting (heads and 3). In Figure 3-8, the possible outcomes are summarized and we can see that the outcome of (heads and 3) is only one of 12 branches. Since the branches correspond to equally

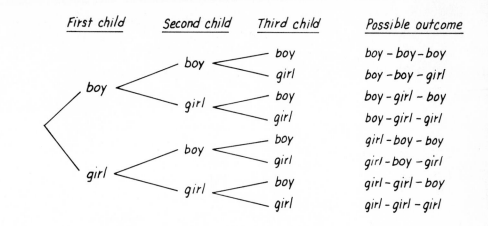

First child	Second child	Third child	Possible outcome

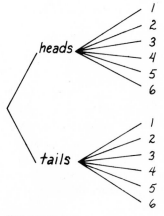

likely outcomes, we get P(heads and 3) = 1/12. Examining Figure 3-8, we see that for each of the 2 outcomes for the coin, there are 6 outcomes for the die. The total number of outcomes is therefore "6 taken 2 times" or 12. Using the tree diagram, we can see the reason for multiplication.

FIGURE 3-7

FIGURE 3-8

EXERCISES A: MULTIPLICATION RULE

3-44.   For each following pair of events, classify the two events as independent or dependent.

(a)   Making a correct guess on the first question of a multiple choice
IND.   quiz.
Making a correct guess on the second question of the same multiple choice quiz.

(b)   Selecting an ace from a deck of cards when 1 card is drawn.
IND.   Selecting another ace from a deck of cards if the first selection is replaced.

(c)   Selecting an ace from a deck of cards when 1 card is drawn.
DEP.   Selecting another ace from a deck of cards if the first selection is not replaced.

(d)   Staying out of jail.
DEP   Cheating on income tax payments.

(e)   Winning the Connecticut lottery in 1980.
IND.   Winning the Connecticut lottery in 1981.

(f)   Rolling 7 on a pair of fair dice.
IND.   Getting 7 on a second roll of a pair of fair dice.

(g)   Getting an unfavorable reaction when a certain drug is adminis-
IND.   tered to a mouse.
Getting an unfavorable reaction when the same drug is administered to another mouse.

(h)   Finding your television inoperable.
IND.   Finding your car inoperable.
(i)   Finding your television inoperable.
DEP.   Finding your kitchen light inoperable.
(j)   Rolling a 7 on a pair of fair dice.
IND.   Flipping a coin and getting heads.

3-45.   If the probability of a dog contracting heartworm is 1/12, find the probability that 2 different dogs selected at random will both have heartworm.

3-46.   Find the probability of flipping a nickel once and a dime once and getting 2 tails.

3-47.   Find the probability of drawing 2 consecutive clubs from a standard shuffled deck if:
(a)   The first card is replaced.
(b)   The first card is not replaced.

3-48.   Find the probability of drawing 2 consecutive red cards from a standard shuffled deck if:
(a)   The first card is replaced.
(b)   The first card is not replaced.

3-49.   Find the probability of drawing 2 consecutive fives from a standard shuffled deck if:
(a)   The first card is replaced.
(b)   The first card is not replaced.

3-50.   Find the probability of getting 10 consecutive heads when a fair coin is flipped 10 times.

3-51.   Find the probability of rolling 3 consecutive threes when a fair die is rolled 3 times.

3-52.   If a death is selected at random, assume that there is a 0.057 probability that it was caused by an accident. Find the probability that 3 randomly selected deaths were all accidental.

3-53.   Find the probability that 2 people selected at random were both born on July 4. (Ignore leap years.) 1/133225

3-54.   If 2 people are randomly selected, find the probability that the second person has the same birthday as the first. 1/365

3-55.   There is a 0.17 probability that a telephone solicitation for a magazine will produce an order. Find the probability that 2 orders are obtained in 2 separate calls.

3-56.   An experiment begins with 4 female mice and 2 male mice in the same cage. One mouse is randomly selected each day and put in a separate cage. Find the probability that the first 3 removals are all females.

3-57. An experiment begins with 6 identical mice in the same cage. Each day 1 mouse is randomly selected, injected with a chemical, and then put back into the same cage. Find the probability of selecting the same mouse on each of the first 3 days.

3-58. A container holds 12 eggs, of which 5 are fertile:
   (a) Find the probability of randomly selecting 3 eggs that are all fertile if each egg is replaced before the next selection is made.
   (b) Find the probability of randomly selecting 3 eggs that are all fertile if the eggs selected are not replaced.

3-59. A car has 8 spark plugs of which 2 are defective. Find the probability of locating both defective spark plugs in only 2 random selections. Assume that the first selection is not replaced.

3-60. You are among a group of 5 students who were arrested in the Banana Republic for jaywalking. You were all put in jail with the stipulation that each day 1 of you would be selected at random and released. Find the probability that you remain in jail after the first 3 days.

3-61. An insurance firm serves 5673 clients of which 4138 are male. There are 2565 clients under 30 years of age, and 1875 of these are male. Find the probability of randomly selecting a female client under 30 years of age. $\frac{690}{5673}$

3-62. Three organizations independently and randomly select a month in which to hold their annual conventions. What is the probability that all 3 months are different?

3-63. Eight defective batteries are present in a bin of 100 batteries. The entire bin is approved for shipment if no defects show up when 3 randomly selected batteries are tested:
   (a) *INP.* Find the probability of approval if the selected batteries are replaced.
   (b) *DEP.* Find the probability of approval if the selected batteries are not replaced.
   (c) Comparing the results to parts (a) and (b), which procedure is more likely to reveal a defective battery? Which procedure do you think is better?

EXERCISES B: MULTIPLICATION RULE

3-64. Sampling problems are of great concern to statisticians. Assume that a random sample of 200 voters is selected from a population of 2 million voters in such a way that precludes the possibility of any subject being chosen more than once. This is sampling without replacement:
   (a) What is the probability of a voter being selected first?

MONKEYS ARE NOT LIKELY TO TYPE "HAMLET"

A classical assertion holds that, if a monkey randomly hits the keys on a typewriter, it will eventually produce the complete works of Shakespeare if it continues to type year after year. Dr. William Bennet used the rules of probability to develop a computer simulation that addressed this problem, and he concluded that it would take a monkey about 1,000,000, 000,000,000,000,000,000, 000,000,000,000 years to reproduce Shakespeare's works.

(b) Assuming that a particular subject was not selected first, what is the probability of being selected second?

(c) Answer part (b) by assuming that selected voters are made available for future selections. (That is, assume replacement.)

(d) Compare the results to parts (b) and (c). Does replacement really make a significant difference in this situation?

3-65. If the probability of a certain pill being effective is 0.95, find the probability that it will be effective in each of 65 independent applications.

3-66. Use a calculator to compute the probability that of 25 people, no 2 share the same birthday.

3-67. Repeat Exercise 3-66 for a group of 50 people.

3-68. A gasoline rationing scheme used in 1973 and again in 1979 mandated that cars with even-numbered license plates may get gas only on even-numbered days of the month. Similarly, cars with odd-numbered plates could refuel only on odd-numbered days of the month. In addition, all gas stations had to close every Sunday. Assume that this plan is now in effect and that you have an even-numbered license plate. Find the probability that you can get gas on a day selected at random from the current year.

3-5 COMPLEMENTARY EVENTS

In this section we begin by defining complementary events and present one last rule of probabilities that relates to these events.

The complement of event A is the event whereby "A does *not* occur." If A represents an outcome of 3 when one die is rolled, then the complementary event $\overline{A}$ is any outcome other than 3.

The definition of complementary events implies that they must be mutually exclusive, since it is impossible for an event and its opposite to occur at the same time. Also, we can be absolutely certain that either A does or does not occur. That is, either A or $\overline{A}$ must occur. These observations enable us to apply the addition rule for mutually exclusive events as follows:

$$P(A \text{ or } \overline{A}) = P(A) + P(\overline{A}) = 1$$

We justify $P(A \text{ or } \overline{A}) = P(A) + P(\overline{A})$ by noting that A and $\overline{A}$ are mutually exclusive, and we justify the total of 1 by our absolute certainty that A either does or does not occur. This result of the addition rule leads to the following three **equivalent** forms:

Rule of complementary events

$$P(A) + P(\overline{A}) = 1$$
$$P(\overline{A}) = 1 - P(A)$$
$$P(A) = 1 - P(\overline{A})$$

The first form comes directly from our original result, while the second (see Figure 3-9) and third variations involve very simple equation manipulations. A major advantage of the rule of complementary events is that it can

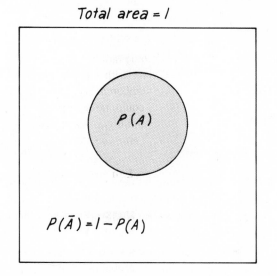

Total area = 1

$P(A)$

$P(\overline{A}) = 1 - P(A)$

FIGURE 3-9

sometimes be used to significantly reduce the workload required to solve certain problems. As an example, let's consider a very ambitious couple planning to have 7 children. We want to determine the probability of getting at least 1 boy among those 7 children. The direct solution to this problem is messy, but a simple indirect approach is made possible by our rule of complementary events. Let's denote the event of getting at least 1 boy in 7 by B. We will begin by finding $P(\overline{B})$, the probability of *not* getting at least 1 boy (which is equivalent to getting 7 girls). Now $P(\overline{B})$ is relatively easy to compute if we make two reasonable assumptions:

1. $P(\text{girl}) = 1/2$.
2. The sexes of successive babies are independent of any younger brothers or sisters.

(Neither of these two assumptions is exactly correct, but they can be used with extremely good results.)

Applying the multiplication rule for independent events we get

$$P(\bar{B}) = P(7 \text{ consecutive girls}) = \frac{1}{2} \cdot \frac{1}{2} \cdot \frac{1}{2} \cdot \frac{1}{2} \cdot \frac{1}{2} \cdot \frac{1}{2} \cdot \frac{1}{2} = \frac{1}{128}$$

However, we are seeking $P(B)$ so that the rule of complementary events can be applied.

$$P(B) = 1 - P(\bar{B})$$

$$= 1 - \frac{1}{128}$$

$$= \frac{127}{128}$$

As complex as this solution may appear, it is trivial in comparison to the alternate solutions that involve a direct approach.

Example

Forty doctors and 10 psychologists are participating in a conference. If 3 different participants are randomly selected for a panel discussion, find the probability that at least 1 is a psychologist.

Solution

Let A represent the event of selecting at least 1 psychologist when 3 participants are randomly selected. Then $\bar{A}$ becomes the event of not getting at least 1 psychologist. That is, $\bar{A}$ signifies that no psychologists are selected, so all 3 participants are doctors. We use the multiplication rule to get

$$P(\bar{A}) = P(\text{all 3 are doctors})$$

$$= \frac{40}{50} \cdot \frac{39}{49} \cdot \frac{38}{48}$$

$$= 0.504$$

We now use the rule of complementary events to find $P(A)$.

$$P(A) = 1 - P(\bar{A})$$

$$= 1 - 0.504$$

$$= 0.496$$

The key concept employed in the two previous examples is equating "at least 1" with the *opposite* of "none." In an informal notation, this means that $P(\text{at least 1}) = 1 - P(\text{none})$.

The rule of complementary events enables us to make many other important conclusions. If, for example, we have a 0.05 probability of an error in a test, we can conclude that the probability of no error is $1 - 0.05$ or 0.95. If we know that the true probability of a baby being a boy is 0.51, we can

BETTORS AT RACES ARE CONNED

One scheme used against bettors at race tracks works this way. One group of bettors is given a hot tip, such as the first horse in the first race. Another group is told that the second horse will win; a third group is told that the third horse will win, and so on. The group that actually ends up with the winner is then subdivided for another round of different tips. The con man has now won the confidence of the small group of two-time winners and offers to sell, for a substantial amount, a third hot tip. The con man collects these fees and then moves on to parts unknown.

conclude that $P(\text{girl}) = 1 - 0.51 = 0.49$. There are many cases when the probability of an event is known and we need the probability of the complementary event.

The concepts and rules of probability theory presented in this chapter consist of elementary and fundamental principles. A more complete study of probability is not necessary at this time since our main objective is to study the elements of statistics, and we have already covered the probability theory that we will need. Hopefully this chapter has generated some interest in probability for its own sake. The importance of probability is continuing to grow as it is used by more and more scientists, economists, politicians, biologists, insurance specialists, executives, and other professionals.

EXERCISES A: COMPLEMENTARY EVENTS

In Exercises 3-69 through 3-78, determine the probability of the given event and the probability of the complementary event:

3-69.   A single fair die is rolled and a 2 turns up.

3-70.   A single fair die is rolled and the result is a number less than 3.

3-71.   A defective capacitor is randomly selected from a box containing 20 capacitors of which 6 are defective. (Only one selection is made.)

3-72.   A letter is randomly selected from the alphabet and the result is a vowel (a, e, i, o, u).

3-73.   A number from 1 through 10 is randomly selected and the result is even.

3-74.   A baby is born and it is a boy.

3-75.   A day of the week is randomly selected and it is a Friday or a Saturday.

3-76.   A tax return is randomly selected from a group of 20, and the return selected shows the largest income in that group.

3-77.   In a class of 15 girls and 10 boys, 1 student is randomly called and it is a girl.

3-78.   Three television stations draw lots to determine which station will cover the local football game, and the station with the lowest budget wins.

3-79.   If a husband and wife plan to have 3 children, find the probability that they have at least 1 boy. Assume that boys and girls are equally likely and that the sex of any successive baby is not affected by the sex of any younger brothers or sisters. $7/8$

$$\tfrac{1}{2} \cdot \tfrac{1}{2} \cdot \tfrac{1}{2} = \tfrac{1}{8} \quad \text{so} \quad 1 - \tfrac{1}{8} = \tfrac{7}{8}$$

3-80. A typing pool is made up of 5 men and 5 women. If 3 different typists are randomly selected from this pool, find the probability that at least one of the 3 is a man.

3-81. A die is shaved so that $P(5)$ becomes 1/4. $\overline{1-\frac{1}{4}}$
   (a) Find the probability of not rolling a 5 with this die. $\frac{3}{4}$
   (b) Find the probability of not getting a 5 in either of two separate rolls of this die. $\frac{3}{4} \cdot \frac{3}{4} = \frac{9}{16}$
   (c) Find the probability of getting at least one 5 in two separate rolls of this die. $1 - \frac{9}{16} = \frac{7}{16}$ or $\frac{1}{4} + \frac{1}{4} - \frac{1}{16} = \frac{7}{16}$

3-82. Three cards are drawn from a standard shuffled deck. Find the probability of getting at least 1 heart if:
   (a) Each card is replaced and the deck is reshuffled before the next selection.
   (b) No cards are replaced. $\frac{20}{120}$

3-83. A pollster must randomly select 5 different residents from a neighborhood containing 40 Catholics, 60 Protestants, and 20 Jews. What is the probability of selecting at least 1 Jew?

$P(\overline{JEW}) = \frac{100}{120}$

$P(\overline{JEW}) = \frac{100}{120} \cdot \frac{99}{119} \cdot \frac{98}{118} \cdot \frac{97}{117} \cdot \frac{96}{116}$

$P(JEW) = 1 - P(\overline{JEW})$

$P(JEW) = .605$

3-84. A certain method of contraception is found to be 95% effective. What is the probability of at least 1 pregnancy among 10 different couples using this method of contraception?

3-85. In a neighborhood of 500 residents there are 4 illegal aliens. A random check is made of 20 of these residents. Find the probability that at least 1 illegal alien will be found among the 20 different residents.

3-86. In a certain state, 4% of all cars on the road are in obvious violation of some law. If a police check involves the random inspection of 15 different cars, what is the probability that at least 1 car will be found in violation of a law?

3-87. When correct procedures are followed, there is a 0.95 probability that a certain test will lead to the correct conclusion. If the correct procedures are used and this test is applied in 5 independent cases, what is the probability that the conclusion will be incorrect at least once?

$\frac{1}{2} \cdot \frac{1}{2} \cdots 10 \, TIME = \frac{1}{1024}$

$1 - \frac{1}{1024} = \frac{1023}{1024}$

3-88. A true–false test of 10 questions is given and an unprepared student must make random guesses in answering each question. What is the probability of at least 1 correct response?

EXERCISES B: COMPLEMENTARY EVENTS

3-89. (a) If $P(A) = 1$ and A and B are complementary events, what is known about event B?

(b) If $P(A) = 0$ and A and B are complementary events, what is known about event B?

(c) If $P(A)$ is at least 0.7 and A and B are complementary events, what is known about event B?

3-90. All pairs of complementary events must be mutually exclusive pairs. Must all mutually exclusive pairs of events also be complementary? Support your response with specific examples.

3-91. Use the addition rule and the rule of complementary events to find a formula for $P(\overline{A \text{ or } B})$, which is the probability of not getting either A or B.

3-92. Find the probability that, of 25 randomly selected people, at least 2 share the same birthday.

3-93. Repeat exercise 3-92 for a group of 50 people.

REVIEW

This chapter began with the basic concept of the **probability** of an event and presented two rules for finding probabilities. Rule 1 represents the **empirical** approach, whereby the probability of an event is approximated by actually conducting or observing the experiment in question:

RULE 1

$$P(A) = \frac{\text{number of times } A \text{ occurred}}{\text{number of times experiment was repeated}}$$

Rule 2 is called the **classical** approach, and it applies only if all of the outcomes are equally likely:

RULE 2

$$P(A) = \frac{s}{n} = \frac{\text{number of ways } A \text{ can occur}}{\text{total number of different outcomes}}$$

We noted that the probability of any impossible event is 0, while the probability of any certain event is 1. Also, for any event A,

$$0 \leq P(A) \leq 1$$

In Section 3-3, we considered the **addition rule** for finding the probability that A **or** B will occur. In evaluating $P(A \text{ or } B)$, it is important to consider whether the events are *mutually exclusive*, that is, they cannot both occur at the same time (see Figure 3-10).

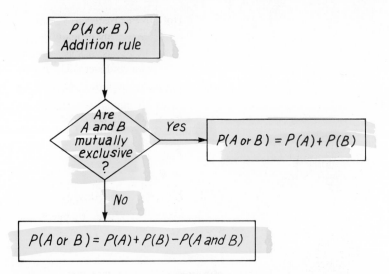

FIGURE 3-10

In Section 3-4 we considered the **multiplication rule** for finding the probability that A **and** B will occur. In evaluating $P(A$ and $B)$, it is important to consider whether the events are *independent*, that is, the occurrence of one event does not affect the probability of the other event (see Figure 3-11).

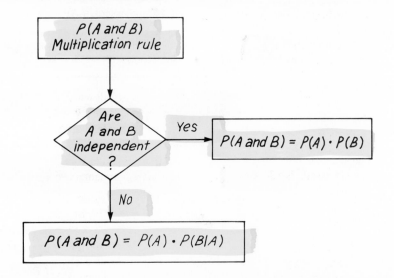

FIGURE 3-11

In Section 3-5 we considered **complementary** events (opposites). Using the addition rule, we were able to develop the **rule of complementary events:** $P(A) + P(\overline{A}) = 1$. We saw that this rule can sometimes be used to simplify probability problems.

Most of the material that follows this chapter deals with statistical inferences based upon probabilities. As an example of the basic approach used, consider a test of someone's claim that his quarter is fair. If we flip the quarter 10 times and get 10 consecutive heads, we can make one of two inferences from these sample results:

1. The coin is actually fair and the string of 10 consecutive heads is a fluke.
2. The coin is not fair.

The statistician's decision is based upon the **probability** of getting 10 consecutive heads which, in this case, is so small (1/1024) that the inference of unfairness is the better choice. The purpose of this example is to emphasize the important role played by probability in the standard methods of statistical inference.

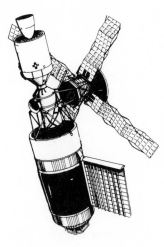

SKYLAB

On July 11, 1979, Skylab left its orbit and 77.5 tons of metal crashed into the earth. No one could predict where Skylab would crash and extensive precautions were taken in areas that NASA judged possible crash sites. On that day, the author was grounded in Gander, Newfoundland along with hundreds of other airline passengers who waited until the danger passed. NASA statisticians calculated that there was a probability of 1/152 that somewhere a human being would be hit. The probability of any particular person being struck was estimated to be 1/600,000,000,000.

REVIEW EXERCISES

3-94. A vaccine is administered to 60 subjects, and 37 unfavorable reactions occurred. What is the approximate probability that this vaccine will produce an unfavorable reaction?

3-95. Of 120 auto ignition circuits, there are 18 defects. If 2 circuits are randomly selected, find the probability that they are both defective if:
 (a) The first selection is replaced before the second selection is made.
 (b) The first selection is not replaced.

3-96. Define the following:
 (a) Mutually exclusive events.
 (b) Independent events.
 (c) Simple events.
 (d) Sample space.
 (e) Complementary events.

3-97. A survey is made in a neighborhood of 65 Democrats and 15 Republicans. 30 of the Democrats are women, while 10 of the Republicans are women. If 1 subject from this group is randomly selected, find the probability of getting:
 (a) A male or a Democrat.
 (b) A male Democrat.
 (c) A Democrat or a Republican.

3-98. An experiment consists of randomly choosing a day of the week and a month of the year. Find the probability of getting ''Saturday-September.''

3-99. Eight men and 7 women have applied for a temporary job. If 3 different applicants are randomly selected from this group, find the probability that:
(a) All 3 are women.
(b) There is at least 1 woman.

3-100. A fair coin is flipped 5 times. Find the probability that at least 1 tail appears.

3-101. A fair coin is flipped 5 times. Find the probability of getting 5 heads or 5 tails.

PROBABILITY
DISTRIBUTIONS

4-1 OVERVIEW

In Chapter 2 we discussed the histogram as a device for showing the frequency distribution of a set of data. In Chapter 3 we discussed the basic principles of probability theory. In this chapter we combine those concepts to develop probability distributions that are basically theoretical models of the frequency distributions we produce when we collect sample data. We construct frequency tables and histograms using *observed* real scores, but we construct probability distributions by presenting possible outcomes along with their *probable* frequencies.

Suppose a casino manager suspects cheating at a dice table. He can compare the frequency distribution of the actual sample outcomes to a theoretical model that describes the frequency distribution likely to occur with fair dice. In this case the probability distribution serves as a model of a theoretically perfect population frequency distribution. In essence, we can determine what the frequency table and histogram would be like for a pair of fair dice rolled an infinite number of times. With this perception of the population of outcomes, we can then determine the values of important parameters such as the mean, variance, and standard deviation.

The concept of a probability distribution is not limited to casino management. In fact, the remainder of this book and the very core of inferential statistics depend on some knowledge of probability distributions. To analyze the effectiveness of a new drug, for example, we must know something about the probability distribution of the symptoms the drug is intended to correct.

This chapter deals mostly with discrete cases while subsequent chapters involve continuous cases. We begin by distinguishing between discrete and continuous random variables.

4-2 RANDOM VARIABLES

We have already stated that an experiment is a process that allows us to obtain observations. Let's assume that we are dealing with experiments of the type that will have some number associated with each outcome. Typically, that number will vary from trial to trial. For example, consider the experiment of launching 2 rockets. If we count the number of successful launches in those 2 attempts, we are associating a number with each outcome, and that number will vary from trial to trial. Since such outcomes vary and are determined by chance, they are called **random variables**. The values of a random variable are the numbers we associate with the different events that comprise the sample space.

Example

A drug is given to 2 sick patients. Let the random variable represent the number of cures that occur. Since there can be 0 cures, 1 cure, or 2 cures, the values of the random variable are 0, 1, and 2.

Example   A quiz consists of 10 multiple-choice questions. Let the random variable represent the number of correct answers. This random variable can take on the values of 0, 1, 2, 3, 4, 5, 6, 7, 8, 9, and 10.

Example   A single fair die is rolled. Let the random variable represent the number of dots that turn up. This random variable can assume any of the values 1, 2, 3, 4, 5, and 6.

Random variables may be discrete or continuous. We know what a finite number of values is (1, or 2, or 3, etc.), but our definition of a discrete random variable also involves the concept of a **countable** number of values. As an example, suppose that a random variable represents the number of times a die must be rolled before a 6 turns up. This random variable can assume any one of the values 1, 2, 3, ... We now have an infinite number of possibilities, but they correspond to the counting numbers. Consequently this type of infinity is called countable. In contrast, the number of points on a continuous scale is not countable and represents a higher degree of infinity. There is no way to count the points on a continuous scale, but we can count the number of times a die is rolled, even if it seems to continue forever.

DEFINITION

> *A **discrete random variable** has either a finite number of values or a countable number of values.*

For example, suppose that a random variable can assume the value which represents the number of United States Senators present for a roll call. That random variable can assume only one of 101 different values (0, 1, 2, . . . , 99, 100) and, since 101 is a finite number, the random variable is discrete.

DEFINITION

> *A **continuous random variable** has infinitely many values, and those values can be associated with points on a continuous scale in such a way that there are no gaps or interruptions.*

Just as count data are usually associated with discrete random variables, measurement data are usually associated with continuous random variables.

As an example, we stipulate that a random variable can assume the value representing the exact speed of a car at a particular instant. For example, the car might be traveling 42.135724 . . . kilometers per hour. This random variable can assume any value that corresponds to a point on the continuous interval shown in Figure 4-1. (We assume that the car never exceeds 80 kilometers per hour.) We now have an infinite number of possible values which are not countable, since there is a correspondence with the continuous scale of Figure 4-1. This random variable is not discrete, but it is continuous.

FIGURE 4-1

|————————————————————————————————|
0 *Speed in kilometers per hour* 80

Random variables that represent heights, weights, times, and temperatures are usually continuous, as are those that represent speeds.

In reality, it is extremely unusual to deal with exact values of a continuous random variable. Instead, we usually convert continuous values to discrete values by rounding off to a limited number of decimal places. If each speed between 0 kilometers per hour and 80 kilometers per hour is rounded off to the nearest integer value, we reduce the total number of possibilities to the finite number of 81. In this way, continuous random variables are made discrete. This chapter is involved almost exclusively with discrete random variables.

HOW DO YOU MEASURE DISOBEDIENCE?

The data you collect are at least as important as the statistical methodology you employ. However, it is often difficult to collect usable or relevant data. How do you collect data that relate to a characteristic that doesn't appear to be measureable, such as the level of disobedience in people? Stanley Milgram is a social psychologist who devised a clever experiment which did just that. A researcher instructed a volunteer to operate a control board which gave increasingly painful "electric shocks" to a third person. Actually, no electric shocks were given and the third person was an actor who feigned increasing levels of pain and anguish. The volunteer began with 15 volts and was instructed to increase the shocks by increments of 15 volts up to a maximum of 450 volts. The disobedience level was the point at which the volunteer refused to follow the researcher's instructions to increase the voltage. Despite the actor's screams of pain and a feigned heart attack, two-thirds of the volunteers continued to obey the researcher. Milgram was surprised by such a high level of obedience. This experiment provides us with some insight into the obedience of Nazis in World War II.

Discrete Random Variables

Inferential statistics is often used to make decisions in a wide variety of different fields. We begin with sample data and attempt to make inferences about the population from which the sample was drawn. If the sample is very large we may be able to develop a good estimate of the population frequency distribution, but samples are often too small for that purpose. The practical approach is to use information about the sample along with general knowledge about population distributions. Much of this general information is included in this and the following chapters. Without a knowledge of probability distributions, users of statistics would be severely limited in the inferences they could make. We intend to develop the ability to work with discrete and continuous probability distributions, and we begin with the discrete case because it is simpler.

As an example, suppose you are analyzing voter preferences for a politician and you have found that of 100 voters surveyed, 57 favor your candidate. You recognize that this constitutes a majority, but you also recognize that samples fluctuate and you want some indication of the significance of the results. To determine whether 57 of 100 voters constitutes a significant majority, you need to know something about the probability distribution of the discrete random variable that this problem involves. You will soon be able to solve problems of this type, but you must first study some preliminaries.

When dealing with a discrete random variable, we can assign probabilities to the individual sample values in such a way that the sum of those probabilities is 1. This cannot be done when we are dealing with a continuous random variable, and therefore we require a different approach which will become apparent when we consider continuous random variables in later chapters.

With the concept of a random variable, we can consider the notion of a probability distribution.

DEFINITION

> *A **probability distribution** is the collection of all values that a random variable can assume, along with the probabilities that correspond to these values.*

TABLE 4-1

x	$P(x)$
1	1/6
2	1/6
3	1/6
4	1/6
5	1/6
6	1/6

TABLE 4-2

x	$P(x)$
0	1/4
1	1/2
2	1/4

For example, when rolling a single fair die, we know that the random variable can assume the values of 1, 2, 3, 4, 5, and 6 and that each of these values is associated with a probability of 1/6. The probability distribution is summarized in Table 4-1.

For the example of the drug administered to 2 sick patients where the random variable is the number of cures, assume that the probability of a cure is 0.5. The probability of 0 cures is 1/4, the probability of exactly 1 cure is 1/2, while the probability of 2 cures is 1/4. Table 4-2 summarizes the probability distribution in this case.

There are various ways of graphing these probability distributions, but we illustrate only the histogram which was introduced in Chapter 2.

The horizontal axis delineates the values of the random variable, while the vertical scale represents probabilities. In Figure 4-2, we show the histogram representing the probability distribution of the last example. Note that, along the horizontal axis, the values of 0, 1, and 2 are located at the centers of the rectangles. This implies that the rectangles are each one unit wide, so the areas of the three rectangles are $1 \cdot 1/4$, $1 \cdot 1/2$, $1 \cdot 1/4$ (or 1/4, 1/2, and 1/4).

FIGURE 4-2

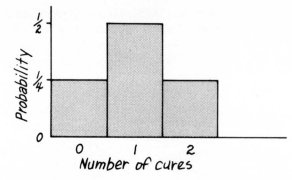

In Figure 4-3 we show the histogram representing the probability distribution for the random variable which assumes the number of dots that turn up when one fair die is rolled. Each rectangle of Figure 4-3 has an area of $1 \cdot 1/6 = 1/6$, and we can again associate area with probability. In general, if we stipulate that each value of the random variable is assigned a width of 1 on the histogram, then the areas of the rectangles will total 1. **We can therefore associate the probability of each numerical outcome with the area of the corresponding rectangle.**

FIGURE 4-3

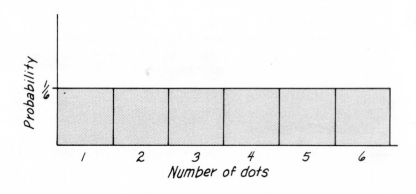

Suppose we have an experiment with an identified discrete random variable. What do we know about any two events that lead to different values of the random variable? They must be mutually exclusive since their simultaneous occurrence would necessarily lead to the same value of the random variable. In the experiment of rolling a fair die, for example, the 6 outcomes which correspond to the 6 different values of the random variable are all mutually exclusive outcomes. (You can't get a 2 and a 5 on one roll of a fair die.) Knowing that all of the values of the random variable will cover all events of the entire sample space, and knowing that events which lead to different values of the random variable are mutually exclusive, we can conclude that the sum of $P(x)$ for all values of x must be 1. Also, $P(x)$ must be between 0 and 1 for any value of x.

RULE 1
RULE 2

$$\Sigma P(x) = 1 \text{ where } x \text{ assumes all possible values.}$$
$$0 \leq P(x) \leq 1 \text{ for any value of } x.$$

These two rules for probability distributions are actually direct descendents of the corresponding rules of probabilities (discussed in Chapter 3).

Example

Does $P(x) = x/5$ (where x can take on the values of 0, 1, 2, 3) determine a probability distribution?

Solution

If a probability distribution is determined, it must conform to the preceding two rules. But

$$\Sigma P(x) = P(0) + P(1) + P(2) + P(3)$$

$$= \frac{0}{5} + \frac{1}{5} + \frac{2}{5} + \frac{3}{5}$$

$$= \frac{6}{5}$$

so that the first rule is violated and a probability distribution is not determined.

Example

Does $P(x) = x/10$ (where x can be 0, 1, 2, 3, or 4) determine a probability distribution?

Solution

From the given function we conclude that

$$P(0) = \frac{0}{10} = 0$$

$$P(1) = \frac{1}{10}$$

$$P(2) = \frac{2}{10}$$

$$P(3) = \frac{3}{10}$$

$$P(4) = \frac{4}{10}$$

The sum of these probabilities is 1, and each $P(x)$ is between 0 and 1, so both requirements are satisfied. Consequently, a probability distribution is determined. The graph of this probability distribution is shown in Figure 4-4. Note that the sum of the areas of the rectangles is 1, and each rectangle has an area between 0 and 1.

FIGURE 4-4

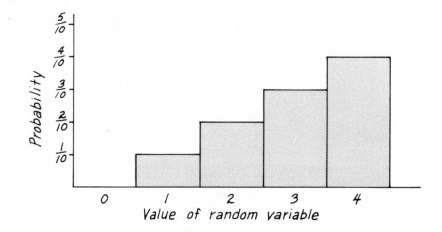

We saw in the Overview that probability distributions are extremely important in the study of statistics. We have just considered probability distributions of discrete random variables, and later chapters will consider fundamental probability distributions of continuous random variables.

EXERCISES A: RANDOM VARIABLES

4-1. Three coins are tossed. Let the random variable x represent the number of tails that turn up.
   (a) List the values that x can assume.
   (b) Determine the probability $P(x)$ for each value of x.
   (c) Construct the histogram that represents the probability distribution for this random variable x.
   (d) Determine the area of each rectangle in the histogram of part (c).

4-2. Do Exercise 4-1 under the assumption that 4 coins are tossed. Compare the shape of the resulting histogram to the histogram obtained in part (c) of Exercise 4-1.

4-3. Two dice are rolled. Let the random variable x represent the total of the dots that turn up.
   (a) List the values that x can assume.
   (b) Determine the probability $P(x)$ for each value of x.

(c) Construct the histogram that represents the probability distribution for this random variable x.

(d) Determine the area of each rectangle in the histogram of part (c).

4-4. A pocket contains 3 pennies, 4 nickels, 6 dimes, and 7 quarters. One coin is randomly selected. Let the random variable x be the value (in cents) of the chosen coin.

(a) List the values that x can assume.

(b) Determine the probability $P(x)$ for each value of x.

(c) Construct the histogram that represents the probability distribution for this random variable x.

(d) Determine the area of each rectangle in the histogram of part (c).

In Exercises 4-5 through 4-10, determine whether a probability distribution is given. Construct a histogram for those that are probability distributions.

4-5. $P(x) = x/10$ for $x = 0, 1, 2, 3, 4, 5$.

4-6. $P(x) = x/20$ for $x = 2, 3, 4, 5, 6$.

4-7.

x	$P(x)$
1	1/12
2	2/12
3	3/12
4	3/12
5	2/12
6	1/12

4-8.

x	$P(x)$
1	1/10
3	2/10
5	3/10
7	2/10
9	1/10

4-9.

x	$P(x)$
0	1/4
1	1/4
2	1/4
3	1/4
4	1/4

4-10. $P(x) = \dfrac{5 - x}{10}$ for $x = 1, 2, 3, 4, 5$.

EXERCISES B: RANDOM VARIABLES

4-11. Toss three coins 100 times and construct a histogram for the number of heads that occurred each time. Compare this histogram to the histogram of the probability distribution obtained in Exercise 4-1.

4-12. Roll a pair of fair dice 100 times and construct a histogram for the totals obtained each time. Compare this histogram to the histogram of the probability distribution obtained in Exercise 4-3.

4-3 MEAN AND VARIANCE FOR A DISCRETE PROBABILITY DISTRIBUTION

We know how to find the mean and variance for a given set of scores (see Chapter 2), but suppose we have scores that are ''conceptualized'' in the sense that we know the probability distribution instead of the actual scores?

Psychologists, biochemists, and medical researchers are among the many professionals who investigate phenomena through experimentation. Analysis of experimental results requires a use and knowledge of standard statistical procedures.

For example, suppose that we have a fair die and we seek the mean and variance for the number of dots that will turn up in future rolls. We have no specific results to work with, but we do know that each outcome (1, 2, 3, 4, 5, 6) has a probability of 1/6, and in the long run each outcome will occur an average of once for each 6 rolls. One way to find the mean and variance for the number rolled is to pretend that theoretically ideal results actually occurred. For our fair die, we can pretend that 6 rolls were made and the results were 1, 2, 3, 4, 5, 6. The mean of these 6 scores is

$$\frac{1 + 2 + 3 + 4 + 5 + 6}{6} = \frac{21}{6} = 3.5$$

while the variance is computed to be 35/12. (The variance is found by applying Formula 2-4. In this use of Formula 2-4 we divide by n instead of $n - 1$ because we assume that we have all scores of the population.)

As another example, consider the problem of determining the mean and variance for the number of boys that will occur in two independent births. (Assume that boys and girls are equally likely to occur.) We list the different possible outcomes:

$$
\begin{array}{ll}
\text{boy–boy} & (2) \\
\text{boy–girl} & (1) \\
\text{girl–boy} & (1) \\
\text{girl–girl} & (0)
\end{array}
$$

Since they are equally likely, we can pretend that the four results actually did occur. The mean of 2, 1, 1, and 0 is 1, while the variance is 0.5. Instead of considering four trials which yield theoretically ideal results, we could pretend that a large number of trials (say 4000) yielded theoretically ideal results as follows:

boy–boy (2 boys 1000 times)	Number of boys	Frequency
boy–girl (1 boy 1000 times)	2	1000
girl–boy (1 boy 1000 times)	1	2000
girl–girl (0 boys 1000 times)	0	1000

Our random variable represents the number of boys in two births, so the preceding results suggest a list of 4000 scores that consist of one thousand 2s, a total of two thousand 1s, and one thousand 0s. The mean of those 4000 scores is again 1, while the variance is again 0.5. In this example, it really makes no difference whether we presume 4 theoretically ideal trials or 4000.

Instead of pretending that theoretically ideal results have actually occurred, we can find the mean of a discrete random variable by Formula 4-1:

Formula 4-1 $MEAN\ \mu = \Sigma\ x \cdot P(x)$

This formula is justified by relating $P(x)$ to its role in describing the relative frequency with which x occurs. Recall that the mean of any list of scores is the sum of those scores divided by the total number of scores n. We usually compute the mean in that order; that is, first sum the scores and then divide the total by n. However, we can obtain the same result by dividing each individual score by n and then summing the quotients. For example, to find the mean of 2, 1, 1, 0 we usually write

$$\frac{2 + 1 + 1 + 0}{4} = 1$$

but we can also compute the mean as

$$\frac{2}{4} + \frac{1}{4} + \frac{1}{4} + \frac{0}{4} = 1$$

Similarly, we can find the mean of the preceding 4000 scores by writing

$$\frac{(2 \cdot 1000) + (1 \cdot 2000) + (0 \cdot 1000)}{4000} = 1$$

but we can also compute that mean as

$$\left(2 \cdot \frac{1000}{4000}\right) + \left(1 \cdot \frac{2000}{4000}\right) + \left(0 \cdot \frac{1000}{4000}\right) = 1$$

The latter approach is employed in Formula 4-1.

For each specific value of x, $P(x)$ can be considered the relative frequency with which x occurs. $P(x)$ effectively accommodates the repetition of specific x scores when we compute the mean. It also incorporates division by n directly into the summation process.

Similar reasoning enables us to take the variance formula from Chapter 2 ($\sigma^2 = \Sigma(x - \mu)^2/n$) and apply it to a random variable of a probability distribution to get $\sigma^2 = \Sigma(x - \mu)^2 \cdot P(x)$. Again the use of $P(x)$ accommodates the repetition of specific x scores and simultaneously accomplishes the division by n. This latter formula for variance is usually manipulated into an equivalent form which facilitates computations. We present this equivalent form in Formula 4-2.

Formula 4-2 $\sigma^2 = \left[\Sigma\, x^2 \cdot P(x)\right] - \mu^2$

To apply Formula 4-2 to a specific case, we square each value of x and multiply that square by the corresponding probability and then sum all of those products. We then subtract the square of the mean. The standard deviation σ can be easily obtained by simply taking the square root of the variance. If the variance is found to be 0.5, the standard deviation is $\sqrt{0.5}$ or about 0.7.

Example

Use Formulas 4-1 and 4-2 to find the mean, variance, and standard deviation of the random variable that represents the number of boys in two independent births. (Assume that a girl or a boy is equally likely to occur.)

Solution

The probability distribution is summarized in the table at left:

We find the mean by applying Formula 4-1.

x	$P(x)$
0	1/4
1	1/2
2	1/4

$$\mu = \Sigma\, x \cdot P(x)$$

$$= \left(0 \cdot \frac{1}{4}\right) + \left(1 \cdot \frac{1}{2}\right) + \left(2 \cdot \frac{1}{4}\right)$$

$$= 0 + \frac{1}{2} + \frac{2}{4}$$

$$= 1$$

We find the variance by applying Formula 4-2.

$$\sigma^2 = \left[\Sigma\, x^2 \cdot P(x)\right] - \mu^2$$

$$= \left(0^2 \cdot \frac{1}{4}\right) + \left(1^2 \cdot \frac{1}{2}\right) + \left(2^2 \cdot \frac{1}{4}\right) - 1^2$$

$$= 0 + \frac{1}{2} + 1 - 1$$

$$= \frac{1}{2} \quad \text{or} \quad 0.5$$

The standard deviation is the square root of the variance so that

$$\sigma = \sqrt{0.5} = 0.7$$

**YOU'RE NOW MORE LIKELY
TO BE MURDERED**

*It is natural to expect the number of murders will increase as the population increases. But there has been an alarming increase in the murder **rate**, which is often given as the number of murders per 100,000 persons. In one decade (1960 to 1970), that rate doubled and caused social scientists to seek an explanation. They noted that the use of the handgun has increased until it is now the weapon used in the majority of murders. A number of social scientists have suggested that the widespread depiction of murders on television and in the movies has encouraged homicides. Recently, a defense attorney claimed that a particular television show was responsible for his client committing murder. A 1966 Gallup poll showed that the majority of Americans were opposed to capital punishment, but a 1978 poll showed about 60% in favor.*

Example | Roll a pair of fair dice. Let the random variable represent the total of the two dice. The probability distribution can be described by the table at left. Find the mean, variance, and standard deviation for the total rolled.

Solution | We first compute the mean by using Formula 4-1.

x	$P(x)$
2	1/36
3	2/36
4	3/36
5	4/36
6	5/36
7	6/36
8	5/36
9	4/36
10	3/36
11	2/36
12	1/36

$$\mu = \Sigma \, x \cdot P(x)$$

$$= \left(2 \cdot \frac{1}{36} \right) + \left(3 \cdot \frac{2}{36} \right) + \left(4 \cdot \frac{3}{36} \right) + \left(5 \cdot \frac{4}{36} \right)$$

$$+ \left(6 \cdot \frac{5}{36} \right) + \left(7 \cdot \frac{6}{36} \right) + \left(8 \cdot \frac{5}{36} \right) + \left(9 \cdot \frac{4}{36} \right)$$

$$+ \left(10 \cdot \frac{3}{36} \right) + \left(11 \cdot \frac{2}{36} \right) + \left(12 \cdot \frac{1}{36} \right)$$

$$= \frac{2}{36} + \frac{6}{36} + \frac{12}{36} + \frac{20}{36} + \frac{30}{36} + \frac{42}{36} + \frac{40}{36}$$

$$+ \frac{36}{36} + \frac{30}{36} + \frac{22}{36} + \frac{12}{36}$$

$$= \frac{252}{36} = 7$$

We now compute the variance by using Formula 4-2.

$$\sigma^2 = \left[\Sigma \, x^2 \cdot P(x) \right] - \mu^2$$

$$= \left(2^2 \cdot \frac{1}{36} \right) + \left(3^2 \cdot \frac{2}{36} \right) + \left(4^2 \cdot \frac{3}{36} \right) + \left(5^2 \cdot \frac{4}{36} \right)$$

$$+ \left(6^2 \cdot \frac{5}{36} \right) + \left(7^2 \cdot \frac{6}{36} \right) + \left(8^2 \cdot \frac{5}{36} \right) + \left(9^2 \cdot \frac{4}{36} \right)$$

$$+ \left(10^2 \cdot \frac{3}{36} \right) + \left(11^2 \cdot \frac{2}{36} \right) + \left(12^2 \cdot \frac{1}{36} \right)$$

$$- 7^2$$

$$= \frac{4}{36} + \frac{18}{36} + \frac{48}{36} + \frac{100}{36} + \frac{180}{36} + \frac{294}{36} + \frac{320}{36} + \frac{324}{36}$$

$$+ \frac{300}{36} + \frac{242}{36} + \frac{144}{36} - 49$$

$$= \frac{1974}{36} - 49 \overset{\circ}{=} 5.8$$

Since the standard deviation σ is the square root of the variance $\sigma^2 \overset{\circ}{=} 5.8$, we get $\sigma \overset{\circ}{=} \sqrt{5.8} \overset{\circ}{=} 2.4$. In summary, we have a mean of 7, a variance of 5.8, and a standard deviation of 2.4. Since the required computations are sometimes lengthy, it can be helpful to organize them as shown in Table 4-3.

TABLE 4-3

x	$P(x)$	$x \cdot P(x)$	x^2	$x^2 \cdot P(x)$
2	1/36	2/36	4	4/36
3	2/36	6/36	9	18/36
4	3/36	12/36	16	48/36
5	4/36	20/36	25	100/36
6	5/36	30/36	36	180/36
7	6/36	42/36	49	294/36
8	5/36	40/36	64	320/36
9	4/36	36/36	81	324/36
10	3/36	30/36	100	300/36
11	2/36	22/36	121	242/36
12	1/36	12/36	144	144/36
TOTAL		7		$\dfrac{1974}{36}$

$$\mu = 7$$
$$\sigma^2 = \frac{1974}{36} - 7^2 = 5.8$$

An important advantage of these techniques is that a probability distribution is actually a model of a theoretically perfect population frequency distribution.

If we could roll a pair of fair dice an infinite number of times and construct a frequency table for the results, it would look essentially like Figure 4-5. Since the probability distribution allows us to perceive the population, we are able to determine the values of important parameters such as the mean,

FIGURE 4-5

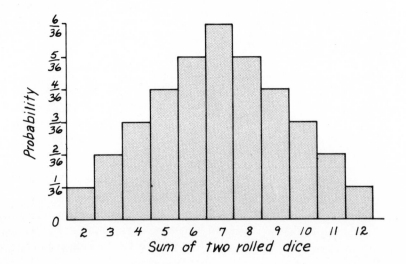

variance, and standard deviation. This in turn allows us to make the inferences that are necessary for decision-making in a multitude of different professions.

EXERCISES A: MEAN AND VARIANCE FOR A DISCRETE PROBABILITY DISTRIBUTION

In Section 4-2 we presented some exercises dealing with probability distributions. In Exercises 4-13 through 4-18, find the mean, variance, and standard deviation of the probability distribution in the exercise named. (The exercises appear on pages 106–107.)

4-13. Exercise 4-1.   4-14. Exercise 4-2.

4-15. Exercise 4-4.   4-16. Exercise 4-6.

4-17. Exercise 4-7.   4-18. Exercise 4-10.

4-19. There is a 0.18 probability that a World Series will last four games, a 0.18 probability that it will last five games, a 0.20 probability that it will last six games, and a 0.44 probability that it will last seven games. Find the mean, variance, and standard deviation for the number of games that World Series' contests last.

4-20. Incoming phone calls in a sheriff's office have the following probability distribution. Find the mean, variance, and standard deviation for the number of incoming calls.

$$4^2(.1) + 5^2(.2) + 6^2(.05) + 7^2(.4) + 8^2(.25) = 6.5$$

$$1.75 \cap 1\frac{3}{4}$$

$$1.3228$$

x	$P(x)$
4	0.10
5	0.20
6	0.05
7	0.40
8	0.25

4-21. A travel agent determines the following probability distribution for the number of customers each day. Find the mean, variance, and standard deviation for the number of daily customers.

x	$P(x)$
0	0.1
1	0.1
2	0.3
3	0.4
4	0.1

4-22. A pilot determines the following probability distribution for the number of radio frequency changes that she must make in a 4-hour flight. Find the mean, variance, and standard deviation for the number of frequency changes required.

x	$P(x)$
6	0.05
7	0.10
8	0.05
9	0.05
10	0.10
11	0.20
12	0.20
13	0.15
14	0.05
15	0.05

4-23. The probability distribution for the number of aircraft arriving at an airport in a given time period follows. Find the mean, variance, and standard deviation for the number of arrivals.

x	$P(x)$
0	0.01
1	0.10
2	0.30
3	0.36
4	0.18
5	0.05

4-24. A telephone company survey studies the number of times a telephone will ring before it is answered. Assume that the probability distribution is as follows. Find the mean, variance, and standard deviation for the number of rings.

x	$P(x)$
1	0.10
2	0.05
3	0.27
4	0.34
5	0.12
6	0.08
7	0.03
8	0.01

4-25. An automatic device is used to set newsprint. The probability distribution for the number of errors per column follows. Find the mean, variance, and standard deviation for the number of errors per column.

x	P(x)
0	0.03
1	0.04
2	0.09
3	0.08
4	0.15
5	0.18
6	0.12
7	0.10
8	0.13
9	0.06
10	0.02

4-26. A couple plans to have five children. Let the random variable be the number of boys that will occur. Assume that a boy or a girl is equally likely to occur and that the sex of any successive child is unaffected by previous brothers or sisters.
(a) List the 32 different possible simple events.
(b) Enter the probabilities in the following table where x represents the number of boys in the five births.

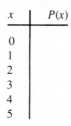

x	P(x)
0	
1	
2	
3	
4	
5	

(c) Graph the histogram for the probability distribution from (b).
(d) Find the mean number of boys that will occur among the five births.
(e) Find the variance for the number of boys that will occur.
(f) Find the standard deviation for the number of boys that will occur.
(g) On the histogram in part (c), identify the location of the mean from part (d). Also identify the location of the value that is exactly one standard deviation above the mean. Then identify the location of the value that is exactly one standard deviation below the mean.

EXERCISES B: MEAN AND VARIANCE FOR A DISCRETE PROBABILITY DISTRIBUTION

4-27. Verify that $\sigma^2 = \left[\Sigma\ x^2 \cdot P(x) \right] - \mu^2$ is equivalent to $\sigma^2 = \Sigma\ (x - \mu)^2 \cdot P(x)$.

4-28. A discrete random variable can assume the values 1, 2, . . . , n and those values are equally likely.
    (a) Show that $\mu = (n + 1)/2$.
    (b) Show that $\sigma^2 = (n^2 - 1)/12$.

        Hint: $1 + 2 + 3 + \cdots + n = n(n + 1)/2$.

               $1^2 + 2^2 + 3^2 + \cdots + n^2 = n(n + 1)(2n + 1)/6$.

4-29. The variance for the discrete random variable x is 1.25.
    (a) Find the variance of the random variable $5x$.
    (b) Find the variance of the random variable $x/5$.
    (c) Find the variance of the random variable $x + 5$.
    (d) Find the variance of the random variable $x - 5$.

4-4 BINOMIAL EXPERIMENTS

Statistics books of all varieties include an abundance of coin-tossing examples. While it's true that the typical American does not require an extensive knowledge of the theory of coin-tossing, this simple event does have certain characteristics that apply to more practical circumstances. For example, tossing coins can be considered a fast and inexpensive simulation of births where the outcomes of boy and girl correspond to heads and tails. We can adjust the standard coin-tossing event to include unfair or biased coins, where occurrences of heads and tails differ from the usual 50–50 distribution. These biased coin experiments can be used to simulate and analyze a multitude of real and practical circumstances such as:

- Elections: voters choosing between two candidates.
- Manufacturing: machines producing acceptable or defective items.
- Education: subjects passing or failing a test.
- Medicine: new drugs being effective or ineffective.
- Psychology: mental health treatment being effective or ineffective.
- Agriculture: crops being profitable or nonprofitable.

All of these situations exhibit an element of "twoness" that induces a particular distribution appropriately called the **binomial distribution**. Specifically, the binomial distribution corresponds to a fixed number of indepen-

DETECTING SYPHILIS

In the 1940s, the United States Army included the Kahn–Wasserman test in its physical examination of inductees. This test, designed to detect the presence of syphilis, required blood samples that were given chemical analyses. Since the testing procedure was time consuming and expensive, a more efficient one was sought. One researcher noted that, if samples from a large number of blood specimens were mixed in pairs and then tested, the total number of chemical analyses would be greatly reduced. Syphilitic inductees could be identified by retesting the few blood samples which were included in the pairs that indicated the presence of syphilis. But if the total number of analyses was reduced by pairing blood specimens, why not put them in groups of three or four or more? Probability theory was used to find the most efficient group size, and a general theory was developed for detecting the defective members of any population. Bell Laboratories, for example, used this approach to identify defective condensers by putting groups of them in a vacuum and testing for leakage.

dent trials where each outcome has only two classifications. The term independent simply means that the outcome of one trial will not affect the probabilities of subsequent trials. If a coin is tossed, the result of heads or tails will not affect the result of any subsequent tosses. The birth of a girl does not affect the sex of the next baby that is born. **A result of this independence of trials is that the probability of either outcome remains constant.** In many cases, the requirement of independence is not strictly satisfied but, for all practical purposes, independence can be assumed.

Pretend, for example, that of 10,000 radios produced, 200 are defective. If we devise a sampling scheme based on the random selection of two radios, we can compute the following:

$$P(\text{first radio is defective}) = 200/10{,}000 = 0.02$$

$$P(\text{second radio is defective}) = \frac{199}{9999} \stackrel{\circ}{=} 0.0199$$

or

$$= \frac{200}{9999} \stackrel{\circ}{=} 0.0200 \quad \textit{depending on whether or not the first radio was defective}$$

Technically, these trials are not independent because the first trial does affect the second, so the probabilities do not remain constant. However, if we assign probabilities the value of 0.02, we have an acceptable error that enables us to use the binomial distribution effectively. If our population consisted of 10 radios instead of 10,000, the error would be gross instead of negligible and the binomial requirements could not be assumed. The binomial requirements are summarized here and, as long as they are satisfied approximately, the binomial distribution can be assumed.

Requirements for an experiment to be binomial:

1. The experiment must have a fixed number of trials.
2. The trials must be independent.
3. Each trial must have all outcomes classified into two categories.
4. The probabilities must remain constant for each trial.

If a real experimental trial has many outcomes, each such outcome must be classified into one of two categories to use the binomial distribution. In a multiple-choice question, there may be five possible selections, but the two categories of "right" and "wrong" are suitable for a binomial model. In randomly selecting a number from 1 to 100, we have 100 possible choices, but we can devise the two categories of "8" and "not 8" for binomial purposes. In short, the property of twoness applies to possible categories and not to specific individual outcomes or simple events.

S and F (success and failure) denote the two possible categories of all outcomes. p and q will denote the probabilities of S and F, respectively, so that

$$P(S) = p$$

$$P(F) = 1 - p = q$$

n will denote the fixed number of trials.

x will denote a specific number of successes in n trials so that x can be any whole number between 0 and n, inclusive.

The word success as used here does not necessarily correspond to a good event. Selecting a defective parachute may be classified a success, even though the results of such a selection may be somewhat less than pleasant. Either of the two possible categories may be called the success S as long as the corresponding probability is identified as p; q can always be found by subtracting p from 1. If $p = 0.4$, then $q = 1 - 0.4$ or 0.6.

Example   A drug given to five patients is effective in 10% of its applications. In this binomial experiment, $n = 5$, $P(S) = p = 0.1$, $P(F) = q = 0.9$, and the number of successes x can be 0, 1, 2, 3, 4, or 5. The trials are independent and the two relevant categories are "effective" and "ineffective."

Example   A baseball player, who has a 31% chance of getting a hit (0.310 batting average), gets four turns at bat during a game. In this binomial experiment, $n = 4$, $P(S) = p = 0.31$, $P(F) = q = 0.69$, and the number of hits x can be 0, 1, 2, 3, or 4. The trials are essentially independent and the two categories are "hit" and "no hit."

Example   Ten special flashbulbs are produced, and five of them are defective. Three bulbs are selected for testing. This does not define a binomial experiment since the small size of the population mandates nonconstant probabilities. That is, the three trials are not independent when the selections are made without replacement as is natural in this sampling experiment.

Let's assume that we do have a binomial experiment in which we seek certain probabilities. For example, a drug with a 10% cure rate is given to five randomly selected patients. We want to determine the probability of getting exactly two cures in the five trials. Letting a cure represent a success, we have $n = 5$, $P(S) = p = 0.1$, $P(F) = q = 0.9$, and we want $P(x)$ for $x = 2$. (That is, we seek $P(2)$.) We can apply the multiplication rule for independent events (see Section 3-4) to find the probability that the first two patients are cured by the drug while the remaining three are not cured.

$$0.1 \cdot 0.1 \cdot 0.9 \cdot 0.9 \cdot 0.9 = 0.00729$$

There are other ways or orders of listing two cures and three failures. Let S represent a cure and F a failure. The probability of 0.00729 represents the particular sequence of S-S-F-F-F, but ten arrangements are possible, and each arrangement has a probability of 0.00729.

$$S\text{-}S\text{-}F\text{-}F\text{-}F$$
$$F\text{-}F\text{-}F\text{-}S\text{-}S$$
$$F\text{-}F\text{-}S\text{-}F\text{-}S$$
$$F\text{-}S\text{-}F\text{-}S\text{-}F$$
$$S\text{-}F\text{-}F\text{-}F\text{-}S$$
$$F\text{-}F\text{-}S\text{-}S\text{-}F$$
$$F\text{-}S\text{-}S\text{-}F\text{-}F$$
$$S\text{-}F\text{-}F\text{-}S\text{-}F$$
$$S\text{-}F\text{-}S\text{-}F\text{-}F$$
$$F\text{-}S\text{-}F\text{-}F\text{-}S$$

Since there are ten ways of getting exactly two cures among the five patients, and each different way has a probability of 0.00729, there is a total probability of $10 \cdot 0.00729$ or 0.0729 that two of the five subjects will be cured by the drug. In general, with n independent trials with $P(S) = p$ and $P(F) = q$, the multiplication rule indicates that the probability of the first x cases being successes while the remaining cases are failures is

$$\underbrace{p \cdot p \cdots \cdots p}_{x \text{ times}} \cdot \underbrace{q \cdot q \cdots \cdots q}_{(n-x) \text{ times}}$$

or

$$p^x \cdot q^{n-x}$$

But $P(x)$ denotes the probability of x successes among n trials in any order, so that $p^x \cdot q^{n-x}$ must be multiplied by the number of ways the x successes and $n - x$ failures can be arranged. The binomial theorem for the expansion of $(x + y)^n$ can be used to show that the number of arrangements we need is given by

$$\frac{n!}{(n-x)! \, x!}$$

where the factorial symbol ! denotes the product of decreasing factors as illustrated in the following examples. (This rationale is not very motivating, but we can see the role of $n!/(n - x)! \, x!$ in the example that follows the binomial probability formula.)

Example

$5! = 5 \cdot 4 \cdot 3 \cdot 2 \cdot 1 = 120$
$4! = 4 \cdot 3 \cdot 2 \cdot 1 = 24$
$3! = 3 \cdot 2 \cdot 1 = 6$
$2! = 2 \cdot 1 = 2$
$1! = 1$
$0! = 1$ (by special definition)
$n! = n \cdot (n - 1) \cdot (n - 2) \cdot \ldots \cdot 1$

We can now combine our result from the binomial theorem with the direct application of the multiplication rule for independent events to get a general formula for computing binomial probabilities.

Binomial proba-bility formula

$$P(x) = \frac{n!}{(n - x)! \, x!} \cdot p^x \cdot q^{n-x}$$

Let's use this formula to repeat the solution to our original problem by finding the probability of exactly two cures among five randomly selected patients where $P(\text{cure}) = P(S) = p = 0.1$. In this problem, recall that $n = 5, p = 0.1$, $q = 0.9$, and $x = 2$ so that

$$P(2) = \frac{5!}{(5 - 2)! \, 2!} \cdot 0.1^2 \cdot 0.9^{5-2}$$

$$= \frac{120}{6 \cdot 2} \cdot 0.1^2 \cdot 0.9^3$$

$$= 10 \cdot 0.01 \cdot 0.729$$

$$= 0.0729$$

This specific example can help us to see the rationale underlying the preceding formula. $n!/(n - x)! \, x!$ counts the number of ways we can get x successes in n trials while $p^x \cdot q^{n-x}$ gives us the probability of getting exactly x successes among n trials in one particular order. The product of those two factors is the total probability representing all ways of getting exactly x successes in n trials. The following examples are intended to illustrate the direct use of the binomial probability formula.

Example

A baseball player bats 0.310 so that $P(\text{hit}) = 0.310$. Find the probability of exactly one hit in four trials at bat.

Solution

With $n = 4, P(S) = p = 0.310$, and $P(F) = q = 1 - 0.310 = 0.690$, we want $P(1)$. Applying the binomial probability formula directly we get

$$P(1) = \frac{4!}{(4-1)! \ 1!} \cdot 0.310^1 \cdot 0.690^3$$

$$= \frac{24}{6 \cdot 1} \cdot 0.310^1 \cdot 0.690^3$$

$$= 4 \cdot 0.310 \cdot 0.3285$$

$$\overset{\circ}{=} 0.407$$

TABLE 4-4

x	$P(x)$
0	0.227
1	0.407
2	0.275
3	0.082
4	0.009

In this example, we can also use the binomial distribution formula to find $P(0)$, $P(2)$, $P(3)$, and $P(4)$ so that the complete probability distribution for the number of hits will be known. These results are shown in Table 4-4 where x denotes the number of hits in four times at bat.

We can depict a table such as 4-4 in the form of a histogram like the one in Figure 4-6. The shape of a probability distribution is often a critically important feature. (We consider that characteristic in more detail in Section 4-6).

FIGURE 4-6

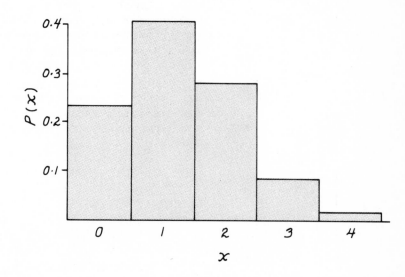

Example

In a certain county, 30% of the voters are Republicans. If ten voters are randomly selected, find the probability that six of them will be Republicans.

Solution

Assuming that the total number of voters in the county is large, the binomial probability model can be used since the necessary conditions are essentially met. From the given information we see that we must find $P(6)$ while $n = 10$, $P(S) = p = 0.3$, and $P(F) = q = 1 - 0.3 = 0.7$. Through direct application of the binomial probability formula we get

$$P(6) = \frac{10!}{(10-6)!\ 6!} \cdot 0.3^6 \cdot 0.7^{10-6}$$

$$= 210 \cdot 0.000729 \cdot 0.2401$$

$$\overset{\circ}{=} 0.037$$

An alternative to computing with the binomial probability formula involves the use of the table of binomial probabilities (see Table A-2 in Appendix A). To use this table, first locate the relevant value of n in the leftmost column and then locate the corresponding value of x that is desired. At this stage, one row of numbers should be isolated. Now align that row with the proper probability of p by using the column across the top. The isolated number represents the desired probability (missing its decimal point at the beginning). A very small probability such as 0.000000345 is indicated by 0+. For example, the table indicates that for $n = 10$ trials, the probability of $x = 2$ successes when $P(S) = p = 0.05$ is 0.075. (A more precise value is 0.0746348, but the table values are approximate.)

Example   Use either the binomial probability formula or the binomial probability table to solve the following problem: A manufacturer has produced 100,000 radios of which 5% are defective. In a random sample of six radios, find the probability of getting exactly two defective radios.

Solution   The large population size and the small number of selections will, in effect, cause the trials to be essentially independent. With $n = 6$, $P(S) = p = 0.05$ and $x = 2$, we compute

$$P(2) = \frac{6!}{(6-2)!\ 2!} \cdot 0.05^2 \cdot 0.95^4$$

$$= \frac{720}{24 \cdot 2} \cdot 0.0025 \cdot 0.8145$$

$$\overset{\circ}{=} 0.031$$

The probability of randomly selecting six radios of which two are defective is therefore 0.031. (Using the table we also get 0.031.)

Example   Find the probability that of five babies, there are exactly four girls. (Assume that a boy or a girl is equally likely to occur.)

Solution   This is a binomial experiment with $n = 5$, $P(S) = p = 0.5$, and $x = 4$. Applying the binomial probability formula we get

$$P(4) = \frac{5!}{(5-4)!\ 4!} \cdot 0.5^4 \cdot 0.5^{5-4}$$

$$= \frac{120}{1 \cdot 24} \cdot 0.0625 \cdot 0.5$$

$$= 0.15625$$

Thus, the probability of getting four girls in five births is 0.156. (Using the table we also get 0.156.)

To keep this section in perspective, remember that the binomial probability formula is but one of many probability formulas that can be used for different situations. It is, however, among the most important and most useful of all probability distributions. In practical cases it is often used in problems such as quality control, voter analysis, medical research, military intelligence, and advertising (see Exercise 4-30).

EXERCISES A:
BINOMIAL EXPERIMENTS

4-30.   Which of the following are binomial experiments?
   (a)   Testing a sample of five batteries from a population of 20 batteries of which 40% are defective.
   (b)   Polling 150 voters on the presidential election from a population of 20,000 voters if 35% are Republicans and 65% are Democrats.
   (c)   Testing a sample of eight drug dosages from a population of 500 of which 2% are contaminated.
   (d)   Polling 1000 voters in the presidential election from a population of 8 million voters of which 40% are Democrats, 35% are Republicans, and 25% are Independents.
   (e)   Firing 20 missiles at a target with a hit rate of 90%.
   (f)   Tossing an unbiased coin 500 times.
   (g)   Tossing a biased coin 500 times.
   (h)   Surveying 500 consumers to find the brands of toothpaste they prefer.
   (i)   Surveying 500 consumers to determine whether their preferred brand of toothpaste is Crest.
   (j)   Administering a driving test to 50 license applicants with a passing rate of 72%.

4-31.   In a binomial experiment, a trial is repeated n times. Find the probability of x successes if $P(S) = p$. (Use the given values of n, x, and p and the table of binomial probabilities (Table A-2).)
   (a) $n = 10$, $x = 3$, $p = 0.5$
   (b) $n = 10$, $x = 3$, $p = 0.4$

(c) $n = 7, x = 0, p = 0.01$
(d) $n = 7, x = 0, p = 0.99$
(e) $n = 7, x = 7, p = 0.01$
(f) $n = 15, x = 5, p = 0.7$
(g) $n = 12, x = 11, p = 0.6$
(h) $n = 9, x = 6, p = 0.1$
(i) $n = 8, x = 5, p = 0.95$
(j) $n = 14, x = 14, p = 0.9$

4-32. For the given values of n (the number of trials), x (the number of successes), and p (the probability of success) in a binomial experiment, why is the table of binomial probabilities inadequate? Give a specific reason for each part.
(a) $n = 30, x = 20, p = 0.3$
(b) $n = 10, x = 8, p = 0.0557$
(c) $n = 4, x = 1, p = 0.99$
(d) $n = 20, x = 10, p = 0.755$
(e) $n = 5, x = 4, p = 0.0375$

In Exercises 4-33 through 4-47, identify the values of n, x, p, *and* q *and find the value requested. (Assume that male and female births are equally likely.)*

4-33. Find the probability of getting exactly four boys in ten births.

4-34. Find the probability of getting exactly six girls in seven births.

4-35. Of 5000 families with four children, how many would be expected to have all girls?

4-36. Of 5000 families with four children, how many would be expected to have three girls and one boy?

4-37. In a certain college, 60% of the entering freshmen graduate. Find the probability that of ten random entering students, exactly six will graduate.

4-38. The probability of a computer component being defective is 0.01. Find the probability of getting exactly two defective components in a sample of ten.

4-39. A company produces batteries in batches of 500. In each batch, a sample of five batteries is tested and, if more than one defect is found, the entire batch is tested. Assume a defect rate of 1%.
(a) Find the probability of getting no defects in the sample of five.
(b) Find the probability of getting exactly one defect in the sample of five.
(c) Use the results of parts (a) and (b) to find the probability of getting more than one defect in the sample of five batteries.

4-40. A quarterback has a 45% pass completion rate. What is the probability of his completing exactly three of the next four passes?

4-41. Thirty percent of the subjects fail a certain civil service examination. Find the probability that of six subjects, exactly four pass the test.

4-42. A question on an I.Q. test is multiple choice with six possible answers of which one is correct. The question is difficult and is worded poorly, so all subjects make random guesses at the solution. Find the probability that of five test subjects, exactly three answer this question correctly.

4-43. A certain seed has a 75% germination rate. If seven seeds are planted, find the probability of having exactly six seeds germinate.

4-44. A jumbo jetliner has four independently operating engines, each having a 0.001 probability of failure. This aircraft can complete a flight if at least two engines are operating. Find the probability of completing a flight.

4-45. A magazine subscription service has found that 20% of telephone solicitations result in an order. Find the probability that of ten telephone contacts, there are fewer than two orders.

4-46. An ESP experiment involves the selection of a randomly chosen integer from 1 through 5. If this test is repeated four times, find the probability of getting at least three correct responses assuming that the subject makes random guesses.

4-47. The rhythm method of contraception is rated as 60% effective. Find the probability that it will be effective in each of five separate trials.

EXERCISES B: BINOMIAL EXPERIMENTS

4-48. The rhythm method of contraception is rated as 60% effective. Find the probability of at least three conceptions in ten separate trials.

4-49. 12,500 couples plan to have four children each. Find the number of couples expected to have:
(a) All boys.
(b) Three boys and one girl.
(c) Two boys and two girls.
(d) One boy and three girls.
(e) All girls.
Make a histogram summarizing these results.

4-50. Use the binomial probability formula to find the probability of x successes in n trials when the probability of success is p and the values of n, x, and p are as follows:
(a) $n = 5$, $x = 4$, $p = 1/3$
(b) $n = 10$, $x = 6$, $p = 3/4$

(c) $n = 6$, $x = 0$, $p = 0.001$

(d) $n = 20$, $x = 19$, $p = 1/2$

(e) $n = 10$, $x = 0$, $p = 0.0001$

4-51. The probability of winning anything in a weekly state lottery is 1/5000. If a person buys one ticket each week for a year, find the probability that he will win at least one prize.

4-5 MEAN AND VARIANCE FOR THE BINOMIAL DISTRIBUTION

The binomial distribution is a probability distribution, so the mean, variance, and standard deviation for the appropriate random variable can be found from the formulas presented in Section 4-3.

$$\mu = \Sigma \, x \cdot P(x)$$

$$\sigma^2 = \left[\Sigma \, x^2 \cdot P(x)\right] - \mu^2$$

However, these formulas which apply to all probability distributions can be significantly simplified for the special case of binomial distributions. Given the binomial probability formula

$$P(x) = \frac{n!}{(n - x)! \, x!} \cdot p^x \cdot q^{n-x}$$

and the preceding general formulas for μ and σ^2, we can pursue a series of somewhat complicated algebraic manipulations that ultimately lead to the following desired result.

> For a *binomial* experiment,
>
> $$\mu = n \cdot p$$
>
> STD. DEV. $\sigma = \sqrt{n \cdot p \cdot q}$
>
> $$\sigma^2 = n \cdot p \cdot q$$

The formula for the mean does make sense intuitively. If we were to toss a fair coin 100 times, we would expect to get about 50 heads, and $n \cdot p$ in this experiment becomes $100 \cdot 1/2$ or 50.

The variance is not so easily justified, and we prefer to omit the complicated algebraic manipulations that lead to the second formula. Instead, we will show that both of these simplified formulas do lead to the same results as the more generalized formulas from Section 4-3.

Example

Of all the mice injected with a certain drug, 30% exhibit unfavorable reactions. If some mice are injected and caged in groups of five, find the mean, variance, and standard deviation for the number of unfavorable reactions in each group.

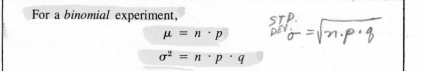

$P(S) = p = .3$

$P(F) = q = .7$

$n = 5$

$\mu = 5 \cdot .3 = 1.5$

$\sigma^2 = 5 \cdot .3 \cdot .7 = 1.05$

$\sigma = 1.024$

Solution   In this binomial experiment we will stipulate that a success is an unfavorable reaction, so we have $n = 5$, $P(S) = p = 0.3$, and $P(F) = q = 1 - 0.3 = 0.7$. We now compute

$$\mu = n \cdot p$$
$$= 5 \cdot 0.3$$
$$= 1.5$$
$$\sigma^2 = n \cdot p \cdot q$$
$$= 5 \cdot 0.3 \cdot 0.7$$
$$= 1.05$$

The standard deviation σ is simply the square root of the variance, so $\sigma = \sqrt{1.05} \doteq 1.02$. We have completed the solution since the values of μ, σ^2, and σ have been determined. However, we want to show that these same values will result from the use of the more general formulas established in Section 4-3. We begin by computing the mean using $\mu = \Sigma x \cdot P(x)$. The possible x values are 0, 1, 2, 3, 4, and 5, but we must first determine $P(0)$, $P(1)$, $P(2)$, $P(3)$, $P(4)$, $P(5)$ by using Table A-2 or the binomial probability formula $P(x) = \left[n!/(n - x)! \, x! \right] \cdot p^x \cdot q^{n-x}$. These probabilities are found to be 0.168, 0.360, 0.309, 0.132, 0.028, and 0.002, respectively.

Now

$$\mu = \Sigma x \cdot P(x)$$
$$= (0 \cdot 0.168) + (1 \cdot 0.360) + (2 \cdot 0.309) + (3 \cdot 0.132)$$
$$+ (4 \cdot 0.028) + (5 \cdot 0.002)$$
$$= 0 + 0.360 + 0.618 + 0.396 + 0.112 + 0.010$$
$$= 1.496$$

(We didn't get exactly 1.5 because of rounding errors in the probabilities.) We compute the variance as follows.

$$\sigma^2 = \left[\Sigma x^2 \cdot P(x) \right] - \mu^2$$
$$= (0^2 \cdot 0.168) + (1^2 \cdot 0.360) + (2^2 \cdot 0.309) + (3^2 \cdot 0.132)$$
$$+ (4^2 \cdot 0.028) + (5^2 \cdot 0.002) - 1.496^2$$
$$= 0 + 0.360 + 1.236 + 1.188 + 0.448 + 0.050 - 2.238$$
$$= 1.044$$

(We didn't get exactly 1.05 because of rounding errors.)

The preceding computations should lead to two conclusions. First, the use of $\mu = n \cdot p$ and $\sigma^2 = n \cdot p \cdot q$ in binomial experiments will produce the same results as the more general formulas of $\mu = \Sigma\, x \cdot P(x)$ and $\sigma^2 = [\Sigma\, x^2 \cdot P(x)] - \mu^2$. Second, the latter formulas tend to be much more complicated, so if we know an experiment is binomial, we should use the simplified formulas. They will conserve time and effort while at the same time reducing the chance for arithmetic errors.

Example   Find the mean and standard deviation for the number of girls that occur in families having exactly four children each. (Assume that a girl or a boy is equally likely to occur.)

Solution   With $n = 4$ and $P(\text{girl}) = p = 0.5$, we compute

$$\mu = n \cdot p$$
$$= 4 \cdot 0.5$$
$$= 2$$

$$\sigma = \sqrt{n \cdot p \cdot q}$$
$$= \sqrt{4 \cdot 0.5 \cdot 0.5}$$
$$= \sqrt{1} = 1$$

Example   On a four-engine aircraft, the probability of any engine failing on a flight is 0.001. Find the mean number of engine failures per flight for such aircraft. Also find the standard deviation. (Assume that the engines operate independently.)

Solution   With $n = 4$ and $P(\text{engine failure}) = p = 0.001$ we get

$$\mu = n \cdot p$$
$$= 4 \cdot 0.001$$
$$= 0.004$$

$$\sigma = \sqrt{n \cdot p \cdot q}$$
$$= \sqrt{4 \cdot 0.001 \cdot 0.999}$$
$$= \sqrt{0.003996}$$
$$\overset{\circ}{=} 0.063$$

Knowledge of the mean and standard deviation can help us to arrive at certain conclusions. For example, we will see later that for many common

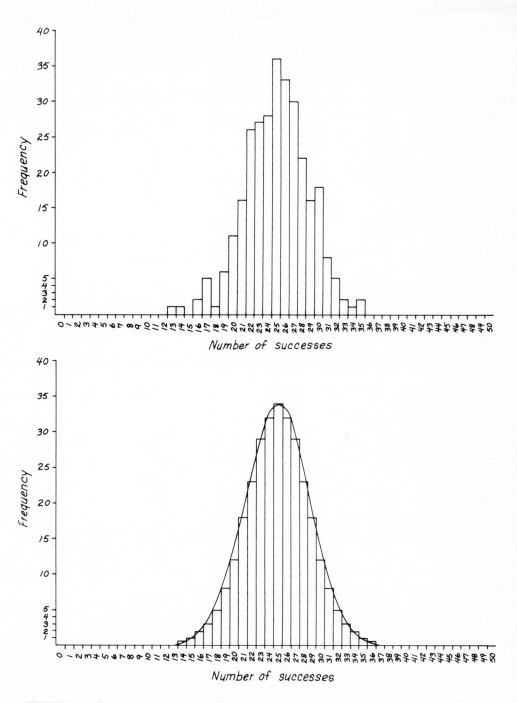

FIGURE 4-7 (top). Summary of 300 real experiments. Each experiment has 50 trials ($n = 50$) with $p = 0.5$. **FIGURE 4-8** (bottom). Summary of 300 ideal theoretical experiments. Each experiment has 50 trials ($n = 50$) with $p = 0.5$.

distributions, it is highly unusual for a score to differ from the mean by more than three standard deviations. In our last example of aircraft engines, it is very unlikely that the number of engine failures will differ from the mean of 0.004 by more than 0.186 (or three standard deviations). However, statistics of this type seem to have negligible impact on those who are afraid of flying. Statistics and reason are ineffective weapons against minds controlled by emotions and prejudices.

When n is large and p is close to 0.5, the binomial distribution tends to resemble the smooth curve which approximates the histograms in Figures 4-7 and 4-8. Note that the data tend to form a bell-shaped curve.

MOTHER HEROINE

EXERCISES A: MEAN AND VARIANCE FOR THE BINOMIAL DISTRIBUTION

In Exercises 4-52 through 4-61, find the mean μ, variance σ^2, and standard deviation σ for the given values of n *and* p. *Assume that the binomial conditions are satisfied in each case.*

4-52. $n = 16$, $p = 0.5$ 4-53. $n = 36$, $p = 0.5$

4-54. $n = 10$, $p = 0.3$ 4-55. $n = 15$, $p = 0.2$

4-56. $n = 20$, $p = 0.9$ 4-57. $n = 11$, $p = 0.1$

4-58. $n = 104$, $p = 0.6$ 4-59. $n = 25$, $p = 0.25$

4-60. $n = 18$, $p = 1/3$ 4-61. $n = 500$, $p = 0.85$

4-62.   Find the mean, variance, and standard deviation for the number of boys in families with five children. (Assume that a boy or a girl is equally likely to occur and that the sex of any child is independent of any brothers or sisters.)

4-63.   A certain computer component is manufactured in lots of 64. If 1% of the components are defective, find the mean and standard deviation for the number of defects in each lot.

4-64.   Thirty percent of the subjects fail a certain civil service examination. If the test is given to groups of 20, find the mean and standard deviation for the number of failures in each group.

4-65.   A certain type of seed has a 75% germination rate. If the seeds are planted in rows of 20, find the mean and standard deviation for the number of seeds that germinate in each row.

4-66.   A magazine subscription service has found that 20% of telephone solicitations result in an order. If each worker makes 80 calls per day, find the mean and standard deviation for the number of orders each worker receives daily.

4-67. A company manufactures batteries and 2% of its products fail before the time specified in the guarantee. If the batteries are produced in lots of 120, find the mean and standard deviation for the number of batteries in each lot that will fail before the guarantee expires.

4-68. An aircraft rated for flight in inclement weather is normally equipped with two independent radios that can be used for communications. If radios of this type have a 1.5% failure rate per flight, find the mean and standard deviation for the number of failures on a flight of an aircraft equipped with two radios.

4-69. A certain pollution-measuring instrument gives accurate readings 95% of the days that it is used. Find the mean and standard deviation for the number of days in one week for which the readings are inaccurate.

4-70. If $n = 30$ and $p = 0.25$, find the mean, variance, and standard deviation assuming that binomial conditions are satisfied.

TABLE 4-5

x	Frequency
0	1
1	17
2	62
3	112
4	142
5	95
6	45
7	22
8	3
9	1
10	0

TABLE 4-6

x	Frequency
0	0
1	5
2	22
3	63
4	96
5	125
6	102
7	64
8	19
9	3
10	1

EXERCISES B:
MEAN AND VARIANCE
FOR THE BINOMIAL DISTRIBUTION

4-71. The most profitable production arrangement for a certain transistor results in a 40% yield, meaning that 40% of the transistors produced are acceptable. The transistors are produced in groups of 10 and Table 4-5 describes the yield for 500 groups. Compute the theoretical mean and standard deviation for the number of acceptable transistors in a group of ten. Compute the mean $\bar{x}$ and standard deviation s based on the sample summarized in the table. Compare the results. (x represents the number of acceptable transistors in a group of 10.)

4-72. Table 4-6 summarizes the number of boys present in 500 families of ten children each. Compute the theoretical population mean μ and standard deviation σ for the number of boys in each family. Compute the sample mean $\bar{x}$ and standard deviation s based on the table. Compare the results. (x represents the number of boys among the ten children.) Assume that a boy or a girl is equally likely to occur and that the sex of any child is independent of any brothers or sisters.

In Exercises 4-73 through 4-76, consider as unusual anything that differs from the mean by more than twice the standard deviation. That is, unusual values are either less than $\mu - 2\sigma$ or greater than $\mu + 2\sigma$.

4-73. Is it unusual to get 450 girls and 550 boys in 1000 independent births?

4-74. Is it unusual for parents of seven children to have exactly two girls?

4-75. Is it unusual to find five defective transistors in a sample of 20 if the defective rate is 20%?

4-76. A company manufactures an appliance, gives a warranty, and 95% of its appliances do not require repair before the warranty expires. Is it unusual for a buyer of ten such appliances to require warranty repairs on two of the items?

4-6 DISTRIBUTION SHAPES

We noted in the first chapter that one of the most fascinating aspects associated with a study of statistics is the surprising regularity and predictability of events that seem to happen by chance. In this section we consider some common collections of data along with their corresponding histograms. An examination of a histogram representing data is extremely helpful in characterizing the way that the data are distributed. An understanding of the distribution may in turn lead to helpful and valuable inferences. Let's begin with a simple example involving the analysis of the last question on an I.Q. test. Assume that this last question is multiple-choice with five possible answers and that the responses of 100 subjects are summarized in Table 4-7. The histogram that corresponds to Table 4-7 is Figure 4-9. You do not have to be an expert statistician to recognize that Figure 4-9 depicts a distribution that is essentially even or uniform.

TABLE 4-7

Response	Frequency
a	20
b	22
c	19
d	18
e	21

FIGURE 4-9

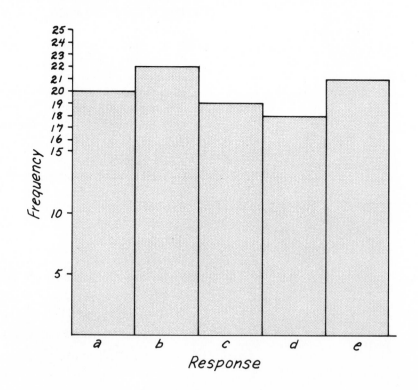

If we determine that the correct response is *b*, we might expect a higher frequency for that correct response. The nature of the distribution should tell us something about the validity or usefulness of that last test question. Because the responses are uniformly distributed, there is good reason to believe that the subjects are guessing. Perhaps there is insufficient time to consider all questions carefully and the last few responses are last minute guesses, or perhaps this last question is so incredibly difficult that everybody makes random guesses. In any event, the distribution of responses suggests that the test can be improved by changing the last question.

Let's consider a second example involving a manufacturer of compact cars who is concerned with the comfort of very tall drivers. The manufacturer seeks information about the distribution of heights and compiles the sample data for adult males summarized in Table 4-8.

This chauvinistic manufacturer isn't concerned with women's heights because most of the purchasers are men. Examination of the sampling distribution leads the manufacturer to conclude that relatively few men (about 3%) are more than 76 inches tall (see Figure 4-10). Since the accommodation of these towering torsos would demand expensive design changes, the car manufacturer elects to sacrifice that tall share of the market that equals 3%. Had the distribution been uniform as in Figure 4-9, then the manufacturer would have proceeded with the necessary design changes, thereby avoiding the sacrifice of about 20% of the potential market.

TABLE 4-8

Height (inches)	Frequency
61–64	4
65–68	27
69–72	46
73–76	20
77–80	3

FIGURE 4-10

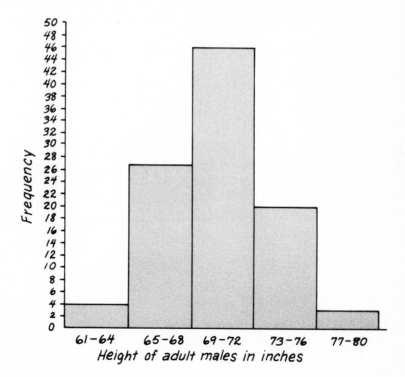

The key observation at this stage is the essential difference between the distributions of Figures 4-9 and 4-10. Both distributions are very real and arise naturally in various circumstances, yet their inherent differences lead to different inferences. Figure 4-11 presents some of the common shapes of histograms that arise in real data problems.

These histograms represent imperfect finite samplings of data. If it were possible to obtain perfect samples of extremely large size, we might represent the data by the smooth curves shown in Figure 4-12. These smooth curves become possible with extremely large data samples if the class intervals are made infinitesimally small so that the corresponding frequency polygon is comprised of connected straight-line segments so small that the net effect is a smooth curve.

FIGURE 4-11

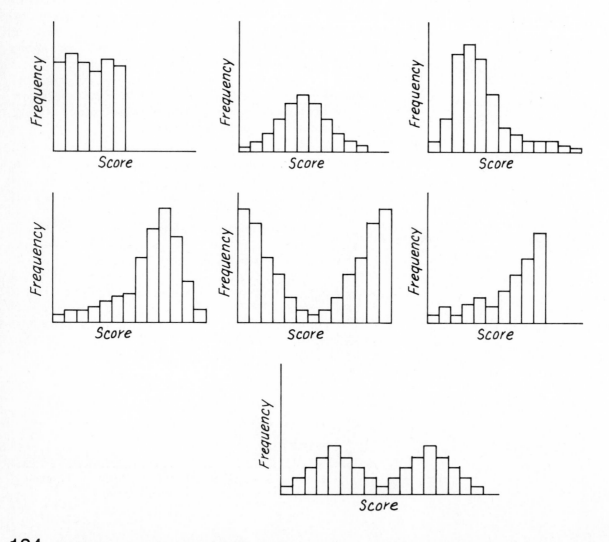

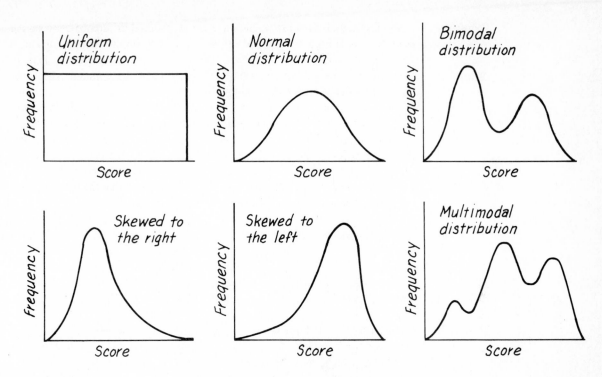

FIGURE 4-12

Armed with a knowledge of a population distribution and the key parameters (like the mean and standard deviation), we can often formulate useful inferences. The following simple example illustrates an important concept which will subsequently be applied to more useful, more realistic, and more difficult situations.

Let's assume that the Curtiss Sugar Company supplies sugar (what else?) in bags labeled 5 pounds. However, the packaging machine isn't perfect and the actual weights are *uniformly* distributed with a mean of 4.98 pounds and a range of 0.12 pound. Using only the mean, range, and type of distribution, we are able to construct the frequency polygon shown in Figure 4-13. The mean will be at the center, while the maximum is above the mean by an amount equal to one-half of the range. The minimum is below the mean by an amount equal to one-half of the range. (If we had been given the values of the minimum and maximum in a uniform distribution, the mean would be one-half of their sum.)

FIGURE 4-13

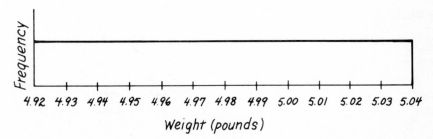

Example Given the uniform distribution of Figure 4-13, find the probability that a randomly selected bag of sugar will weigh less than 4.98 pounds.

Solution Since all bags weigh between 4.92 and 5.04 pounds, we can state that for the weight X of one randomly selected bag, $P(4.92 \le X \le 5.04) = 1$. That is, we assign 1 to the area under the line in Figure 4-13 so that we can establish a natural and usable correspondence between area and probability. Since exactly one-half of the area is associated with the stipulated weight of less than 4.98 pounds, we get a probability of 1/2 as a solution.

Example Given the same distribution as in Figure 4-13, find the probability of randomly selecting a weight between 4.93 and 5.03 pounds.

Solution The area associated with the interval between 4.93 and 5.03 is 10/12 of the total area, so the probability we seek is 10/12 or 5/6.

Example If weights less than 4.96 pounds are unacceptable, find the probability that a randomly selected bag of sugar is acceptable.

Solution Again using Figure 4-13, we see that 4/12 or 1/3 of the total area corresponds to unacceptable weights (less than 4.96), so the probability of selecting an acceptable weight must be 8/12 or 2/3.

This concept of assigning 1 to the area under a curve works well for uniform distributions, since the associated rectangles are easily partitioned into smaller units. Normal distributions similar to the one depicted in Figure 4-14 occur often in reality, but the area computations are much more difficult because of the curve involved. (Methods for normal distribution computations are pursued in Chapter 5.)

FIGURE 4-14

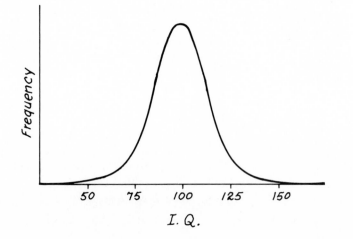

CLUSTER OF LEUKEMIA CASES IN RUTHERFORD, NEW JERSEY

In a recent 5-year period, there were 13 cases of leukemia in Rutherford, New Jersey. The expected leukemia rate for a typical community of equivalent size is about one-half of a case over a 5-year span. A New Jersey Health Department official declared that the incidence of leukemia in Rutherford is significantly greater than the national average.

Research studies of such a cluster can lead to valuable conclusions. For example, researchers discovered that asbestos fibers can be carcinogenic after studying a cluster of cancer cases near African asbestos mines. But care must be taken to avoid the statistical mistake of creating artificial clusters by placing the boundaries very close to death locations.

EXERCISES A: DISTRIBUTION SHAPES

4-77. A computer simulation involves the random generation of whole numbers between 5001 and 10000, inclusive. If the distribution of these numbers is uniform, find the probability that a randomly selected number is:
   (a)   Less than 3000.
   (b)   Greater than 8000.
   (c)   Less than 7501.
   (d)   Between 6000 and 7508.
   (e)   Greater than 8000 but less than 9000.

4-78. A set of sample data represents salaries of sales personnel employed by a vacuum cleaner manufacturer. Assuming that those salaries are uniformly distributed with a mean of $13,000 and a range of $5,000, find:
   (a)   The maximum salary.
   (b)   The minimum salary.
   (c)   The probability of randomly selecting a salary between $9,500 and $14,000.

4-79. The heights of three-year-old pine trees in New York state follow a normal bell-shaped distribution, but an experiment begins with the

formation of a uniformly distributed population having a mean of 25 inches and a range of 3 inches. For the experimental group, find:
(a) The height of the tallest tree.
(b) The height of the shortest tree.
(c) The probability of randomly selecting a tree taller than 24.5 inches.

4-80. Cholesterol levels are artificially introduced into blood samples so that the resulting mixtures are uniformly distributed with a mean of 295 and a range of 110. Find:
(a) The highest cholesterol level.
(b) The lowest cholesterol level.
(c) The probability of randomly selecting a mixed sample and getting a cholesterol level between 253 and 387, inclusive.

4-81. A machine is designed to pour 50 cc of a drug into a bottle. Because of a flaw, the machine pours amounts which are uniformly distributed with a mean of 51.5 cc and a range of 1.2 cc.
(a) What is the maximum volume of the drug that is poured by the machine?
(b) What is the minimum volume of the drug that is poured by the machine?
(c) If a poured sample is randomly selected, find the probability that it is between 51.0 cc and 52.0 cc.

4-82. Pollution levels are uniformly distributed over a city. If the mean pollution index is 53.0 with a range of 7.5, find:
(a) The maximum pollution level.
(b) The minimum pollution level.
(c) The probability of randomly selecting a place where the pollution index is between 46.0 and 50.0.

In Exercises 4-83 through 4-92, construct the frequency polygon representing the given frequency table and classify the distribution according to the following categories (see Figure 4-12): Uniform distribution; normal distribution; distribution skewed to the right; distribution skewed to the left; bimodal; and multimodal.

4-83.

Score	Frequency
0–9	1
10–19	0
20–29	2
30–39	1
40–49	3
50–59	1
60–69	4
70–79	8
80–89	9
90–99	1

4-84.

Score	Frequency
59–60	1
61–62	2
63–64	6
65–66	3
67–68	4
69–70	7
71–72	3
73–74	0
75–76	1

4-85.

Score	Frequency
60–61	1
62–63	0
64–65	6
66–67	14
68–69	23
70–71	25
72–73	17
74–75	8
76–77	3
78–79	3

4-86.

Score	Frequency
57–58	1
59–60	1
61–62	3
63–64	7
65–66	12
67–68	11
69–70	8
71–72	4
73–74	1
75–76	2

4-87.

Score	Frequency
1	73
2	69
3	70
4	74
5	68
6	72

4-88.

Score	Frequency
1	0
2	3
3	6
4	8
5	11
6	14
7	17
8	14
9	11
10	9
11	5
12	2

4-89.

Score	Frequency
0–3	94
4–7	103
8–11	101
12–15	100
16–19	96
20–23	99
24–27	105
28–31	103
32–35	107
36–39	92

4-90.

Score	Frequency
0–499	37
500–999	75
1000–1499	98
1500–1999	89
2000–2499	41
2500–2999	27
3000–3499	16
3500–3999	8
4000–4499	6
4500–4999	3

4-91.

Score	Frequency
0.0– 4.9	2
5.0– 9.9	0
10.0–14.9	5
15.0–19.9	9
20.0–24.9	12
25.0–29.9	4
30.0–34.9	3
35.0–39.9	10
40.0–44.9	14
45.0–49.9	8
50.0–54.9	6
55.0–59.9	1

4-92.

Score	Frequency
0–0.065	11
0.066–0.131	32
0.132–0.197	97
0.198–0.263	45
0.264–0.329	15
0.330–0.395	69
0.396–0.461	81
0.462–0.527	89
0.528–0.593	75
0.594–0.659	52
0.660–0.725	40
0.726–0.791	21
0.792–0.857	76
0.858–0.923	98
0.924–0.989	67
0.990–1.055	43

EXERCISES B:
DISTRIBUTION SHAPES

4-93. Given the triangular distribution, find the probability of randomly selecting a score that is:

(a) Less than 3.

(b) Between 2 and 4.

(*Hint:* The area of any triangle = 1/2 × base × height.)

4-94. Given the accompanying distribution, find the probability of randomly selecting a score that is:

(a) Less than 12.

(b) Between 5 and 15.

(c) Between 5 and 19.

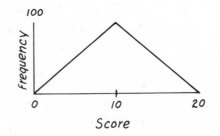

4-95. (a) If a distribution is perfectly uniform, what would you choose for the mean, median, mode, and midrange?

(b) If a distribution is perfectly normal (as in Figure 4-14), what would you guess the mean, median, mode, and midrange to be?

4-96. How is wealth distributed in the United States? Does this distribution resemble the distribution of the traits that should affect wealth?

REVIEW

The central concern of this chapter was the concept of a probability distribution. Here we dealt with **discrete** probability distributions, while successive chapters deal with continuous probability distributions.

In an experiment yielding numerical results, the **random variable** can take on those different numerical values. A **probability distribution** consists of all values of a random variable, along with their corresponding probabilities. By constructing a histogram of a probability distribution, we can see a useful correspondence between those probabilities and the areas of the rectangles in the histogram. The fundamental rules governing any probability distribution are:

1. $\Sigma P(x) = 1$ for all possible values of x.
2. $0 \leq P(x) \leq 1$ for any particular value of x.

where $P(x)$ denotes the probability of the random variable x.

Also, for any discrete probability distribution, the mean and variance are as follows:

$$\mu = \Sigma \, x \cdot P(x)$$

$$\sigma^2 = \left[\Sigma \, x^2 \cdot P(x)\right] - \mu^2$$

Of the infinite number of different probability distributions, special attention is given to the important and useful **binomial probability distribution,** which is characterized by these properties:

1. There is a fixed number of trials (denoted by n).
2. The trials must be independent.
3. Each trial must have outcomes that can be classified into *two* categories.
4. The probabilities involved must remain constant for each trial.

We saw that probabilities for the binomial distribution can be computed by reference to Table A-2 or by using the formula

$$P(x) = \frac{n!}{(n - x)! \, x!} \cdot p^x \cdot q^{n-x}$$

where
    n is the number of trials
    x is the number of successes
    p is the probability of a success
    q is the probability of a failure

For the special case of the binomial probability distribution, the mean and variance of the random variable can be easily computed by using the formulas

$$\mu = n \cdot p$$

$$\sigma^2 = n \cdot p \cdot q$$

We concluded the chapter with a section emphasizing the importance of determining the shape of a distribution and the usefulness of making a correspondence between probability and area in a histogram.

REVIEW EXERCISES

4-97. Does $P(x) = (x + 1)/5$ (for $x = -1, 0, 1, 2$) determine a probability distribution? Explain.

4-98. One coin is randomly selected from a bag containing three pennies, four nickels, six dimes, and two quarters. Let the random variable x represent the value (in cents) of the selected coin.
   (a) List the values that x can assume.
   (b) Determine the probability $P(x)$ for each value of x.
   (c) Construct the histogram which represents the probability distribution for this random variable x.
   (d) Determine the area of each rectangle in the histogram of part (c).
   (e) Find the mean μ for the value of the random variable x.
   (f) Find the variance σ^2 for the value of the random variable x.
   (g) Find the standard deviation σ for the value of the random variable x.

4-99. Forty percent of the voters in a certain region are Democrats, and five of these voters are randomly selected.
   (a) Find the probability that all five are Democrats.
   (b) Find the probability that exactly three of the five are Democrats.
   (c) Find the mean number of Democrats that would be selected in such groups of five.
   (d) Find the variance for the number of Democrats that would be selected in such groups of five.
   (e) Find the standard deviation for the number of Democrats that would be selected in such groups of five.

4-100. (a) Evaluate $P(3)$ if $P(x) = \dfrac{n!}{(n-x)!\,x!} \cdot p^x \cdot q^{n-x}$ and $n = 7$ and $p = 0.2$.
   (b) What is the mean number of heads that will turn up if we continue to toss 400 fair coins?
   (c) What is the standard deviation for the number of heads that will turn up if we continue to toss 400 fair coins?
   (d) A certain binomial experiment consists of ten trials and the mean for the random variable is zero. What is known about these trials?
   (e) Thirteen cards are dealt from a deck without any being replaced. Do these 13 trials comprise a binomial experiment? Why or why not?

4-101. A uniform distribution of weights has a mean of 60 and a range of 20.
   (a) What is the median?
   (b) Determine the minimum and maximum values.
   (c) If a score is randomly selected, what is the probability that it is less than 64?

NORMAL PROBABILITY DISTRIBUTIONS

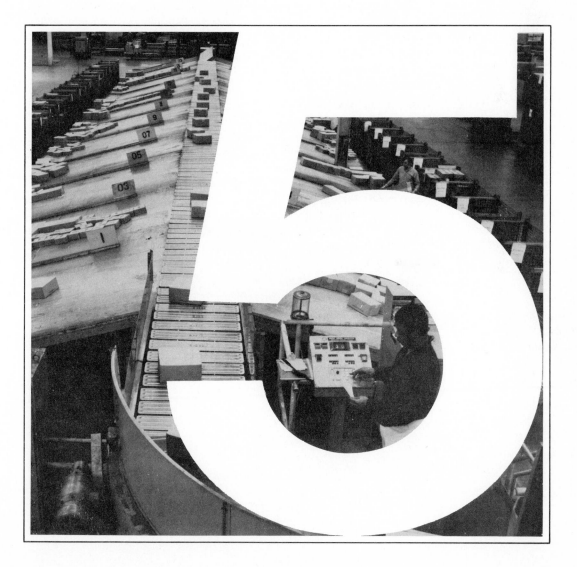

5-1 OVERVIEW

Chapter 4 explained the difference between discrete random variables and continuous random variables and dealt with discrete probability distributions. This chapter and subsequent ones consider continuous probability distributions. Specifically, this chapter involves the normal distribution, which is the most important of all continuous probability distributions because it is so widely used.

The normal distribution was originally developed as a result of studies dealing with errors that occur in various experiments. We now recognize that many real and natural occurrences, as well as many physical measurements, have frequency distributions that are approximately normal. Blood cholesterol levels, heights of adult women, weights of 5-year-old boys, diameters of New York McIntosh apples, and the lengths of newborn sharks are all examples of collections of values whose frequency polygons will closely resemble the normal probability distribution (see Figure 5-1).

Scores on standardized tests, such as I.Q. tests or college entrance examinations, also tend to be normally distributed. We will see in this chapter that sample means tend to be normally distributed, despite the distribution of the population from which the samples are drawn. Also, the normal distribution is often used as a good approximation to other types of distributions — both discrete and continuous.

The smooth bell-shaped curve shown in Figure 5-1 can be algebraically described by the equation

$$y = \frac{e^{-(x - \mu)^2/2\sigma^2}}{\sigma\sqrt{2\pi}}$$

(5-1)

where

μ represents the mean score of the entire population
σ is the standard deviation of the population
π is approximately 3.142
e is approximately 2.718

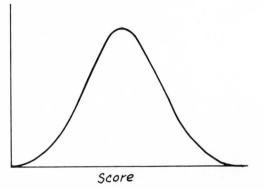

FIGURE 5-1
The normal distribution.

ARE GENIUSES PECULIAR?

A 35-year study of children with extremely high I.Q.'s showed, through statistical analysis, that a disproportionately large number of these gifted children came from families of high social class, while relatively few came from lower social and economic classes. Also, the gifted children in this sample achieved greater than average success in later years. This study further showed that the appearances and personalities of the geniuses did not tend to be peculiar. In fact, these people exhibited a lower incidence of maladjustment and unusual appearance than those having average I.Q. scores.

Equation 5-1 relates the horizontal scale of x values to the vertical scale of y values. (We will not use this equation in actual computations since we would need some knowledge of calculus to do so.)

In reality, collections of scores may have frequency polygons that approximate the normal distribution illustrated in Figure 5-1, but we cannot expect to find perfect normal distributions that conform to the precise relationship of equation (5-1). As a realistic example, consider the 150 I.Q. scores summarized in Table 5-1 and illustrated in Figure 5-2. The histogram of Figure 5-2 closely resembles the smooth bell-shaped curve of Figure 5-1, but it does contain some imperfections. In a theoretically ideal normal distribution, the tails extend infinitely far in both directions as they get closer to the horizontal axis. But this property is usually inconsistent with the limitations of reality. I.Q. scores, for example, cannot be less than zero, so no histogram of I.Q. scores can extend indefinitely far to the left. Similarly, objects cannot have negative weights nor can they have negative distances. Limitations such as these require that frequency distributions of real data can only approximate a normal distribution.

TABLE 5-1

Frequency table.

I.Q.	Frequency
55.0–64.9	1
65.0–74.9	5
75.0–84.9	15
85.0–94.9	31
95.0–104.9	39
105.0–114.9	36
115.0–124.9	15
125.0–134.9	4
135.0–144.9	3
145.0–154.9	1

FIGURE 5-2

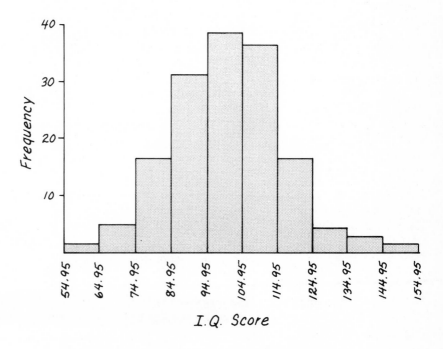

This chapter presents the standard methods used to analyze normally distributed scores, and it includes applications. In addition to the importance of the normal distribution itself, the methods are important in establishing basic patterns and concepts that will apply to other continuous probability distributions.

Original I.Q. Scores for Table 5-1 and Figure 5-2

96	107	112	101	72	147
108	94	97	104	99	75
104	84	80	95	90	115
88	99	119	75	122	109
103	84	102	130	102	99
86	102	93	87	114	105
109	93	112	89	113	68
101	121	111	101	88	105
97	136	119	110	117	80
122	87	96	115	103	98
83	79	107	117	86	88
87	74	107	113	96	95
93	80	101	114	86	88
93	84	88	107	103	74
107	104	102	83	84	105
127	77	108	107	117	135
94	103	112	101	138	104
89	114	109	106	96	76
118	121	119	80	79	101
111	98	114	107	83	94
90	99	90	108	92	109
107	76	96	117	107	130
110	128	91	100	108	102
109	104	93	88	103	85
95	118	113	74	89	98

5-2 THE STANDARD NORMAL DISTRIBUTION

There are actually many different normal probability distributions, and each one depends on only two parameters: the population mean μ and the population standard deviation σ. Figure 5-3 shows four different normal distributions of I.Q. scores where the differences are due to the changes in the mean and standard deviation. Among the infinite possibilities, one particular normal distribution is of special interest.

DEFINITION

> *The **standard normal distribution** is a normal probability distribution which has a mean of zero and a standard deviation of one.*

The choices of $\mu = 0$ and $\sigma = 1$ enable us to simplify the general algebraic equation of normal distributions as follows:

$$y = \frac{e^{-(x-\mu)^2/2\sigma^2}}{\sigma\sqrt{2\pi}} = \frac{e^{-x^2/2}}{\sqrt{2\pi}} \qquad (5\text{-}2)$$

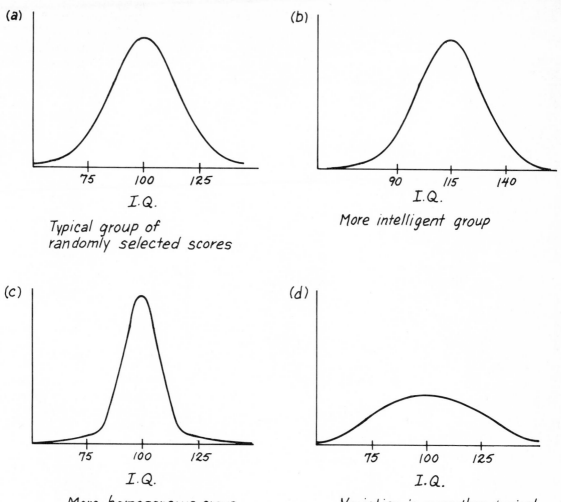

FIGURE 5-3
(a) Normally distributed I.Q. scores. Mean is 100 and standard deviation is 15. (b) Mean is 115 and standard deviation is 15. (c) Mean is 100 and standard deviation is 7. Overall intelligence is average, but the scores vary less than in a typical group. (d) Mean is 100 and standard deviation is 22. Overall intelligence is average, but the scores vary less than in a typical group.

These particular values of μ and σ therefore lead to an equation which cannot be simplified more. $y = e^{-x^2/2} / \sqrt{2\pi}$ is the easiest form of the general equation. Working with this simplified form, mathematicians are able to perform various analyses and computations. Figure 5-4 shows a graph of the standard normal distribution with some of the computed results. For example, the area under the curve and bounded by scores of 0 and 1 is 0.3413. The area under the curve and bounded by scores of −1 and −2 is 0.1359. The sum of the six known areas in Figure 5-4 is 0.9974, but if we include the small areas in the two tails we get a total of 1. (In any probability distribution, the total area under the curve is 1.)

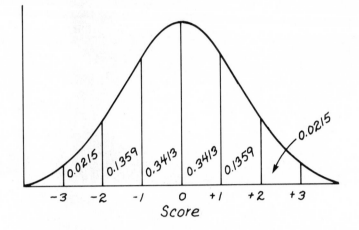

FIGURE 5-4
Standard normal distribution.

Figure 5-4 illustrates only six probability values, but a more complete table has been compiled to provide more precise data. Table A-3 (in Appendix A) gives the probability corresponding to the area under the curve bounded on the left by a vertical line above the mean of zero and bounded on the right by a vertical line above any specific positive score (denoted by z) (see Figure 5-5). Note that when you use Table A-3, the hundredths part of the z score is found across the top row. To find the probability associated with a score between 0 and 1.23, for example, begin with the z score of 1.23 by locating 1.2 in the left column. Then survey the adjoining row of probabilities to find the value that is directly below 0.03. There is a probability of 0.3907 of randomly

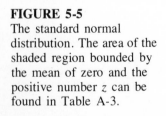

FIGURE 5-5
The standard normal distribution. The area of the shaded region bounded by the mean of zero and the positive number z can be found in Table A-3.

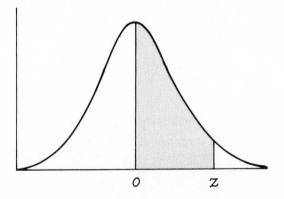

HOW VALID ARE UNEMPLOYMENT FIGURES?

The actions of presidents, economists, budget directors, corporation heads, and many other key decision-makers are often based on the monthly unemployment figures. Yet these figures have been criticized as being exaggerated or understated. Julius Shickin, a commissioner of the Bureau of Labor Statistics, says that "if you're thinking of unemployment in terms of economic potential, then the answer is no, the figure doesn't overstate the problem. But if you're thinking of unemployment in terms of hardships then the answer is yes it does. It all depends upon your value judgments."

Critics point out that, in developing unemployment figures, families are asked whether anyone out of work and aged 16 years or older has actually looked for work sometime within the last 4 weeks. However, some people have been out of work so long that they have abandoned efforts to find a job, while others do ask about work, but they do it in very casual ways. At the other end of the scale is a chemical engineer who is counted as employed even though she is driving a cab 35 hours a week while she looks for meaningful employment. Such people distort the unemployment figure, which is arrived at through a survey involving about 50,000 families each month. The survey data are sent to Washington, processed by the Census Bureau, and then given to the Bureau of Labor Statistics. This procedure, in effect since 1940, costs about $5 million per year.

selecting a score between zero and 1.23. It is essential to remember that this table is designed only for the standard normal distribution which has a mean of zero and a standard deviation of one. Nonstandard cases will be pursued later.

Since normal distributions originally resulted from studies of experimental errors, the following examples that deal with errors in measurements should be helpful.

Example A manufacturer of scientific instruments produces thermometers that are supposed to give readings of 0 degrees Celsius at the freezing point of water. Tests on a large sample of these instruments reveal that some readings are too low (denoted by negative numbers) and that some readings are too high (denoted by positive numbers). Assume that the mean reading is 0 degrees while the standard deviation of the readings is 1 degree. Also assume that the frequency distribution of errors closely resembles the normal distribution. If one thermometer is randomly selected, find the probability that, at the freezing level of water, the reading is between 0 degrees and $+1.58$ degrees.

Solution We are dealing with a standard normal distribution and we are looking for the area of the shaded region in Figure 5.5 with $z = 1.58$. We find from Table A-3 that the shaded area is 0.4429. The probability of randomly selecting a thermometer with an error between 0 degrees and $+1.58$ degrees is therefore 0.4429.

The solutions to this and the following examples are contingent on the values listed in Table A-3. But these values did not appear spontaneously. They were arrived at through calculations that relate directly to equation (5-2). Table A-3 therefore serves as a convenient means of circumventing difficult computations involving that equation.

Example With the thermometers from the preceding example, find the probability of randomly selecting one thermometer which reads (at the freezing point of water) between 0 degrees and -2.43 degrees.

Solution We are looking for the region shaded in Figure 5-6(a), but Table A-3 is designed to apply only to regions to the right of the mean (zero) as in Figure 5-5. However, by observing that the normal probability distribution possesses symmetry about zero, we see that the shaded regions in parts (a) and (b) of Figure 5-6 have the same area. Referring to Table A-3 we can easily determine that the shaded area of Figure 5-6(b) is 0.4925, so the shaded area of Figure 5-6(a) must also be 0.4925. That is, the probability of randomly selecting a thermometer with an error between 0 degrees and -2.43 degrees is 0.4925.

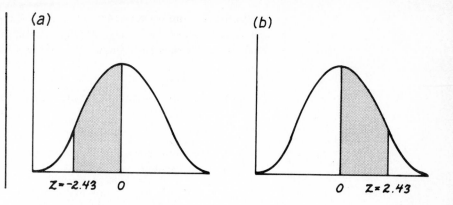

FIGURE 5-6

(a)   (b)

$Z = -2.43$ O O $Z = 2.43$

We incorporate an obvious but useful observation in the following example, but first go back for a minute to Figure 5-5. A vertical line directly above the mean of zero divides the area under the curve into two equal parts, each containing an area of 0.5. (Remember that since we are dealing with a probability distribution, the total area under the curve must be 1.)

Example With these same thermometers, we again make a random selection. Find the probability that the chosen thermometer reads (at the freezing point of water) greater than +1.27 degrees.

Solution We are again dealing with normally distributed values having a mean of 0 degrees and a standard deviation of 1 degree. The probability of selecting a thermometer which reads above +1.27 degrees corresponds to the shaded area of Figure 5-7. Table A-3 cannot be used to find that area directly, but we can use the table to find the adjacent area of 0.3980. We can now reason that

FIGURE 5-7

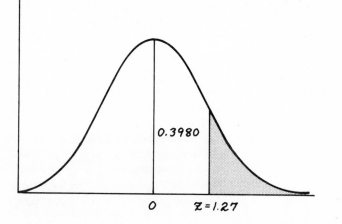

0.3980

O $Z = 1.27$

since the total area to the right of zero is 0.5, the shaded area is 0.5 − 0.3980 or 0.1020. Thus there is a 0.1020 probability of randomly selecting one of the thermometers with a reading greater than +1.27 degrees.

We are able to determine the area of the shaded region in Figure 5-7 by an *indirect* application of Table A-3. The following example illustrates yet another indirect use.

Example Back to the same thermometers. Assume that one thermometer is randomly selected and find the probability that it reads (at the freezing point of water) between 1.20 degrees and 2.30 degrees.

Solution The probability of selecting a thermometer which reads between +1.20 degrees and +2.30 degrees corresponds to the shaded area of Figure 5-8. However, Table A-3 is designed to provide only for regions bounded on the left by the vertical line above zero. We can use the table to find the areas of 0.3849 and 0.4893 as shown in this figure. If we denote the area of the shaded region by A, we can see from the figure that

$$0.3849 + A = 0.4893$$

so that
$$A = 0.4893 − 0.3849$$
$$= 0.1044$$

The probability we seek is therefore 0.1044.

FIGURE 5-8

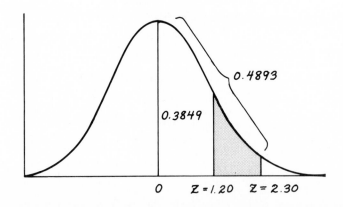

The previous examples were pat in the sense that the mean of zero and the standard deviation of one coincided exactly with the values of the standard normal distribution described in Table A-3. In reality, it would be unusual to find such a nice relationship, since typical normal distributions involve means different from zero and standard deviations different from one.

These nonstandard normal distributions introduce another problem. What table of probabilities can be used since Table A-3 is designed around a mean of zero and a standard deviation of one? For example, I.Q. scores are normally distributed with a mean of 100 and a standard deviation of 15. Scores in this range are far beyond the scope of Table A-3. Section 5-3 examines these nonstandard normal distributions and the methods used in dealing with them.

EXERCISES A: THE STANDARD NORMAL DISTRIBUTION

In Exercises 5-1 through 5-25, assume that the readings on the thermometers are normally distributed with a mean of 0 degrees and a standard deviation of 1 degree. A thermometer is randomly selected and tested. Find the probability that the reading in degrees is:

5-1. Between 0 and 0.89

5-2. Between 0 and 2.17

5-3. Less than 0

5-4. Greater than 0

5-5. Between 0 and -1.62

5-6. Between 0 and -2.7

5-7. Between -1.8 and 1.65

5-8. Between -0.35 and 0.82

5-9. Between -2.1 and 2.9

5-10. Between -0.2 and 2.57

5-11. Between 1.6 and 2.4

5-12. Between 0.3 and 1.8

5-13. Between -0.8 and -0.1

5-14. Between -1.3 and -1.7

5-15. Greater than 1.5

5-16. Greater than 1.99

5-17. Greater than 2.93

5-18. Less than 2.93

5-19. Less than -1.08

5-20. Less than -2.08

5-21. Less than 1.72

5-22. Greater than -0.55

5-23. Greater than -1.00

5-24. Either less than -2.00 or greater than 0.68

5-25. Either less than -0.50 or greater than 0.75

EXERCISES B: THE STANDARD NORMAL DISTRIBUTION

5-26. (a) In a normally distributed collection of scores with mean zero and standard deviation one, find the area under the curve bounded by the lines above $z = 1.40$ and $z = 1.80$.

(b) Suppose we let $z = 1.80 - 1.40 = 0.40$. We then find the corresponding area from Table A-3. Does this result agree with the answer to part (a)? Why or why not?

5-27. The equation

$$y = \frac{e^{-x^2/2}}{\sqrt{2\pi}}$$

can be approximated by

$$y = \frac{2.7^{-x^2/2}}{2.5}$$

Graph the last equation after finding the y coordinates that correspond to the following x coordinates: $-4, -3, -2, -1, 0, 1, 2, 3,$ and 4. (A calculator capable of dealing with exponents will be helpful.) Attempt to determine the approximate area bounded by the curve, the x axis, the vertical line passing through 0 on the x axis, and the vertical line passing through 1 on the x axis. Compare this result to Table A-3.

5-3 NONSTANDARD NORMAL DISTRIBUTIONS

Section 5-2 considered the standard normal distribution, and in this section we extend the same basic concepts to include nonstandard normal distributions. This inclusion will greatly expand the variety of practical applications we can make since, in reality, most normally distributed populations will have either a nonzero mean or a standard deviation different from 1.

We continue to use Table A-3, but we require a way of standardizing these nonstandard cases. This is done by letting

$$z = \frac{x - \mu}{\sigma} \tag{5-3}$$

JANITOR HAS I.Q. OF 173

A Portland, Oregon janitor with an I.Q. of 173 says that "the only way you don't get used is by having a job that people consider worthless." The mean I.Q. is 100 and the genius level is considered by many to be 140 or higher.

so that z is the number of standard deviations that a particular score x is away from the mean. z is called the *z score* or *standard score* and it is used in Table A-3 (see Figure 5-9).

FIGURE 5-9

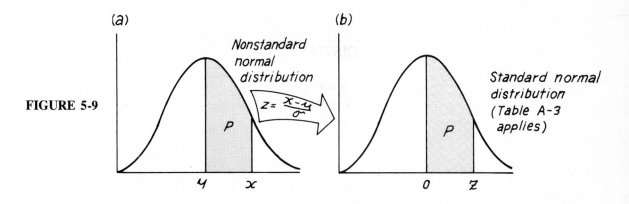

(a)

Nonstandard normal distribution

$z = \frac{x - \mu}{\sigma}$

P

μ x

(b)

Standard normal distribution (Table A-3 applies)

P

0 z

Suppose, for example, that we are considering a normally distributed collection of I.Q. scores known to have a mean of 100 and a standard deviation of 15. If we seek the probability of randomly selecting one I.Q. score that is between 100 and 130, we are concerned with the area shown in Figure 5-10. The difference between 130 and the mean of 100 is 30 I.Q.

FIGURE 5-10

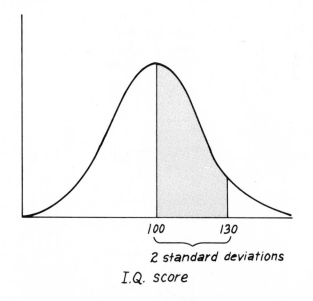

100   130

2 standard deviations

I.Q. score

points or exactly two standard deviations. The shaded area in Figure 5-10 will therefore correspond to the shaded area of Figure 5-5 where $z = 2$. We get

$z = 2$ either by reasoning that 130 is two standard deviations above the mean of 100, or we can compute

$$z = \frac{x - \mu}{\sigma} = \frac{130 - 100}{15} = \frac{30}{15} = 2$$

With $z = 2$, Table A-3 indicates that the shaded region we seek has an area of 0.4772 so that the probability of randomly selecting an I.Q. score between 100 and 130 is 0.4772. Thus Table A-3 can be indirectly applied to any normal probability distribution if we use equation (5-3) as the algebraic way of recognizing that the z score is actually the number of standard deviations that z is away from the mean. The following examples are intended to illustrate this observation.

Example   If I.Q. scores are normally distributed with a mean of 100 and a standard deviation of 15, find the probability of randomly selecting a subject with an I.Q. betwen 100 and 133.

Solution   Referring to Figure 5-11, we seek the probability associated with the shaded region. In order to use Table A-3, the nonstandard data must be standardized by applying equation (5-3).

$$z = \frac{x - \mu}{\sigma} = \frac{133 - 100}{15} = \frac{33}{15} = 2.20$$

The score of 133 therefore differs from the mean of 100 by 2.20 standard deviations. Corresponding to a z score of 2.20, Table A-3 indicates a probability of 0.4861. There is therefore a probability of 0.4861 of randomly selecting a subject having an I.Q. between 100 and 133.

FIGURE 5-11

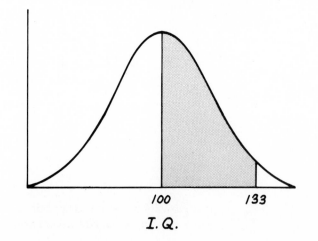

Note that when Table A-3 is used in conjunction with equation (5-3), the nonstandard population mean corresponds to the standard mean of zero. As a result, probabilities extracted directly from Table A-3 must represent regions whose left boundary is the line denoting the mean.

Example   In one region, trees grow in such a way that the maximum heights are normally distributed with a mean of 12.0 meters and a standard deviation of 2.0 meters. Find the probability of randomly selecting a tree with a maximum height of 8.4 meters or less.

Solution   In Figure 5-12, the shaded region corresponds to the area representing trees with maximum heights of 8.4 meters or less. We cannot find the area of the shaded region directly, but we can use equation (5-3) to find the adjacent region immediately to the right of the region we seek.

$$z = \frac{x - \mu}{\sigma} = \frac{8.4 - 12.0}{2.0}$$

$$= \frac{-3.6}{2.0} = -1.80$$

Noting the symmetry about the vertical centerline, we conclude that a z score of -1.80 will produce the same value as a z score of 1.80, and we proceed to Table A-3 to get an area of 0.4641. The area bounded by heights of 8.4 meters and 12.0 meters is 0.4641, and the total area to the left of 12.0 is 0.5, so the shaded region must be $0.5 - 0.4641$ or 0.0359. That is, there is a probability of 0.0359 of randomly selecting a tree that reaches a maximum height of 8.4 meters or less.

FIGURE 5-12

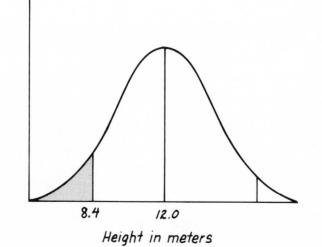

8.4   12.0

Height in meters

Example

A college entrance examination is designed so that the scores are normally distributed with a mean of 500 and a standard deviation of 100. Find the percentage of subjects with scores between 550 and 675.

Solution

Figure 5-13 depicts the relevant region (B), but we cannot obtain the desired probability directly since the left boundary is not the mean. Instead we use equation (5-3) to find the total probability for regions A and B combined and then proceed to subtract the probability of region A. To get the probability for regions A and B combined, we let $\mu = 500$, $\sigma = 100$, and $x = 675$ to get

$$z = \frac{x - \mu}{\sigma} = \frac{675 - 500}{100} = \frac{175}{100} = 1.75$$

FIGURE 5-13

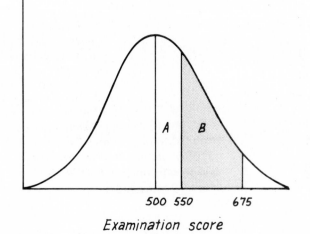

Examination score

From Table A-3 with $z = 1.75$ we get a probability of 0.4599. To get the probability representing region A, we let $\mu = 500$, $\sigma = 100$, and $x = 550$ to get

$$z = \frac{x - \mu}{\sigma} = \frac{550 - 500}{100} = \frac{50}{100} = 0.50$$

From Table A-3 with $z = 0.50$ we get a probability of 0.1915.

$$P(\text{regions } A \text{ and } B \text{ combined}) - P(\text{region } A)$$
$$= 0.4599 - 0.1915$$
$$= 0.2684$$

Since there is a probability of 0.2684 that a randomly selected subject will score between 550 and 675, we conclude that 26.84% of the subjects will score in that range.

Here we extended the concept of Section 5-2 to include more realistic nonstandard normal probability distributions. We noted that the formula $z = (x - \mu)/\sigma$ algebraically describes the observation that the z score is the number of standard deviations that a particular score x is away from the mean. However, all of the examples we have considered so far are of the same general type: a probability (or percentage) determined by using the normal distribution (described in Table A-3) and given the values of the mean, standard deviation, and relevant score(s). In some practical cases, the probability (or percentage) is known and we must determine the relevant score(s). Problems of this type are considered in Section 5-4.

RALPH NADER QUESTIONS
APTITUDE TESTS

Public consumer advocate Ralph Nader criticizes the way the Educational Testing Service's (ETS) College Board scores are used. He points out that the tests are designed to predict performance the first year in college, and that the tests are not really broad "aptitude" tests. Bowdoin College began to disregard College Board scores when it found a low correlation between SAT scores and the academic achievement of graduates with high honors. A common criticism of College Board scores is that they are biased in favor of whites and high income families. Also, these tests do not measure such important characteristics as creativity and drive. Furthermore, the importance placed on these tests tends to downplay the importance of the characteristics they can't measure.

EXERCISES A: NONSTANDARD NORMAL DISTRIBUTIONS

5-28. Assume that daily industrial waste amounts in the United States are normally distributed with a mean of 300,000 tons and a standard deviation of 20,000 tons. Find the probability that, on a day selected at random, the industrial waste is between:
   (a)   300,000 tons and 350,000 tons.
   (b)   300,000 tons and 345,000 tons.

5-29. Assume that daily shipments of glass containers of beverages are normally distributed with a mean of 74 million containers and a standard deviation of 5 million containers. Find the probability that on a day selected at random, the number of beverage containers shipped is between:
   (a)   74 million containers and 82 million containers.
   (b)   74 million containers and 75 million containers.

5-30. Assume that weights of newborn children are normally distributed with a mean of 116 ounces and a standard deviation of 12 ounces. Find the probability of randomly selecting a newborn child whose weight is between:
(a) 108 ounces and 124 ounces.
(b) 110 ounces and 122 ounces.

5-31. A standard I.Q. test produces normally distributed results with a mean of 100 and a standard deviation of 15. If an average I.Q. is defined to be any I.Q. between 90 and 109, find the probability of randomly selecting an I.Q. which is average.

5-32. A machine fills sugar boxes in such a way that the weights of the contents are normally distributed with a mean of 2260 grams and a standard deviation of 20 grams. If a box is randomly selected, find the probability that the weight of the contents is between:
(a) 2250 grams and 2275 grams.
(b) 2225 grams and 2300 grams.

5-33. Certain trees grow in such a way that their maximum heights are normally distributed with a mean of 12.0 meters and a standard deviation of 2.0 meters. Find the percentage of trees having maximum heights:
(a) Between 10.0 meters and 15.0 meters.
(b) Above 13.0 meters.

5-34. A test for reaction times produces normally distributed results with a mean of 0.78 second and a standard deviation of 0.06 second. Find the percentage of subjects with reaction times:
(a) Between 0.68 second and 0.76 second.
(b) Slower than 0.75 second.
(c) Faster than 0.80 second.

5-35. A scientist repeats an experiment and records the times required for a certain chemical reaction. The times are normally distributed with a mean of 44.35 seconds and a standard deviation of 0.30 second. Find the percentage of times:
(a) Between 44.00 seconds and 45.00 seconds.
(b) Below 44.00 seconds.
(c) Above 44.00 seconds.

5-36. Scores on a college entrance examination are normally distributed with a mean of 500 and a standard deviation of 100. One college gives priority acceptance to subjects scoring above 650. What percentage of subjects are eligible for priority acceptance?

5-37. In a certain country, heights of adult males are normally distributed with a mean of 70.2 inches and a standard deviation of 4.2 inches. If cars are designed to accommodate a maximum height of 6 feet 4 inches, what percentage of adult males will find these cars unsuitable?

5-38. A manufacturer of bulbs for movie projectors advertises a life of 50 hours. A study of these bulbs indicates that their lives are normally distributed with a mean of 61 hours and a standard deviation of 6.3 hours. What is the percentage of bulbs that fail to last as long as the manufacturer claims?

5-39. A manufacturer contracts to supply ball bearings having diameters between 24.6 millimeters and 25.4 millimeters. Product analysis indicates that the ball bearings manufactured have diameters that are normally distributed with a mean of 25.1 millimeters and a standard deviation of 0.2 millimeter. What percentage of ball bearings satisfy the contract specifications?

EXERCISES B: NONSTANDARD NORMAL DISTRIBUTIONS

5-40. Assume that the following scores are representative of a normally distributed population.
   (a) Find the mean $\bar{x}$ of this sample.
   (b) Find the standard deviation s of this sample.
   (c) Find the percentage of these sample scores that are between 51 and 54 inclusive.
   (d) Find the percentage of *population* scores between 51 and 54. Use the sample values of $\bar{x}$ and s as estimates of μ and σ.

50	51	52	51	47	57
51	50	50	51	50	47
51	48	48	50	49	52
49	50	53	47	53	52
51	48	51	54	51	50

5-41. Is wealth in the United States normally distributed? What individual characteristics affect personal wealth? Are these characteristics normally distributed? What social implications follow?

5-4 FINDING SCORES WHEN GIVEN PROBABILITIES

All of the examples and problems in Sections 5-2 and 5-3 involved the determination of a probability or percentage based on the normal distribution data of Table A-3, a given mean, standard deviation, and relevant score(s). Here we consider the same types of circumstances, but we will alter the known data so the computational procedure will change. These techniques will closely parallel some of the important procedures that will be introduced later in the book.

Let's begin with a practical problem. We have an intelligence test that produces normally distributed scores with a mean of 500 and a standard deviation of 100. Assuming that we want to identify the top 33% so that we can develop a specialized experimental learning program, what specific score serves as a cutoff that separates the top 33% from the lower 67%? Figure 5-14 depicts the relevant normal distribution.

FIGURE 5-14

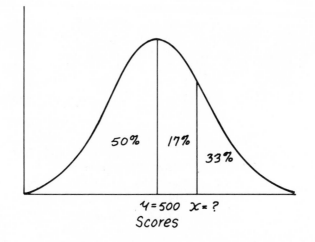

We can find the z score that corresponds to the x value we seek after first noting that the region containing 17% corresponds to a probability of 0.1700. Looking at Table A-3 we can see that a probability of 0.1700 corresponds to a z score of 0.44. This means that the value of x that we want is 0.44 standard deviation away from the mean. Since the standard deviation is given as 100, we can conclude that 0.44 standard deviation is equivalent to 44. The score x is above the mean and 44 units away. Since the mean is given as 500, the score x must be 500 + 44 or 544. That is, the test score of 544 separates the top 33% from the lower 67%.

We could have achieved the same results by noting that

$$z = \frac{x - \mu}{\sigma} \quad \text{becomes} \quad 0.44 = \frac{x - 500}{100}$$

when we substitute the given values for the mean μ, the standard deviation σ, and the z score corresponding to a probability of 0.1700. We can solve this last equation for x by multiplying both sides by 100 and then adding 500 to both sides. We get $x = 544$.

Example   A clothing manufacturer finds it unprofitable to service very tall or very short adult males. The executives decide to discontinue production of goods for the tallest 7.5% and the shortest 7.5% of the adult male population. Find the minimum and maximum heights they will continue to serve.

Solution Assume that heights of adult males are normally distributed with a mean of 70.2 inches and a standard deviation of 4.2 inches (see Figure 5-15). We note that the two outer regions total 15%, so the two equal inner regions must comprise the remaining 85%. This implies that each of the two inner regions must represent 42.5%, which is equivalent to a probability of 0.425. We can use Table A-3 to find the z score that yields a probability of 0.425. With $z = 1.44$, the corresponding probability of 0.4251 is close enough so we conclude that the upper and lower cutoff scores will each differ from the mean by 1.44 standard deviations. With $\sigma = 4.2$ inches, 1.44 standard deviations is $1.44 \cdot 4.2$ inches or 6.048 inches. Thus, the upper cutoff height is 70.2 inches + 6.048 inches or 76.248 inches. By similar reasoning, the lower cutoff height is 70.2 inches − 6.048 inches or 64.152 inches. That is, the clothing company will serve only adult males between 64.152 inches and 76.248 inches in height.

FIGURE 5-15

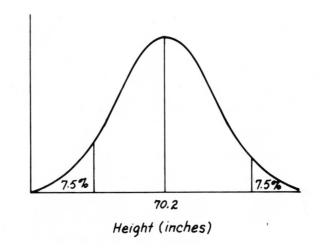

In the preceding example we can see that thoughtless use of equation 5-3 will produce the maximum height of 76.248 inches, but determination of the minimum height requires an adjustment that comes only with an understanding of the whole situation. The moral is clear: don't blindly plug in numbers. Instead, develop an understanding of the underlying meaning. Always draw a graph of the normal distribution with the relevant labels and apply common sense to guarantee that the results are reasonable. The graph and the common-sense check help reduce the incidence of errors.

The following example shows how blind application of equation (5-3) leads to serious errors in the case of a negative z score.

Example As one of its admissions' criteria, a college requires an entrance examination score that is among the top 70% of all scores. Assuming a normal distribution with $\mu = 500$ and $\sigma = 100$, find the minimum acceptable score.

Solution   From Figure 5-16 we see that the area bounded by the unknown relevant score x and the mean comprises 20% of the total area. If that 20% region were to the right of the mean, we could use Table A-3 directly with no difficulty. Let's proceed by pretending that this is the case. Twenty percent or 0.20 is approximated by a z score of 0.52 (see Table A-3). We can now reason as follows. The value of x differs from the mean of 500 by 0.52 standard deviation. With $\sigma = 100$, 0.52 standard deviation becomes $0.52 \cdot 100 = 52$, so x must be 52 *below* 500. That is, x must be 448. We can see from Figure 5-16 that x is less than 500, so the difference of 52 must be subtracted from 500 instead of added to 500. If this solution were attempted through a superficial application of equation (5-3), we would get the erroneous answer of 552. But 552 is unreasonable because the given data require a result less than 500.

FIGURE 5-16

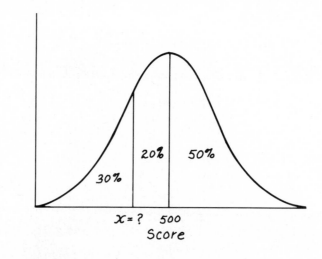

EXERCISES A: FINDING SCORES WHEN GIVEN PROBABILITIES

In Exercises 5-42 through 5-51, assume that the readings on a scale are normally distributed with a mean of 0 meters and a standard deviation of 1 meter.

5-42.   Ninety-five percent of the errors are below what value?

5-43.   Ninety-nine percent of the errors are below what value?

5-44.   Ninety-five percent of the errors are above what value?

5-45.   Ninety-nine percent of the errors are above what value?

5-46.   If the top 0.5% and the bottom 0.5% of all errors are unacceptable, find the minimum and maximum acceptable errors.

DECLINING STUDENT SCORES

Much concern has been generated in recent years over the fact that the mean scores for the Scholastic Aptitude Tests (SAT) have been steadily declining. The first few years of the downward trend were explained by the fact that the pool of college applicants included a large proportion of poor students who would not have considered college in the past. This explanation became obsolete as the downward trend continued even with a more stable pool of test takers. Many people now blame the elementary schools for deteriorating levels of education. Others suggest that the rising cost of tuition is preventing many good students from applying to colleges that require SAT scores, thereby lowering the mean. In any event, we see an attempt to use statistics for a better understanding of an apparent problem that may seriously affect our society.

5-47. If the top 5% and the bottom 5% of all errors are unacceptable, find the minimum and maximum acceptable errors.

5-48. If the top 10% and the bottom 5% of all errors are unacceptable, find the minimum and maximum acceptable errors.

5-49. If the top 15% and the bottom 20% of all errors are unacceptable, find the minimum and maximum acceptable errors.

5-50. Find the value which separates the top 40% of all errors from the bottom 60%.

5-51. Find the value which separates the top 82% of all errors from the bottom 18%.

5-52. Assume that daily industrial waste amounts in the United States are normally distributed with a mean of 300,000 tons and a standard deviation of 20,000 tons. If this waste overloads the system 6% of all days, find the maximum capacity that the system can accommodate.

5-53. Assume the daily shipments of glass containers of beverages are normally distributed with a mean of 74 million containers and a standard deviation of 5 million containers. Find the number of containers separating the top 12% from the bottom 88%.

5-54. Assume that weights of newborn children are normally distributed with a mean of 116 ounces and a standard deviation of 12 ounces. Find the upper and lower limits that separate the top 5% and the bottom 5%.

5-55. A standard I.Q. test produces normally distributed results with a mean of 100 and a standard deviation of 15. If a superior intelligence score is associated with the top 10% of all scores, find the lowest superior score.

5-56. A soda machine fills cups in such a way that the weights of the contents are normally distributed with a mean of 7.9 ounces and a standard deviation of 0.7 ounce. If 12% of the weights are insufficient, find the lowest sufficient weight.

5-57. Certain trees grow in such a way that their maximum heights are normally distributed with a mean height of 12 meters and a standard deviation of 2.0 meters. The tallest 5% of these trees are used for flagpoles in summer camps. Find the height of the shortest tree used for a flagpole.

5-58. A test for reaction times produces normally distributed results with a mean of 0.78 second and a standard deviation of 0.06 second. If the slowest 8% of all reaction times are unsatisfactory for car racing, find the slowest acceptable reaction time.

5-59. A scientist repeats an experiment and records the times required for a given chemical reaction. The times are normally distributed with a mean of 44.35 seconds and a standard deviation of 0.30 second. If the scientist decides to exclude the top 15% and the bottom 15% of these times, find the minimum and maximum times that remain.

5-60. Scores on a college entrance examination are normally distributed with a mean of 500 and a standard deviation of 100. One college automatically rejects the bottom 30% and takes a further look at the remainder. Find the minimum score of the group that will receive additional consideration.

5-61. In a certain country, heights of adult males are normally distributed with a mean of 70.2 inches and a standard deviation of 4.2 inches. Doorways are designed so that 99% of all adult males have at least a 6-inch clearance. Find this standard doorway height.

5-62. A manufacturer of bulbs for movie projectors finds that the lives of the bulbs are normally distributed with a mean of 61.0 hours and a standard deviation of 6.3 hours. The manufacturer will guarantee the bulbs so that only 3% will be replaced because of failure before the guaranteed number of hours. For how many hours should the bulb be guaranteed?

5-63. A manufacturer has contracted to supply ball bearings. Product analysis reveals that the diameters are normally distributed with a mean of 25.1 millimeters and a standard deviation of 0.2 millimeter. The largest 7% of the diameters and the smallest 13% of the diameters are unacceptable. Find the limits for the diameters of the acceptable ball bearings.

EXERCISES B: FINDING SCORES WHEN GIVEN PROBABILITIES

5-64. Assume that the following scores are representative of a normally distributed list of test scores.
(a) Find the mean $\bar{x}$ of this sample.
(b) Find the standard deviation s of this sample.
(c) If a grade of A is given to the top 5%, find the minimum numerical score that corresponds to A in this sample.
(d) Find the theoretical score that separates the top 5% by using the sample mean and standard deviation as estimates for the population mean and standard deviation.

59	69	64	60	77	72	74	64
76	61	77	47	69	82	76	69
60	72	66	92	80	74	78	54
82	59	64	77	69	66	56	53
72	79	58	59	59	50	72	76

5-65. A teacher gives a test and gets normally distributed results with a mean of 50 and a standard deviation of 10. Grades are to be assigned according to the following scheme.
A: Top 10%.
B: Scores above the bottom 70% and below the top 10%.
C: Scores above the bottom 30% and below the top 30%.
D: Scores above the bottom 10% and below the top 70%.
F: Bottom 10%.

Find the numerical limits for each letter grade.

5-66. There are 5000 entrants for the 1-mile run in a large track tournament that can accommodate only 500 runners in that event. Regional qualifying meets are held and the times are normally distributed with a mean of 290 seconds and a standard deviation of 25 seconds. What is the slowest time a runner can have and still be among the 500 finalists?

5-5 NORMAL AS APPROXIMATION TO BINOMIAL

"In a recent presidential election, 55% of the Massachusetts voters favored the Democratic candidate, while 45% favored the Republican candidate. Assuming that these percentages remain constant, find the probability that, in a poll of 150 Massachusetts voters, at least 100 indicate a Democratic preference."

The preceding problem, like many sampling experiments, can be properly classified as a binomial experiment with $n = 150, p = 0.55, q = 0.45$, and x assuming the values of 100, 101, 102, . . . , 149, 150. The formula

$$P(x) = \frac{n!}{(n - x)! \, x!} p^x q^{n-x}$$

can be applied 51 times beginning with

$$P(100) = \frac{150!}{(150 - 100)! \, 100!} \cdot 0.55^{100} \cdot 0.45^{150-100}$$

The resulting 51 probabilities can be totalled to produce the correct result. However, to call these computations formidable would be like calling King Kong a house pest. Fortunately, there is another approach that can be used.

In some cases, which will be described soon, the normal distribution serves as a good approximation to the binomial distribution. Reexamine the graph of Figure 4-8 where we illustrate the binomial distribution for $n = 50$ and $p = 0.5$. The graph of that figure strongly resembles a normal distribution. Similar observations have led mathematicians to recognize that the normal distribution may be used as an approximation to the binomial distribution. Although results will be approximate, they are usually good and we can circumvent the lengthy computations of the binomial probability function and use normal distribution computations instead. This explains why we bother with an approximation when results can be exact. The price of exactness (namely, very lengthy computations) is too great. We settle for the normal distribution approximation for reasons of time and effort.

Thus far we have justified the use of the normal distribution as an approximation to the binomial distribution simply because of a strong resemblance between the two graphs as in Figure 4-8. Actually, this is not a sound justification. There are other distributions (such as the t-distribution to be examined later) with the same basic bell shape yet they cannot be approximated by the normal distribution since unacceptable errors result. The true justification that allows us to use the normal distribution as an approximation to the binomial distribution results from more advanced mathematics. Specifically, the Central Limit Theorem (discussed in Section 5-6) tells us that if $np \geq 5$ and $nq \geq 5$, then the binomial random variable is approximately normally distributed with the mean and standard deviation given as

$$\mu = np$$
$$\sigma = \sqrt{npq}$$

Results of higher mathematics are used to show that the sampling distribution of sample means from *any* population tend to be normally distributed as long as the sample size is large enough. This will be discussed in more detail later. Relative to binomial experiments, "large enough" is equivalent to having both $np \geq 5$ and $nq \geq 5$. In effect, we are stating that there is a formal theorem that justifies our approximation. Unfortunately (or fortunately, depending upon your perspective) it is not practical to outline here the details of the proof for that theorem. For the present, you should accept the intuitive evidence of the strong resemblence between the binomial and normal distributions and take the more rigorous evidence on faith.

Figure 5-17 depicts the normal distribution which approximates the binomial experiment involving the survey of 150 voters. The mean of 82.5 responses by Democrats was obtained by applying the formula $\mu = np$ which describes the mean for any binomial experiment. (This was discussed in Section 4-5.)

$$\mu = np = 150 \cdot 0.55 = 82.5$$

Similarly, the standard deviation can be computed as follows:

$$\sigma = \sqrt{npq} = \sqrt{150 \cdot 0.55 \cdot 0.45} = \sqrt{37.125} \overset{\circ}{=} 6.093$$

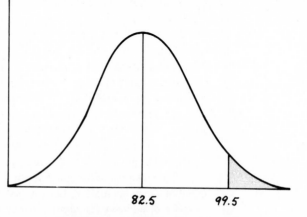

FIGURE 5-17

82.5   99.5

Number of voters favoring the Democrat

We seek the probability of getting *at least* 100 Democrat preferences, so the probability of *exactly* 100 Democrat preferences must be included. But the *discrete* value of 100 is approximated in the *continuous* normal distribution by the interval from 99.5 to 100.5 Such conversions from a discrete to a continuous distribution are called **continuity corrections** (see Figure 5-18). If we ignore or forget the continuity correction, the additional error will be very small as long as *n* is large.

FIGURE 5-18

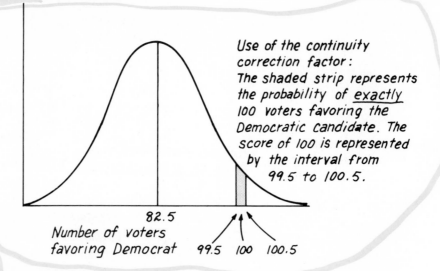

Use of the continuity correction factor:
The shaded strip represents the probability of <u>exactly</u> 100 voters favoring the Democratic candidate. The score of 100 is represented by the interval from 99.5 to 100.5.

82.5
Number of voters
favoring Democrat   99.5 100 100.5

TABLE 5-2 Minimum sample required to approximate a binomial distribution by a normal distribution.

p	n must be at least
0.001	5001
0.01	501
0.1	51
0.2	26
0.3	17
0.4	13
0.5	11
0.6	13
0.7	17
0.8	26
0.9	51
0.99	501
0.999	5001

Reverting to standard procedures associated with normal distributions and Table A-3, we recognize that the probability corresponding to the shaded region of Figure 5-17 cannot be obtained directly. Our strategy will be to find the probability representing the area immediately to the left of the shaded region and subtract that value from 0.5.

$$z = \frac{x - \mu}{\sigma} = \frac{99.5 - 82.5}{6.093} = \frac{17}{6.093} \stackrel{\circ}{=} 2.79$$

Table A-3 indicates that a z score of 2.79 yields a probability of 0.4974 so that the shaded region has a probability of $0.5 - 0.4974$ or 0.0026. There is a probability of 0.0026 that the 150 responses will consist of at least 100 Democrat preferences. Without the continuity correction, we would have used 100 in place of 99.5 and an answer of 0.0021 would have resulted.

In this example, the normal distribution does serve as a good approximation to the binomial experiment, but this is not always the case. For large values of *n* with *p* close to 0.5, the normal distribution approximates the binomial fairly accurately, but for *small values of* n *with* p *near 0 or 1, the binomial distribution is approximated very poorly by the normal distribution.* The approximation is suitable only when $np \geq 5$ and $nq \geq 5$; both conditions must be met. Table 5-2 lists some specific values of p along with the corresponding minimum values of n.

Suppose, for example, that a couple plans to have four children and they seek the probability of having three girls and one boy. With $n = 4$, $p = 0.5$, $q = 0.5$ and $x = 3$, we see that $np = 4 \cdot 0.5 = 2$ and $nq = 4 \cdot 0.5 = 2$, so the normal distribution should not be used. Table 5-2 indicates that they should plan on at least 11 children if they really want to use the normal distribution as an approximation to the binomial. (Admittedly, this is not much of an inducement for a large family.) The binomial probability formula can be applied easily to the given situation.

The following example illustrates another successful use of the normal distribution as an approximation to the binomial distribution.

Example
I

The conviction rate for auto theft is 37%. Find the probability that of the next 100 arrests, there are *exactly* 37 convictions.

Solution

The conditions described satisfy the criteria for the binomial distribution with $n = 100$, $p = 0.37$, $q = 0.63$, and $x = 37$. The binomial probability formula applies, but

$$P(37) = \frac{100!}{(100 - 37)! \, 37!} \cdot 0.37^{37} \cdot 0.63^{100-37}$$

is too difficult to compute. (Most calculators cannot evaluate anything above 70!, but the availability of calculators or a computer would simplify this approach.) Let's proceed with the binomial distribution approximated by the normal distribution. We begin by checking the suitability of the approximation.

$$np = 100 \cdot 0.37 = 37 \geq 5$$

$$nq = 100 \cdot 0.63 = 63 \geq 5$$

Since np and nq are both at least 5, we conclude that the normal approximation to the binomial is satisfactory. We now obtain μ and σ as follows:

$$\mu = np = 100 \cdot 0.37 = 37$$
$$\sigma = \sqrt{npq} = \sqrt{100 \cdot 0.37 \cdot 0.63} = \sqrt{23.21} \doteq 4.828$$

The shaded region of Figure 5-19 represents the probability we want. Use of the continuity correction results in the representation of "37" by the region extending from 36.5 to 37.5.

The format of Table A-3 requires that we first find the probability corresponding to the region bounded on the left by the vertical line through the mean of 37 and on the right by the vertical line through 37.5.

$$z = \frac{x - \mu}{\sigma} = \frac{37.5 - 37}{4.828} = \frac{0.5}{4.828} = 0.10$$

FIGURE 5-19

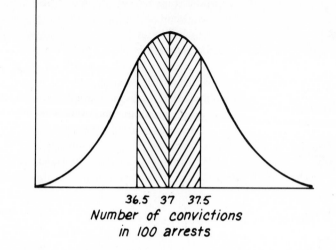

36.5 37 37.5
Number of convictions
in 100 arrests

In Table A-3 we find that a probability of 0.0398 corresponds to $z = 0.10$. Consequently, the entire shaded region of Figure 5-19 depicts a probability of $0.0398 + 0.0398$ or 0.0796. (A computer computation using the binomial probability formula produced a result of 0.0824; the discrepancy of 0.0028 is very small.)

Let's assume that an experiment consists of randomly selecting 16 births and recording the number of boys. With $n = 16$, $p = 0.5$, and $q = 0.5$, we could let x assume the values 0, 1, 2, . . . , 15, 16 and apply the binomial probability formula to obtain the results summarized in Table 5-3. The graph

FIGURE 5-20
Normal approximation compared to binomial result.

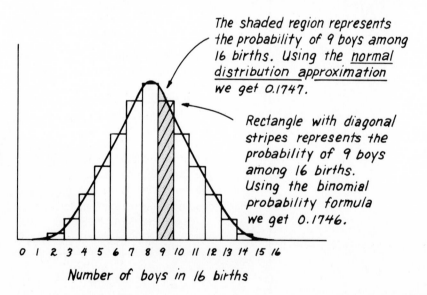

The shaded region represents the probability of 9 boys among 16 births. Using the _normal distribution approximation_ we get 0.1747.

Rectangle with diagonal stripes represents the probability of 9 boys among 16 births. Using the binomial probability formula we get 0.1746.

0 1 2 3 4 5 6 7 8 9 10 11 12 13 14 15 16
Number of boys in 16 births

of Table 5-3 consists of the rectangles shown in Figure 5-20. (Figure 5-20 provides a visual comparison of a binomial distribution and the approximating normal distribution.)

The approximating normal curve is characterized by $\mu = np = 16 \cdot 0.5 = 8$ and $\sigma = \sqrt{npq} = \sqrt{16 \cdot 0.5 \cdot 0.5} = \sqrt{4} = 2$. As a basis for comparison, we find the probability of getting exactly nine boys using the binomial distribution and the normal distribution. Using the binomial distribution we see that $P(9) = 0.1746$ from Table 5-3. Using the normal distribution with the continuity correction, $P(9)$ is computed to be 0.1747 and the discrepancy of 0.0001 is negligible.

We now have three different methods for determining probabilities in binomial experiments:

1. Refer to Table A-2 if $n \leq 25$ and p is one of the following values: 0.01, 0.05, 0.10, 0.20, 0.30, 0.40, 0.50, 0.60, 0.70, 0.80, 0.90, 0.95, 0.99.

2. Use the binomial probability formula

$$P(x) = \frac{n!}{(n - x)! \, x!} \, p^x q^{n-x}$$

if it is not too time consuming and difficult (see Section 4-4).

3. Use the normal distribution approximation with $\mu = np$ and $\sigma = \sqrt{n \cdot p \cdot q}$ if $np \geq 5$ and $nq \geq 5$.

TABLE 5-3 The values of $P(x)$ are computed with the binomial probability formula.

x	$P(x)$
0	0.00002
1	0.0002
2	0.0018
3	0.0085
4	0.0278
5	0.0667
6	0.1222
7	0.1746
8	0.1964
9	0.1746
10	0.1222
11	0.0667
12	0.0278
13	0.0085
14	0.0018
15	0.0002
16	0.00002

$$
\begin{aligned}
P(x) &= \frac{n!}{(n - x)! \, x!} \, p^x q^{n-x} \\
&= \frac{16!}{(16 - x)! \, x!} \cdot 0.5^x \cdot 0.5^{16-x} \\
&= \frac{16!}{(16 - x)! \, x!} \cdot 0.5^{16}
\end{aligned}
$$

EXERCISES A: NORMAL AS APPROXIMATION TO BINOMIAL

5-67. For the given values associated with binomial experiments, determine whether the normal distribution is a suitable approximation.
   (a) $n = 6, p = 0.4$
   (b) $n = 500, p = 0.01$
   (c) $n = 8, p = 0.1$
   (d) $n = 12, p = 0.4$
   (e) $n = 8, p = 0.9$
   (f) $n = 100, p = 0.99$
   (g) $n = 1000, p = 0.006$
   (h) $n = 1,000,000, p = 0.999$
   (i) $n = 75, p = 0.67$
   (j) $n = 15, p = 0.7$

5-68. In a binomial experiment with $n = 16$ and $p = 0.5$, find P(at least 8) using:
   (a) Table 5-3.
   (b) The normal distribution approximation.

5-69. In a binomial experiment with $n = 16$ and $p = 0.5$, find P(at least 10) using:
   (a) Table 5-3.
   (b) The normal distribution approximation.

5-70. In a binomial experiment with $n = 16$ and $p = 0.5$, find $P(8)$ using:
   (a) Table 5-3.
   (b) The normal distribution approximation. (Use the continuity correction to represent 8 by the interval from 7.5 to 8.5.)

5-71. In a binomial experiment with $n = 16$ and $p = 0.5$, find $P(10)$ using:
   (a) Table 5-3.
   (b) The normal distribution approximation. (Use the continuity correction to represent 10 by the interval from 9.5 to 10.5.)

5-72. Find the probability of getting exactly 50 boys in 100 births. (Hint: Use the continuity correction.)

5-73. Find the probability of getting at least 60 boys in 100 births.

5-74. Find the probability of passing a true-false test of 100 questions if 65% is passing and all responses are random guesses.

5-75. In an eastern college, 60% of the entering freshmen graduate. Find the probability that, of 2000 entering freshmen, there will be at most 1150 graduates.

5-76. The probability of a particular computer component being defective is 0.01. Find the probability of getting at most 60 defects in a random sample of 5000 components.

5-77. A quarterback completes 35% of his passes. What is the probability of his completing fewer than 165 of his next 400 passes?

5-78. The failure rate on a civil service examination is 30%. Find the probability of getting more than 325 failures on the next 1000 exams.

5-79. On a question in an I.Q. test, 25% of the respondents answer correctly. Find the probability that, of 250 responses, fewer than 70 are correct.

5-80. A certain seed has a 75% germination rate. If 50,000 seeds are planted, find the probability that more than 37,400 will germinate.

5-81. An engine on a light aircraft has a 99.9% chance of completing a flight without failure. Find the probability that of 6000 flights, there are more than 10 engine failures. (Assume that all aircraft have one engine.)

5-82. A magazine subscription service has found that 8% of its telephone solicitations result in orders. Find the probability that of 120 contacts, there are fewer than 10 orders.

5-83. An experiment in parapsychology involves the guessing of a selected number from 1 through 5. If this test is repeated on 250 subjects, find the probability of getting more than 60 correct responses. Would it be unusual to get more than 70 correct responses?

5-84. The probability of winning anything in the weekly New York State lottery is about 1/500. Three thousand tickets are sold in a small village. What is the probability that no prizes are won in that village?

5-85. Experience has shown that 12% of all subjects fail a visual perception test. Find the probability that of 200 randomly selected subjects, there are at least 20 failures.

5-86. An airline company experiences a 7% rate of no shows on advance reservations. Find the probability that of 250 randomly selected advance reservations, there will be at least 10 no shows.

5-87. A certain genetic characteristic appears in one-quarter of all offspring. Find the probability that of 40 randomly selected offspring, less than 5 exhibit the characteristic in question.

5-88. Of all taxpayers whose returns are audited, 70% end up paying additional taxes. Find the probability that of 500 randomly selected returns, at least 400 end up paying additional taxes.

EXERCISES B: NORMAL AS APPROXIMATION TO BINOMIAL

5-89. In a binomial experiment with $n = 25$ and $p = 0.4$, find P (at least 18) using:
  (a) The table of binomial probabilities (Table A-2).
  (b) The binomial probability formula.
  (c) The normal distribution approximation.

5-90. The cause of death is related to heart disease in 52% of the cases studied. Find the probability that in 500 randomly selected cases, the number of heart disease-related deaths differs from the mean by more than two standard deviations.

5-91. An airline company works only with advance reservations and experiences a 7% rate of no shows. How many reservations could be accepted for an airliner with a capacity of 250, if there is at least a 0.95 probability that all reservation holders who show will be accommodated?

5-6 THE CENTRAL LIMIT THEOREM

We have already stated that the normal distribution is extremely important in the study of statistics because so many natural phenomena yield scores that are normally distributed. In Section 5-5 we showed how the normal distribution can be used to approximate the binomial distribution. Here we support the importance of the normal distribution by indirectly extending it to nonnormal distributions by way of the **Central Limit Theorem.**

Let's assume that the variable x represents scores which may or may not be normally distributed, and that the mean of the x values is μ while the standard deviation is σ. Suppose we collect a sample of size n and calculate the sample mean $\bar{x}$. What do we know about the collection of all sample means that we produce by repeating this experiment, collecting a sample of size n to get the sample mean? The Central Limit Theorem tells us that as the sample size n increases, the sample means will tend to approach a normal distribution with mean μ and standard deviation $\sigma/\sqrt{n}$. This conclusion is not intuitively obvious, and it was arrived at through extensive research and analysis. The formal rigorous proof requires advanced mathematics and is beyond the scope of this text, so we can only illustrate the theorem and give examples of its practical use.

CENTRAL LIMIT THEOREM

> *If the variable* x *has some distribution so that the mean value is* μ *and the standard deviation is* σ, *the sample means* $\bar{x}$ *(based on random samples of size* n) *will, as* n *increases, approach a **normal** distribution with mean* μ *and standard deviation* $\sigma/\sqrt{n}$.

In random samples of size n, the mean of the sample means is denoted by $\mu_{\bar{x}}$ so that $\mu_{\bar{x}} = \mu$.

Also, the standard deviation of the sample means is denoted by $\sigma_{\bar{x}}$ so that $\sigma_{\bar{x}} = \sigma/\sqrt{n}$.

$\sigma_{\bar{x}}$ is often called the **standard error of the mean.**

For the moment pretend that we have never heard of the Central Limit Theorem and that we wish to investigate sample means extracted from some nonnormal population. We can begin with a *uniform* distribution which is nonnormal and relatively simple. Take the infinite population consisting of the random selection (with replacement) of numbers between 1 and 9 inclusive. From this population we select samples of size $n = 10$. (Thirty such samples are listed in Table 5-4.) In Figure 5-21, the original 300 random numbers between 1 and 9 are depicted in a histogram that does support the uniformity of the distribution. Figure 5-22 consists of a histogram representing the 30 sample means shown in Table 5-4. We can see that Figure 5-22

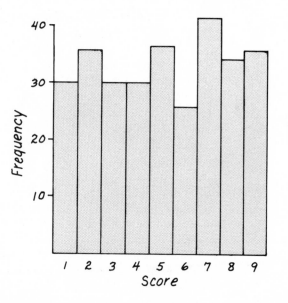

FIGURE 5-21
Histogram of the 300 *original scores* randomly selected (between 1 and 9).

roughly approximates the shape of a normal distribution. Observations exactly like this eventually led to the formulation of the Central Limit Theorem. If our sample means were based on samples of a size larger than 10, Figure 5-22 would more closely resemble a normal distribution.

Statisticians have found that for samples of size 31 or larger, the sample means can be approximated reasonably well by a normal distribution with mean μ and standard deviation $\sigma/\sqrt{n}$. Sometimes even smaller sample sizes

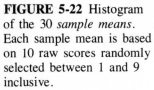

FIGURE 5-22 Histogram of the 30 *sample means*. Each sample mean is based on 10 raw scores randomly selected between 1 and 9 inclusive.

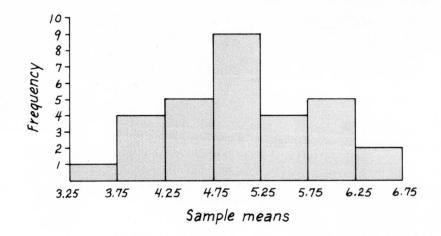

TABLE 5-4 Thirty collections of samples with ten random numbers between 1 and 9 in each sample. The right column consists of the corresponding sample means.

Sample	Data										Sample mean
1.	2	7	5	5	2	1	7	7	9	4	4.9
2.	5	8	1	1	5	7	1	4	1	4	3.7
3.	7	6	9	8	5	1	6	4	7	9	6.2
4.	7	3	1	7	3	6	7	9	4	3	5.0
5.	9	7	7	6	1	6	8	3	4	7	5.8
6.	5	3	3	4	2	5	9	9	1	9	5.0
7.	5	5	3	9	5	3	1	9	1	5	4.6
8.	4	3	9	5	5	9	1	7	7	8	5.8
9.	2	1	7	8	6	7	7	9	8	3	5.8
10.	3	4	5	6	8	4	8	3	4	5	5.0
11.	5	3	2	2	6	8	1	5	5	9	4.6
12.	7	5	9	6	8	2	2	7	2	1	4.9
13.	3	1	4	1	7	9	3	2	3	8	4.1
14.	6	2	7	4	4	5	2	6	8	6	5.0
15.	9	6	2	9	4	2	6	3	5	5	5.1
16.	9	2	2	3	6	2	6	6	8	3	4.7
17.	5	4	2	1	9	4	2	9	4	2	4.2
18.	8	1	2	1	4	3	2	8	5	4	3.8
19.	5	8	9	6	2	7	9	3	8	5	6.2
20.	5	6	8	7	5	9	6	4	8	7	6.5
21.	7	9	9	8	3	5	5	1	4	6	5.7
22.	8	4	7	8	7	8	7	7	1	8	6.5
23.	5	5	1	7	5	7	7	2	9	8	5.6
24.	9	5	2	5	9	2	5	3	5	8	5.3
25.	4	5	8	4	2	9	2	6	6	1	4.7
26.	1	7	7	3	4	7	7	2	8	7	5.3
27.	8	1	1	7	6	2	2	1	4	9	4.1
28.	9	4	3	7	3	7	8	4	3	2	5.0
29.	1	2	9	3	8	2	4	6	2	8	4.5
30.	2	9	3	3	1	2	6	7	8	7	4.8

RELIABILITY AND VALIDITY

The reliability of data refers to the consistency with which the same results occur, but the validity of data refers to how well the data measure what they are supposed to measure. The reliability of an I.Q. test can be tested by comparing scores for the test given on one date to the scores for the same test given on another date. Many critics claim that the popular I.Q. tests are reliable but not valid. That is, they produce consistent results, but they do not really measure intelligence levels. It is much easier to use statistics to analyze reliability than validity, but validity is the more important characteristic. To test the validity of an intelligence test, we can compare the test scores to another indicator of intelligence, such as academic performance.

can produce means that approximate a normal distribution. If the original population is itself normally distributed, then samples of size $n = 1$ are large enough. In this case, the sample means are actually the original normally distributed scores.

Comparison of Figures 5-21 and 5-22 should confirm that the original numbers have a nonnormal distribution while the sample means approximate a normal distribution. The Central Limit Theorem also indicates that the mean of all such sample means should be μ (the mean of the original population) and the standard deviation of all such sample means should be $\sigma/\sqrt{n}$ (where σ is the standard deviation of the original population and n is the sample size of 10). We can find μ and σ for the original population of numbers between 1 and 9 by noting that, if those numbers occur with equal frequency as they should, then the population mean μ is given by

$$\mu = \frac{1+2+3+4+5+6+7+8+9}{9} = 5.0$$

Similarly, we can find σ by again using 1, 2, 3, 4, 5, 6, 7, 8, 9 as an ideal or theoretical representation of the population. Following this course, σ is computed to be 2.58. The mean and standard deviation of the sample means can now be found as follows:

$$\mu_{\bar{x}} = \mu = 5.0$$

$$\sigma_{\bar{x}} = \frac{\sigma}{\sqrt{n}} = \frac{2.58}{\sqrt{10}} = \frac{2.58}{3.16} \overset{\circ}{=} 0.82$$

The preceding results represent *all* sample means of size $n = 10$. For the 30 sample means shown in Table 5-4 we have a mean of 5.08 and a standard deviation of 0.75. We can see that our real data conform quite well to the theoretically predicted values for $\mu_{\bar{x}}$ and $\sigma_{\bar{x}}$.

We can use a simple random selection process beginning with a uniform distribution to artificially create normally distributed data. Realistic and practical simulations become possible through this use of the Central Limit Theorem. More mundane applications of the Central Limit Theorem are illustrated in the following examples.

Example

A study of the time required for airplanes to land at a certain airport shows that the mean is 280 seconds with a standard deviation of 20 seconds. (The times are not normally distributed.) For a sample of 36 randomly selected incoming airplanes, find the probability that their mean landing time is between 280 seconds and 284 seconds.

Solution

Even though the landing times are not normally distributed, the sample size is large enough (it is larger than 30) so that the sample means approximate a normal distribution. Given the population mean (280 seconds) and standard deviation (20 seconds), we can compute $\mu_{\bar{x}}$ and $\sigma_{\bar{x}}$ for a sample of 36 as follows:

$$\mu_{\bar{x}} = \mu = 280 \text{ seconds}$$

$$\sigma_{\bar{x}} = \frac{\sigma}{\sqrt{n}} = \frac{20 \text{ seconds}}{\sqrt{36}} = \frac{20 \text{ seconds}}{6} \stackrel{\circ}{=} 3.33 \text{ seconds}$$

Figure 5-23 depicts the normal distribution applicable to samples of size 36 for the situation under consideration.

FIGURE 5-23
This is the approximate distribution of means computed from samples of size 36 where the original values are the landing times (in seconds) of airplanes at a certain airport.

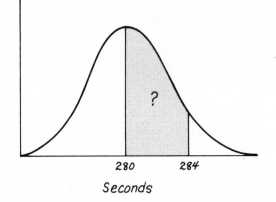

Proceeding as usual with a normal distribution, we find the probability in question by first obtaining the relevant z score.

$$z = \frac{\bar{x} - \mu_{\bar{x}}}{\sigma_{\bar{x}}} = \frac{\bar{x} - \mu}{\sigma/\sqrt{n}} = \frac{284 - 280}{20/\sqrt{36}} = 1.20$$

From Table A-3 we see that $z = 1.20$ corresponds to a probability of 0.3849, so P (mean time is between 280 seconds and 284 seconds) $= 0.3849$.

Example   A strobe light is designed so that the mean time between flashes is 10.0 seconds while the standard deviation is 0.4 second. If 30 strobe lights are randomly selected from this population, find the probability that their mean time between flashes is longer than 10.1 seconds.

Solution   By the Central Limit Theorem and the size of the sample, we conclude that the distribution of mean times between flashes is approximately normal. The mean and standard deviation of sample means (based on samples of size 30) are as follows:

$$\mu_{\bar{x}} = \mu = 10.0 \text{ seconds}$$

$$\sigma_{\bar{x}} = \frac{\sigma}{\sqrt{n}} = \frac{0.4 \text{ second}}{\sqrt{30}} \stackrel{\circ}{=} 0.073 \text{ second}$$

Figure 5-24 depicts the approximating normal distribution and we are looking for the shaded region. We determine the shaded area by first finding the area bounded by 10.0 and 10.1. We then subtract that result from 0.5 to get the shaded area:

$$z = \frac{\bar{x} - \mu_{\bar{x}}}{\sigma_{\bar{x}}} = \frac{10.1 - 10.0}{0.073} \stackrel{\circ}{=} 1.37$$

From Table A-3 we see that a z score of 1.37 corresponds to an area of 0.4147, so the shaded region is $0.5 - 0.4147 = 0.0853$. There is a probability of 0.0853 that the sample of 30 lights will have a mean time between flashes that is longer than 10.1 seconds.

FIGURE 5-24

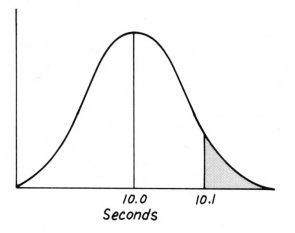

10.0   10.1
Seconds

It is interesting to note that, as the sample size increases, the sample means tend to vary less, since $\sigma_{\bar{x}} = \sigma/\sqrt{n}$ gets smaller as n gets larger. For example, assume that I.Q. scores have a mean of 100 and a standard deviation of 15. Samples of 36 will produce means with $\sigma_{\bar{x}} = 15/\sqrt{36} = 2.5$ so that 99% of all such samples will have means between 93.6 and 106.4. If the sample size is increased to 100, $\sigma_{\bar{x}}$ becomes $15/\sqrt{100}$ or 1.5 so that 99% of the samples will have means between 96.1 and 103.9. (These particular computations and concepts are explained in Section 7.2.)

These results are supported by common sense: as the sample size increases, the corresponding sample mean will tend to be closer to the true population mean. The effect of an unusual or outstanding score tends to be dampened as it is averaged in as part of a sample.

Our use of $\sigma_{\bar{x}} = \sigma/\sqrt{n}$ assumes that the population is infinite. When we sample with replacement of selected data, for example, the population is effectively infinite. Yet realistic applications involve sampling without replacement, so successive samples are dependent upon previous outcomes.

Just as n denotes the *sample* size, N denotes the size of a *population*.

For finite populations of size N, we should incorporate the **finite population correction factor** $\sqrt{(N - n)/(N - 1)}$ so that $\sigma_{\bar{x}}$ is found as follows:

$$\sigma_{\bar{x}} = \frac{\sigma}{\sqrt{n}} \sqrt{\frac{N - n}{N - 1}}$$

If the sample size n is small in comparison to the population size N, the finite population correction factor will be close to 1. Consequently its impact will be negligible and it can therefore be ignored.

Statisticians have devised the following rule of thumb: use the finite population correction factor when computing $\sigma_{\bar{x}}$ if the population is finite and $n > 0.05N$. That is, use the correction factor only if the sample size is greater than 5% of the population size.

Example For 1,000 fuses, the breaking points have a mean of 7.5 amperes and a standard deviation of 1.0 ampere. If a sample of 150 fuses is obtained without replacement, find the probability that the sample yields a mean between 7.5 amperes and 7.6 amperes.

Solution With $\mu = 7.5$ amperes, $\sigma = 1.0$ ampere, $N = 1000$ and $n = 150$, we compute

$$\mu_{\bar{x}} = \mu = 7.5 \text{ amperes}$$

$$\sigma_{\bar{x}} = \frac{\sigma}{\sqrt{n}} \sqrt{\frac{N - n}{N - 1}} = \frac{1.0 \text{ ampere}}{\sqrt{150}} \sqrt{\frac{1000 - 150}{1000 - 1}} \doteq 0.075 \text{ ampere}$$

FIGURE 5-25

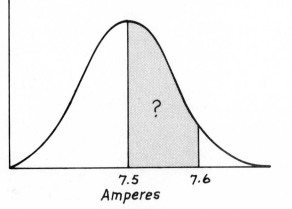

7.5   7.6
Amperes

The finite population correction factor was used because $n > 0.05N$, or $150 > 0.05(1000)$. This means that the sample size is large in comparison to the population size. Having determined the values of $\mu_{\bar{x}}$ and $\sigma_{\bar{x}}$, we can apply the Central Limit Theorem by proceeding with the methods associated with the normal distribution. Figure 5-25 illustrates the normal distribution and the region representing the desired probability.

$$z = \frac{\bar{x} - \mu_{\bar{x}}}{\sigma_{\bar{x}}} = \frac{7.6 - 7.5}{0.075} = \frac{0.1}{0.075} \overset{\circ}{=} 1.33$$

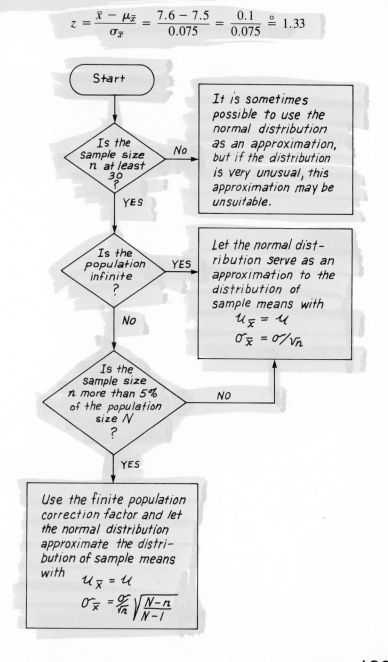

FIGURE 5-26
Flow chart summarizing the decisions to be made when considering a distribution of sample means.

From Table A-3 we see that $z = 1.33$ corresponds to a probability of 0.4082 so that P(mean is between 7.5 and 7.6 amperes) = 0.4082. If we did not use the finite population correction factor, our answer would have been 0.3888 instead of the better result of 0.4082. Figure 5-26 (page 183) outlines most of the key points presented in this section.

One final provocative query: can it be that normally distributed populations actually reflect a composite total of factors that are not themselves normally distributed? The weights of newborn whales, for example, may be affected by genetic and environmental factors such as the weights and diets of predecessors and the random selection of certain genes. Perhaps the contributing factors were not themselves normally distributed, but when the total net effect is realized, the Central Limit Theorem applies by suggesting that the weight of the newborn represents a mean of the contributing sample factors. The normal distribution of the sample means is then reflected by the normal distribution of weights of newborn whales. This may help to explain the frequent occurrence of the normal distribution among various physical, sociological, and psychological measurements.

EXERCISES A: THE CENTRAL LIMIT THEOREM

5-92.   If samples of size n are selected from populations of size N, identify those cases in which the finite population correction factor can be ignored. If it cannot be ignored, find its value. Assume that sampling is done without replacement.

(a) $N = 5000$, $n = 200$ (b) $N = 12,000$, $n = 1000$
(c) $N = 4000$, $n = 500$ (d) $N = 8000$, $n = 3000$
(e) $N = 1500$, $n = 50$ (f) $N = 750$, $n = 50$
(g) $N = 673$, $n = 32$ (h) $N = 866$, $n = 73$
(i) $N = 50,000$, $n = 10,000$ (j) $N = 8362$, $n = 935$

5-93.   If $\sigma = 15$, find the standard error of the mean $\sigma_{\bar{x}}$ for each part of Exercise 5-92. Incorporate the finite population correction factor whenever necessary.

5-94.   A population has a standard deviation of 20. Samples of size n are taken randomly and the means of the samples are computed. What happens to the standard error of the mean if the sample size is increased from 100 to 400?

5-95.   A population has a standard deviation of 20. Samples of size n are randomly selected and the means of the samples are computed. What happens to the standard error of the mean if the sample size is decreased from 64 to 16?

5-96.   A sample of $n = 49$ is taken from a very large population that has a mean of 50 and a standard deviation of 14. Find the probability that the sample mean $\bar{x}$ will be between 50 and 54.

5-97.   A sample of $n = 100$ is taken from a very large population that has a mean of 50 and a standard deviation of 14. Find the probability that the sample mean $\bar{x}$ will be between 50 and 54.

5-98.   A sample of $n = 64$ is taken from a population of 2000. (Samples are taken without replacement.) The population mean μ is 200 and the standard deviation σ is 40. Find the probability that the sample mean $\bar{x}$ will be between 200 and 209.

5-99.   A sample of $n = 144$ is taken from a population of 5000. (Samples are taken without replacement.) The population mean μ is 400 and the population standard deviation σ is 60. Find the probability that the sample mean $\bar{x}$ will be more than 404.

5-100.   A sample of $n = 70$ is taken from a population of 1573. The population mean μ is 78.4 and the population standard deviation σ is 8.2. Find the probability that the sample mean $\bar{x}$ will be less than 77.1.

5-101.   Assume that the mean height of adult males is 70 inches while the standard deviation is 2.6 inches. If a sample of 100 adult males is selected from a population of 5736, find the probability that the mean height of the sample is between 69.5 inches and 70.5 inches.

5-102.   On one I.Q. test, the mean score is 100 and the standard deviation is 15. A sample of 50 scores is selected from a very large population. Find the probability that the mean of the sample group is more than 103.

5-103.   A test in extrasensory perception is given to 500 subjects and the mean and standard deviation are computed to be 48 and 8, respectively. If a random sample of 25 is selected without replacement, find the probability that the mean of the sample group is more than 50.

5-104.   A study of 10,000 males of age 25 who smoke at least two packs of cigarettes daily shows that the mean life span is 65.3 years while the standard deviation is 3.4 years. If a sample of 16 is selected, find the probability that the mean life span of the sample is less than the retirement age of 65 years.

5-105.   A strobe light is designed so that the mean time between flashes is 10.00 seconds with a standard deviation of 0.40 second. A sample of 81 lights is selected from the 2000 lights produced in one week. Find the probability that for the sample, the mean time between flashes is between 10.00 seconds and 10.10 seconds.

5-106.   A battery is designed to last for 25 hours of operation during normal use. The batteries are produced in batches of 1000 and 20 batteries from each batch are tested. If the mean life of the sample is less than 24 hours, the entire batch is rejected. Assuming that $\mu = 25$ hours and $\sigma = 2$ hours, find the probability that a batch will be rejected.

5-107. A college entrance examination produces a mean score of 500 while the standard deviation is 100. Find the probability that, for a sample of 750 randomly selected subjects, the mean is between 495 and 505.

5-108. A traffic study shows that the mean number of occupants in a car is 1.85 people while the standard deviation is 0.31 people. For a sample of 30 cars, find the probability that the mean number of occupants is more than 2.00.

5-109. Four hundred subjects are given a test for reaction times and the mean is 0.820 second with a standard deviation of 0.180 second. If a sample of 30 is selected from this population of 400 subjects, find the probability that the sample mean is less than 0.825 second.

5-110. In a county of 230,762 households, the mean annual income per household is $11,387 and the standard deviation is $3,419. Find the probability that, in a sample of 3% of the households, the mean annual income is between $10,500 and $11,500.

5-111. In a county of 9,850 households, the mean annual income per household is $11,387 and the standard deviation is $3,419. Find the probability that, in a sample of 1500 households, the mean annual income is between $10,500 and $11,500.

5-112. A visual perception test is given to children about to enter kindergarten and the mean score of 1000 subjects is 4.85 while the standard deviation is 0.45. If a sample of 120 subjects is randomly selected from this population, find the probability that the sample mean is more than 4.90.

EXERCISES B: THE CENTRAL LIMIT THEOREM

5-113. Assume that a population of 10,000 has $\sigma = 15$. If a sample of 56 is selected, find the probability that a random sample mean will differ from the population mean by more than 4.

5-114. Assume that a population is infinite. Find the probability that the mean of a sample of 100 differs from the population mean by more than $\sigma/4$.

5-115. To halve the standard error of the mean, the sample size must be _____. (Prove your answer.)

5-116. Using a telephone book, select 25 telephone numbers at random and find the sum of the last four digits. Determine μ and σ for the sum of the last four digits where the population consists of all possible telephone numbers. Determine your sample mean and standard error of the mean.

The main concern of this chapter is the concept of a normal probability distribution, the most important of all continuous probability distributions. Many real and natural occurrences yield data that are normally distributed or can be approximated by a normal distribution. The normal distribution, which tends to appear bell-shaped when graphed, can be described algebraically by an equation, but the complexity of that equation usually forces us to use a table of values instead.

Table A-3 represents the **standard normal distribution** which has a mean of zero and a standard deviation of one. This table relates deviations away from the mean with areas under the curve. Since the total area under the curve is 1, those areas correspond to probability values.

In the early sections of this chapter, we work with the standard procedures used in applying Table A-3 to a variety of different situations. We see that Table A-3 can be applied indirectly to normal distributions that are nonstandard. (That is, μ and σ are not 0 and 1, respectively.) We are able to find the number of standard deviations that a score x is away from the mean μ by computing $z = (x - \mu)/\sigma$.

In sections 5-3 and 5-4 we consider real and practical examples as we convert from a nonstandard to a standard normal distribution. In Section 5-5 we see that we can sometimes approximate a binomial probability distribution by a normal distribution. (If $np \geq 5$ and $nq \geq 5$, the binomial random variable x is approximately normally distributed with the mean and standard deviation given as $\mu = np$ and $\sigma = \sqrt{npq}$.) Since the binomial probability distribution deals with discrete data while the normal distribution deals with continuous data, we introduce the **continuity correction** which should be used in normal approximations to binomial distributions if n is small. Finally, in Section 5-6, we consider the distribution of sample means that can come from normal or nonnormal populations. The **Central Limit Theorem** asserts that sample means $\bar{x}$ (based on random samples of size n) will, as n increases, approach a normal distribution with mean μ and standard deviation $\sigma/\sqrt{n}$. This means that if samples are of size n where $n \geq 30$, we can approximate the distribution of those sample means by a normal distribution. The **standard error of the mean** is $\sigma/\sqrt{n}$ as long as the population is infinite or the sample size is not more than 5% of the population. But if the sample n exceeds 5% of the population N, then the standard error of the mean must be adjusted by the **finite population correction factor** with $\sigma/\sqrt{n}$ multiplied by $\sqrt{(N - n)/(N - 1)}$. Figure 5-26 summarizes these concepts.

This chapter represents our first plunge into inferential statistics. Chapter 6 applies many of these concepts to the extremely important process of testing hypotheses. Since basic concepts of this chapter serve as critical prerequisites for the following material, it would be wise for you to master these ideas and methods now.

REVIEW EXERCISES

5-117. Errors on a scale are normally distributed with a mean of 0 kilograms and a standard deviation of 1 kilogram. One item is randomly selected and weighed. (The errors can be positive or negative.)

(a) Find the probability that the error in kilograms is between 0 and 0.74.

(b) Find the probability that the error is greater than 1.76 kilograms.

(c) Find the probability that the error is greater than −1.08 kilograms.

(d) Find the probability that the error in kilograms is between −1.89 and 2.14.

(e) If the top 5% of all errors are unacceptable, find the weights that separate acceptable and unacceptable errors.

5-118. Scores on a hearing test are normally distributed with a mean of 600 and a standard deviation of 100.

(a) If one subject is randomly selected, find the probability that his score is between 600 and 735.

(b) If one subject is randomly selected, find the probability that his score is more than 450.

(c) If one subject is randomly selected, find the probability that his score is between 500 and 800.

(d) If a job requires a score in the top 80%, find the lowest acceptable score.

(e) If 50 subjects are randomly selected, find the probability that their mean score is between 600 and 635.

5-119. The probability of a particular computer component being defective is 0.2. Five hundred of these components are randomly selected and tested. Find the probability of getting at least 120 defective components.

5-120. Scores on a standard I.Q. test are normally distributed with a mean of 100 and a standard deviation of 15.

(a) Find the probability that a randomly selected subject will achieve a score above 105.

(b) Thirty subjects are randomly selected and tested. Find the probability that their mean I.Q. score is above 105.

(c) Compare the results of parts (a) and (b) and justify any difference.

5-121. Eight hundred scores appear to have a very unusual distribution with $\mu = 120.0$ and $\sigma = 4.0$. If 50 of these scores are randomly selected, find the probability that their mean is above 121.0.

TESTING HYPOTHESES

6-1 OVERVIEW

WAS THIS POLL MISLEADING?

The following is an excerpt from an article analyzing the results of a teacher survey.

"Poll results showed the NYSUT leaders failed to get a clear membership mandate to disaffiliate from NEA. There were 72,208 votes in favor of disaffiliation and 28,527 against. The point is, however, that more than half of the teachers didn't vote, and the teachers voting represented only 35% of NYSUT's total membership. Further, Al Shanker's United Federation of Teachers staff campaigned hard in New York City to turn out the disaffiliation vote, and it is likely that many of the city's 60,000-plus members were among the majority of the pro-disaffiliation voters. NEA doubts that the disaffiliation campaign had strong support outside New York City."

In this chapter our main objective is to develop an understanding of the concept that underlies hypothesis testing. We also want you to develop the skills required to execute the test of a hypothesis relating to a mean, proportion, or variance successfully.

A hypothesis is a statement that something is true. The following statements are examples of hypotheses that can be tested by the procedures that we develop in this chapter:

- "The Republican nominee will win the election in November."
- "Our brand of tires lasts an average of 50,000 kilometers."
- "The students in this statistics class have an average I.Q. that is greater than the average for all other statistics classes."
- "People who exercise daily have less cholesterol than those who do not."
- "The average weight of a car in New York State is 1800 kilograms."

Suppose you take a dime from your pocket and then claim that it favors heads when it is flipped. You have formulated a hypothesis. Most reasonable people wishing to test your claim would begin by examining the coin to see if it is two-headed or possesses any other gross irregularities. If the dime seems normal, the next reasonable step is to flip the coin several times to see what happens. Let's suppose that heads occurs 94 times out of 100 tosses. On the basis of those sample data, most people would conclude that your claim is correct.

Perhaps the dime is actually fair and does not favor heads or tails. Perhaps the occurrence of 94 heads out of 100 tosses is simply a very unusual manifestation for a coin that usually exhibits normal behavior. But since the probability of flipping a fair dime 100 times and getting 94 heads is so incredibly small, it is more reasonable to conclude that the coin does favor heads. This is the way statisticians think when they are testing hypotheses.

We operate under the assumption that when events which appear to be very unusual occur, they probably occur not by chance but by the influence of another factor. If a gambler started a friendly dice game and began by rolling ten consecutive 7s, it is not likely that he would be congratulated for such a streak of good fortune. A more likely reaction would be a request to examine the dice. The suspicious players would be behaving like statisticians as they tended to attribute very unusual events to causes other than pure chance.

This intuitive discussion should reveal a fundamental concept that underlies the method of testing hypotheses. That method involves a variety of standard terms and conditions in the context of an organized procedure. We suggest that you begin the study of this chapter by first reading Section 6-2 casually to obtain a very general idea of its concepts. Then read the material more carefully to gain a familiarity with the terminology. Subsequent readings should incorporate the details and refinements into the basic procedure. You are not expected to master the principles of hypothesis testing in one reading, since a text description of the statistical method of testing hypotheses bears no resemblance to a novel by Truman Capote. (This undoubtedly pleases Truman Capote.)

6-2 TESTING A CLAIM ABOUT A MEAN

The coin and dice examples mentioned in the introduction are largely intuitive and devoid of some of the components that are required in a formal statistical test of a hypothesis. For example, we stated that the occurrence of 94 heads in 100 tosses is a very unusual event, but we made no attempt to specify the exact criteria used to identify unusual events. We will begin this section with another illustrative example, but we will include more of the necessary details. We will then proceed to define the terms, conditions, and procedures that constitute the formal method of a standard hypothesis test.

Assume that you have just theorized that blondes have above average intelligence. The only way to prove or disprove this contention is to measure the I.Q. of all blondes, but that is clearly impractical. Instead, you must rely upon sample data to form your conclusion. You decide to recruit 100 randomly selected blondes and administer an I.Q. test. You know that this test has been designed so that scores are normally distributed with a mean of 100 and a standard deviation of 15. After compiling the I.Q. scores of 100 blondes, you compute the mean and obtain $\bar{x} = 101.8$. At this point you may be tempted to conclude that the hypothesis of greater intelligence for blondes is correct simply because the sample mean of 101.8 is greater than the population mean of 100. But let's analyze this I.Q. experiment critically. First, we recall that tests are essentially measurements that may fluctuate and display errors of various sizes. A person taking I.Q. tests on different days is likely to achieve different scores on the tests. Similarly, 100 randomly selected volunteers cannot be expected to achieve a mean I.Q. score of exactly 100. Their actual mean score will differ from the theoretically predicted score of 100 because of chance fluctuations. Recognizing this, we formulate the key question: Does the sample mean score of 101.8 represent a statistically **significant** increase above the population mean of 100, or is the difference more likely due to **chance** variations in I.Q. scores? Obviously, if the 100 blondes achieved a mean I.Q. score below 100, we would immediately conclude that the hypothesis of greater intelligence for blondes is unfounded. Also, if the 100 blondes achieved a mean score such as 160, the increase would obviously be significant and not due to chance. However, many values are not quite so obvious in their implications. Where do we draw the line? What critical value separates **significant increases** from those due to **chance fluctuations**?

Let's pursue the claim that blondes have greater intelligence by examining the data in relation to what we know about sample means. Specifically, we know from Section 5-6 that sample means tend to be normally distributed with a mean equal to the population mean and with a standard deviation equal to $\sigma/\sqrt{n}$.

$$\mu_{\bar{x}} = \mu$$

$$\sigma_{\bar{x}} = \frac{\sigma}{\sqrt{n}}$$

We can use that knowledge to determine how unusual a sample mean of 101.8 is for a sample of 100 blondes randomly selected from the population. If 101.8 could easily occur in a sample of normal people, then we would conclude that the claim of greater intelligence for blondes is unwarranted. But if it is very unlikely that a random sample of 100 normal subjects will produce a mean I.Q. score as high as 101.8, then we would be inclined to agree with the hypothesis of greater intelligence.

What exactly do we mean by unusual or very unlikely? Let's arbitrarily select 5% (or a probability of 0.05) as a level which separates a significant difference from a chance fluctuation. That is, we are defining unusual to mean that the event has a 5% chance (or less) of occurring. If the probability of getting a sample mean I.Q. score of 101.8 or greater is 0.05 or less, we will conclude that the sample mean represents a significant increase over the population mean of 100 and we will therefore agree with the claim of greater intelligence for blondes. Otherwise, we will reject that claim.

The test of the claim now centers on the probability of getting a sample mean of 101.8. Sample means have distributions that can be approximated by normal distributions and since $\mu = 100$ and $\sigma = 15$ we get

$$\mu_{\bar{x}} = \mu = 100$$

$$\sigma_{\bar{x}} = \frac{\sigma}{\sqrt{n}} = \frac{15}{\sqrt{100}} = \frac{15}{10} = 1.5$$

(see Section 5-6). We are applying the Central Limit Theorem and our knowledge about the sample size ($n = 100$) and the design of the I.Q. test ($\mu = 100$, $\sigma = 15$) to get these results. Figure 6-1 illustrates these results and our 5% choice as the level separating chance fluctuations from significant differences.

We find that $\bar{x} = 101.8$ is equivalent to $z = 1.20$ by computing

$$z = \frac{\bar{x} - \mu_{\bar{x}}}{\sigma_{\bar{x}}} = \frac{101.8 - 100}{15/\sqrt{100}} = \frac{1.8}{1.5} = 1.20$$

We determine that $z = 1.645$ is the cutoff value separating unusual results from chance fluctuations by observing that, if the shaded region in the right tail represents 5% or 0.05 of the total area, then the left limit of that region and $z = 0$ must encompass 45% or 0.45 of the total. From Table A-3, we see that 0.4500 is between $z = 1.64$ and $z = 1.65$, so we split the difference (or interpolate) to get $z = 1.645$.

FIGURE 6-1

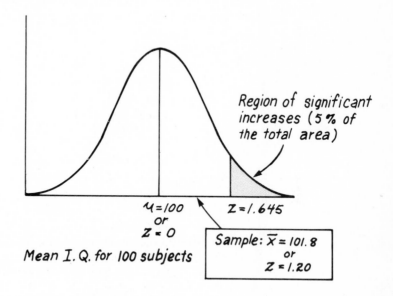

In effect, Figure 6-1 shows that a sample mean of 101.8 is *not* unusually high. Unusually high sample means have a 0.05 probability or less, and they fall within the shaded region of Figure 6-1. We therefore withdraw support for the claim that blondes tend to be more intelligent; that is, the experimental results do not warrant such support. Because this kind of procedure is used in decision-making and because the consequences of decisions can often be severe, we will examine this method closely.

A **hypothesis test** or **test of significance** involves procedures that allow us to make inferences about whole populations by analyzing samples. In this decision-making process, we begin by hypothesizing (sometimes just guessing) about the population. After gathering sample data, we try to determine whether the data support the hypothesis and whether they are statistically significant. If we hypothesize that "blondes have higher I.Q.'s" and then find that the sample mean score is 101.8, we question the *significance* of that increase. We have already indicated that such an increase is not statistically significant and that the hypothesis should not be accepted.

We need to formalize this decision-making process. The following are some of the standard terms and components required to do this (see also Table 6-1).

TABLE 6-1

	The null hypothesis is true	The null hypothesis is false
We decide to reject the null hypothesis	Type I error	Correct decision
We fail to reject the null hypothesis	Correct decision	Type II error

- *Null hypothesis:* (denoted by H_0) The statement of a zero or null difference that is directly tested. The null hypothesis corresponds to the original claim only if that claim includes the condition of no change or difference. However, the null hypothesis is the negation of an original claim which asserts a significant change or difference. The null hypothesis is directly tested in the sense that the final conclusion will be either the rejection of H_0 or failure to reject H_0.

- *Alternative hypothesis:* (denoted by H_1) The statement that must be true if the null hypothesis is false.

- *Type I error:* The mistake of rejecting the null hypothesis when it is true.

- *Type II error:* The mistake of failing to reject the null hypothesis when it is false.

- α (*alpha*): Symbol used to represent the probability of a type I error.

- β (*beta*): Symbol used to represent the probability of a type II error.

- *Test statistic:* A sample statistic or a value based on the sample data. It is used in making the decision to accept the null hypothesis or to reject it.

- *Critical region:* The set of all values of the test statistic that would cause us to reject the null hypothesis.

- *Critical value(s):* The value(s) that separates the critical region from the values of the test statistic that would not lead to rejection of the null hypothesis. The critical value(s) depends on the nature of the null hypothesis, the relevant sampling distribution, and the level of significance α.

- *Elation:* The feeling experienced when the techniques of hypothesis testing are mastered.

FAMILY PLANNING PROGRAM LOWERS INFANT MORTALITY RATE

A study conducted by John Hopkins University shows that family planning programs appear to lower the mortality rate among infants born to nonwhite parents. There were 27 deaths per 1000 infants for a 2-year base period before the family planning program was expanded, but the postexpansion rate was 19 deaths per 1000 infants. Researchers felt that the mortality rate was lowered through a decrease in high-risk births, and that the decrease was made possible by the use of birth control methods.

Most of our hypothesis testing will be concerned with values of *parameters* — numerical characteristics of a population such as the mean, variance, or proportion. The following rules will be helpful in these tests:

1. For the sake of simplicity, always arrange the original claim and its alternative so that the null hypothesis contains the condition of equality. The alternative hypothesis will then involve exactly one of the three signs, $>$, $<$, or $\neq$.

2. H_0 is presumed true until significant sample evidence suggests that it be rejected. As a result of the test, we either reject H_0 or we fail to reject H_0. These methods of statistically testing H_0 *never prove* that H_0 is true.

3. The value of α should be selected before the test is conducted. If a type I error is very serious and has dire consequences, α should be a small value such as 0.01. Less serious type I errors usually involve larger values of α. For the significance level α, the values of 0.05 and 0.01 are very common.

The essential steps for testing hypotheses are as follows:

1. Based upon some claim, formulate the null hypothesis H_0 and the alternative hypothesis H_1.

2. Based on the seriousness of a type I error, select α (the probability of making a type I error). The probability of a type II error (β) will be determined when the sample size is fixed. (We will not deal with β.)

3. Determine which sample statistic is appropriate. Also determine the appropriate sampling distribution.

4. Using the sample data, compute the test statistic.

5. Using the computed test statistic and the corresponding critical value, either reject H_0 or assert that the sample data do not warrant a rejection of H_0. (At this point, it is usually wise to graph the appropriate sampling distribution with the test statistic, critical region(s), and critical value(s) identified.)

6. In simple nontechnical terms, state what the results suggest. Figure 6-2 summarizes these key steps. To the right of the diagram are the actual results of the steps applied to the example at the beginning of this section.

It is easy to become entangled in a web of cookbook-type steps without ever understanding the underlying rationale for the procedure. The key to that understanding lies with recognition of this concept: **if an event can easily occur, we attribute it to chance; but if the event appears to be unusual**, **we attribute that significant departure to the presence of different characteristics**. If we keep this idea in mind as we examine various examples, the method of hypothesis testing will become meaningful instead of a rote mechanical process.

Our first example of hypothesis testing involves the claim that "blondes have above average intelligence." The test used is called a **right-tailed** test because the critical region of Figure 6-1 is in the extreme right region under the curve. We reject the null hypothesis H_0 if our test statistic is in the critical region, because that indicates a significant conflict between the null

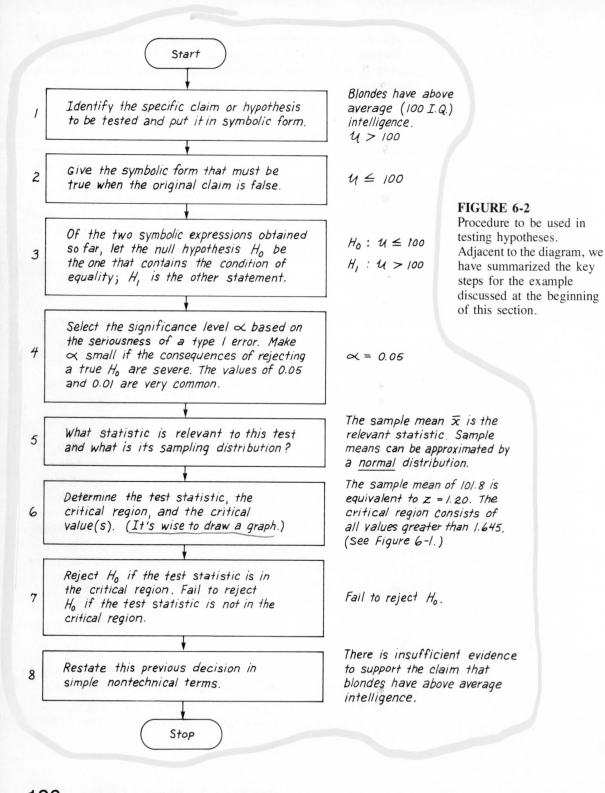

FIGURE 6-2
Procedure to be used in testing hypotheses. Adjacent to the diagram, we have summarized the key steps for the example discussed at the beginning of this section.

Start

1. Identify the specific claim or hypothesis to be tested and put it in symbolic form.

Blondes have above average (100 I.Q.) intelligence.
$u > 100$

2. Give the symbolic form that must be true when the original claim is false.

$u \leq 100$

3. Of the two symbolic expressions obtained so far, let the null hypothesis H_0 be the one that contains the condition of equality; H_1 is the other statement.

$H_0 : u \leq 100$
$H_1 : u > 100$

4. Select the significance level α based on the seriousness of a type I error. Make α small if the consequences of rejecting a true H_0 are severe. The values of 0.05 and 0.01 are very common.

$\alpha = 0.05$

5. What statistic is relevant to this test and what is its sampling distribution?

The sample mean $\bar{x}$ is the relevant statistic. Sample means can be approximated by a <u>normal</u> distribution.

6. Determine the test statistic, the critical region, and the critical value(s). (<u>It's wise to draw a graph.</u>)

The sample mean of 101.8 is equivalent to $z = 1.20$. The critical region consists of all values greater than 1.645. (See Figure 6-1.)

7. Reject H_0 if the test statistic is in the critical region. Fail to reject H_0 if the test statistic is not in the critical region.

Fail to reject H_0.

8. Restate this previous decision in simple nontechnical terms.

There is insufficient evidence to support the claim that blondes have above average intelligence.

Stop

FIGURE 6-3
Critical regions are shaded.

hypothesis and the sample data. Some tests will be **left-tailed** with the critical region located in the extreme left region under the curve. Other tests may be **two-tailed** because there are two critical regions located in the two extreme regions under the curve. Figure 6-3 illustrates these three possibilities. In each

Two-tailed test

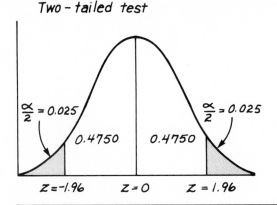

Claim: Blondes have a mean I.Q. equal to 100.
$H_0 : \mu = 100$
$H_1 : \mu \neq 100$
$\alpha = 0.05$
Reject H_0 if the sample mean $\bar{x}$ is significantly above 100 (right tail) or below 100 (left tail).

Left-tailed test

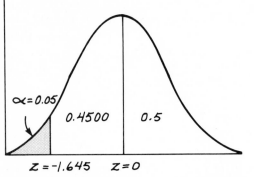

Claim: Blondes have a mean I.Q. less than 100
$H_0 : \mu \geq 100$
$H_1 : \mu < 100$
$\alpha = .05$
Reject H_0 if the sample mean $\bar{x}$ is significantly below 100 (left tail)

Right-tailed test

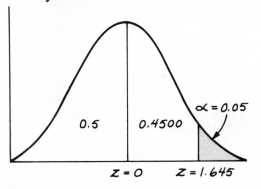

Claim: Blondes have a mean I.Q. greater than 100.
$H_0 : \mu \leq 100$
$H_1 : \mu > 100$
$\alpha = 0.05$
Reject H_0 if the sample mean $\bar{x}$ is significantly greater than 100 (right tail).

case, we convert the claim into symbolic form and then determine the symbolic alternative. The null hypothesis H_0 becomes the symbolic statement containing the condition of equality. The alternative hypothesis becomes the other symbolic statement. We reject H_0 if there is significant evidence supporting H_1. For this reason, critical regions correspond to the extremes indicated by H_1. In Figure 6-3, the inequality sign points to the critical region.

Sign used in H_1	Type of test
$>$	Right-tailed
$<$	Left-tailed
$\neq$	Two-tailed

Example

Test a car manufacturer's claim that the Gasmiser model has a mean fuel consumption rate equal to 35 miles per gallon. The quality control group suggests that $\sigma = 4$ miles per gallon and a sample of 50 Gasmisers yields $\bar{x} = 33.6$ miles per gallon.

Solution

We outline our test of the manufacturer's claim by following the scheme in Figure 6-2. The result is shown in Figures 6-4 and 6-5. The test is two-tailed because a sample mean significantly greater than 35 miles per gallon (right tail) or less than 35 miles per gallon (left tail) is strong evidence against the null hypothesis that $\mu = 35$ miles per gallon. Our sample mean of 33.6 miles per gallon is found to be equivalent to $z = -2.47$ through the following computation:

$$z = \frac{\bar{x} - \mu_{\bar{x}}}{\sigma_{\bar{x}}} = \frac{33.6 - 35}{4/\sqrt{50}} \stackrel{\circ}{=} -2.47$$

The critical z values are found by distributing $\alpha = 0.05$ equally between the two tails to get 0.025 in each tail. We then refer to Table A-3 (since we are assuming a normal distribution) to find the z value corresponding to $0.5 - 0.025$ or 0.4750. After finding $z = 1.96$, we use the property of symmetry to conclude that the left critical value is -1.96.

Example

A brewery distributes beer in bottles labeled 32 ounces. The local Bureau of Weights and Measures randomly selects 50 of these bottles, measures their contents, and obtains a sample mean of 31.0 ounces. Assuming that σ is known to be 0.75 ounces, is it valid at the 0.01 significance level to conclude that the brewery is cheating the consumer? That is, can we conclude that the true mean value is less than 32 ounces?

Solution

We outline the test of the claim that "the mean is less than 32 ounces" by

FIGURE 6-4

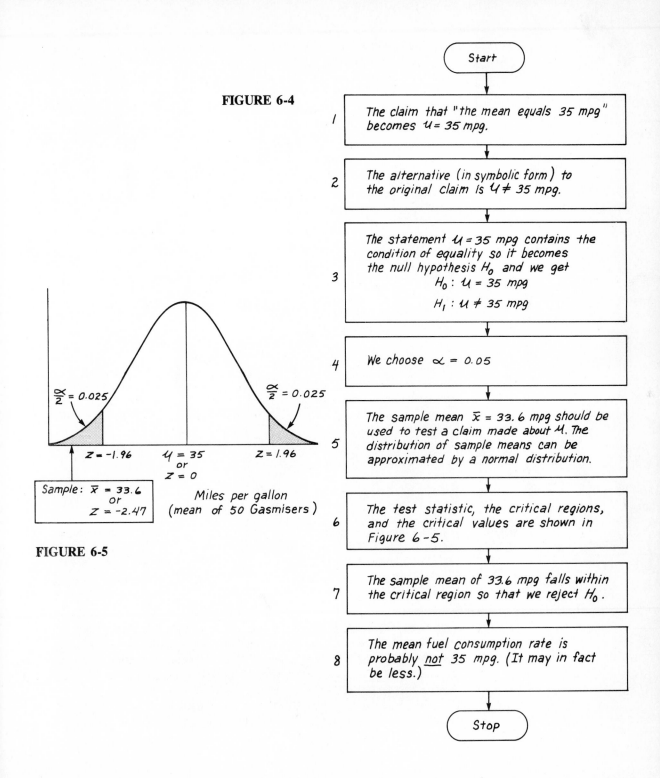

Start

1. The claim that "the mean equals 35 mpg" becomes $\mu = 35$ mpg.

2. The alternative (in symbolic form) to the original claim is $\mu \neq 35$ mpg.

3. The statement $\mu = 35$ mpg contains the condition of equality so it becomes the null hypothesis H_0 and we get
$$H_0 : \mu = 35 \text{ mpg}$$
$$H_1 : \mu \neq 35 \text{ mpg}$$

4. We choose $\alpha = 0.05$

5. The sample mean $\bar{x} = 33.6$ mpg should be used to test a claim made about μ. The distribution of sample means can be approximated by a normal distribution.

6. The test statistic, the critical regions, and the critical values are shown in Figure 6-5.

7. The sample mean of 33.6 mpg falls within the critical region so that we reject H_0.

8. The mean fuel consumption rate is probably <u>not</u> 35 mpg. (It may in fact be less.)

Stop

$\frac{\alpha}{2} = 0.025$

$\frac{\alpha}{2} = 0.025$

$z = -1.96$

$\mu = 35$
or
$z = 0$

$z = 1.96$

Sample: $\bar{x} = 33.6$
or
$z = -2.47$

Miles per gallon
(mean of 50 Gasmisers)

FIGURE 6-5

again following the model of Figure 6-2. The results are presented in Figures 6-6 and 6-7. The z value of -9.43 is computed as follows:

$$z = \frac{\bar{x} - \mu_{\bar{x}}}{\sigma_{\bar{x}}} = \frac{31 - 32}{0.75/\sqrt{50}} \doteq -9.43$$

The critical z value is found in Table A-3 as the z value corresponding to an area of 0.4900.

FIGURE 6-6

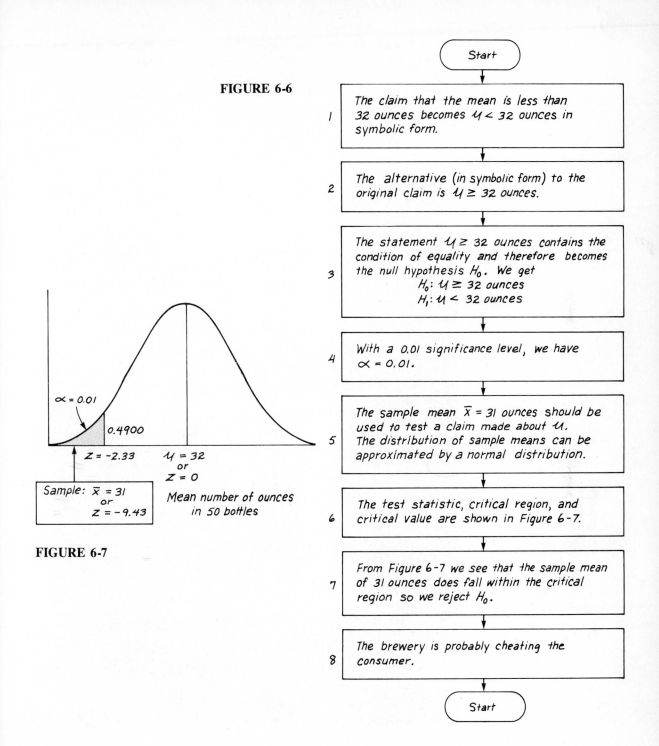

∝ = 0.01

0.4900

Z = -2.33 $\mathcal{U}$ = 32
             or
             Z = 0

Sample: $\bar{x}$ = 31
        or
        Z = -9.43

Mean number of ounces
in 50 bottles

FIGURE 6-7

Start

1 | The claim that the mean is less than 32 ounces becomes $\mathcal{U} < 32$ ounces in symbolic form.

2 | The alternative (in symbolic form) to the original claim is $\mathcal{U} \geq 32$ ounces.

3 | The statement $\mathcal{U} \geq 32$ ounces contains the condition of equality and therefore becomes the null hypothesis H_0. We get
$H_0: \mathcal{U} \geq 32$ ounces
$H_1: \mathcal{U} < 32$ ounces

4 | With a 0.01 significance level, we have $\propto$ = 0.01.

5 | The sample mean $\bar{x}$ = 31 ounces should be used to test a claim made about $\mathcal{U}$. The distribution of sample means can be approximated by a normal distribution.

6 | The test statistic, critical region, and critical value are shown in Figure 6-7.

7 | From Figure 6-7 we see that the sample mean of 31 ounces does fall within the critical region so we reject H_0.

8 | The brewery is probably cheating the consumer.

Start

A potentially unrealistic feature of the preceding example is the assumption that $\sigma = 0.75$. Realistic tests of hypotheses must often be made without knowledge of the population standard deviation. If the sample is large enough (30 or larger), we can compute the sample standard deviation, and we may be able to use that value of s as an estimate for σ. For smaller samples with unknown standard deviations, we may be able to use a t statistic (discussed in Section 6-3).

EXERCISES A: TESTING A CLAIM ABOUT A MEAN

6-1. The null hypothesis should include the condition of equality, while the alternative hypothesis should involve one of the three signs $>$, $<$, or $\neq$. For each of the following claims, identify H_0 and H_1 as in the following example: "The mean I.Q. of doctors is greater than 110."

$$H_0: \mu \leq 110$$

$$H_1: \mu > 110$$

(a) The mean age of professors is more than 30 years.
(b) The mean I.Q. of criminals is below 100.
(c) The mean I.Q. of college students is at least 100.
(d) The mean annual salary of police officers is at least $16,000.
(e) The mean annual salary of police officers is at most $16,000. $\leq$
(f) The mean annual salary of police officers is less than $16,000. $\geq$
(g) The mean annual salary of police officers is more than $16,000. $\leq$
(h) The mean annual salary of police officers is $16,000. $\neq$
(i) The mean weight of girls at birth is 3.2 kilograms. $=$
(j) The mean life of a car battery is longer than 30 months. $\leq$
(k) The reaction time of normal adult drivers is 0.73 seconds. $=$
(l) The mean net weight of containers of sugar is at least 1 pound. $\geq$
(m) The mean monthly maintenance cost of an aircraft is $3,271. $=$
(n) The mean height of females is 1.6 meters. $\neq$
(o) The mean I.Q. of television programming executives is below $\geq$ 75.

6-2. Identify the type I error that corresponds to each claim in Exercise 6-1.

6-3. Identify the type II error that corresponds to each claim in Exercise 6-1.

6-4. For each claim in Exercise 6-1, categorize the hypothesis test as a right-tailed test, a left-tailed test, or a two-tailed test.

6-5. Find the critical z value for the following conditions. In each case, assume that the normal distribution applies so that Table A-3 can be used.

(a) Right-tailed test; $\alpha = 0.05$.
(b) Right-tailed test; $\alpha = 0.01$.
(c) Two-tailed test; $\alpha = 0.05$.
(d) Two-tailed test; $\alpha = 0.01$.
(e) Left-tailed test; $\alpha = 0.05$.
(f) Left-tailed test; $\alpha = 0.02$.
(g) Two-tailed test; $\alpha = 0.10$.
(h) Right-tailed test; $\alpha = 0.005$.
(i) Right-tailed test; $\alpha = 0.025$.
(j) Left-tailed test; $\alpha = 0.025$.

In each of the following exercises, test the given hypotheses by following the procedure suggested by Figure 6-2. Draw the appropriate graph.

6-6.   Test the claim that $\mu \geq 20$ given a sample of $n = 100$ for which $\bar{x} = 18.7$. Assume that $\sigma = 3$ and test at the $\alpha = 0.05$ significance level.

6-7.   Test the claim that $\mu \leq 100$ given a sample of $n = 81$ for which $\bar{x} = 100.8$. Assume that $\sigma = 5$ and test at the $\alpha = 0.01$ significance level.

6-8.   Test the claim that $\mu \leq 40$ given a sample of $n = 150$ for which $\bar{x} = 41.6$. Assume that $\sigma = 9$ and test at the $\alpha = 0.01$ significance level.

6-9.   Test the claim that $\mu > 40$ given a sample of $n = 50$ for which $\bar{x} = 42$. Assume that $\sigma = 8$ and test at the $\alpha = 0.05$ significance level.

6-10.   Test the claim that $\mu = 65$ given a sample of $n = 50$ for which $\bar{x} = 66.1$. Assume that $\sigma = 4$ and test at the $\alpha = 0.05$ significance level.

6-11.   Test the claim that $\mu = 500$ given a sample of $n = 300$ for which $\bar{x} = 510$. Assume that $\sigma = 100$ and test at the $\alpha = 0.10$ significance level.

6-12.   In 1950, a study of fuel consumption by cars showed that the mean mileage per gallon was 14.95. In 1972, a sample of 100 randomly selected cars produced a mean of 13.67 miles per gallon. Test the claim that in 1972, cars achieved a lower mile-per-gallon rating. Assume that $\sigma = 4$ and use $\alpha = 0.005$.

6-13.   Triple X sugar is packaged in boxes labeled 5 pounds. One hundred packages are randomly selected and measured. The sample mean and standard deviation are found to be 4.95 pounds and 0.15 pound, respectively. At the 0.01 significance level, test the claim that the sample comes from a population in which the mean weight is less than 5 pounds. (Assume that the sample standard deviation can be used for σ.)

6-14.   On a college entrance examination, the mean score is 500 while the standard deviation is 100. A high school principal boasts better than average scores for her graduates. To support this claim, she randomly selects 75 graduates and determines that the mean score for this sample is 515. Is her claim actually supported by the sample data? Let $\alpha = 0.05$.

6-15. The mean time between failures for a certain type of radio used in light aircraft is 420 hours. Thirty-five new radios have been modified for more reliability, and tests show that the mean time between failures for this sample is 385 hours. Assume that σ is known to be 24 hours and let $\alpha = 0.05$. Test the claim that the modifications improved reliability. (Be careful with this one!)

6-16. A certain nighttime cold medicine bears a label indicating the presence of 600 milligrams of acetaminophelan in each fluid ounce of the drug. The Food and Drug Administration randomly selects 65 1-ounce samples and finds that the mean acetaminophelan content is 589 milligrams while the standard deviation is 21 milligrams. With $\alpha = 0.01$, test the claim that the population mean is equal to 600 milligrams. (Assume that the sample standard deviation can be used for σ.)

6-17. Excluding Monday mornings, a secretary can type an average of 50 words per minute (with a standard deviation of 4 words). The secretary's boss will invest in a new typewriter if the new model can be shown to increase productivity. Forty separate tests are conducted on the new model with a resulting mean of 53 words per minute. At the $\alpha = 0.01$ significance level, test the claim that the new model increases productivity.

6-18. A paint is applied to tin panels and baked for 1 hour so that the mean index of hardness is 35.2. Thirty-eight test panels are painted and baked for 3 hours producing a sample mean index of hardness equal to 35.9. Assuming that $\sigma = 2.7$, test (at the $\alpha = 0.05$ significance level) the claim that longer baking increases hardness of the paint.

6-19. A psychological test of motor skills requires the subject to sort cards into different boxes. The mean time for this task is 125.3 seconds, with a standard deviation of 15.0 seconds. Fifty pianists were given the test and their mean time was 130.1 seconds. At the $\alpha = 0.01$ significance level, test the claim that pianists are *faster*.

6-20. Tests on automobile braking reaction times for normal young men have produced a mean and standard deviation of 0.610 second and 0.123 second, respectively. Forty young male graduates of a driving school were randomly selected and tested for their braking reaction times and a mean of 0.587 second resulted. At the $\alpha = 0.10$ significance level, test the claim of the driving instructor that his graduates had faster reaction times.

6-21. A dairy produces milk with an average of 18.24 grams of fat in each quart. (σ = 3.88 grams.) The dairy supervisor hypothesizes that skinny cows would produce milk with less fat. Fifty quarts of milk are randomly selected from 50 different skinny cows, and the fat contents are measured with a resulting mean of 17.98 grams. At the α = 0.05 significance level, test the supervisor's hypothesis.

6-22. The mean and standard deviation for the fuel consumption in miles per gallon for a given car are 22.6 and 3.4, respectively. A revolutionary new spark plug is used and a sample of 33 tests yields a mean of 20.1 miles per gallon. At the α = 0.05 significance level, test the claim that the new spark plug lowered fuel consumption.

EXERCISES B: TESTING A CLAIM ABOUT A MEAN

6-23. At the α = 0.03 significance level, test the claim that the following I.Q. scores come from a special group in which the mean is above 100. Use the sample standard deviation as an estimate for σ.

101	110	114	105	79	144	111	99	101	107	103	82
107	90	91	99	95	117	93	103	120	82	123	112
107	89	105	130	106	103	100	118	98	101		

6-24. A brewery claims that consumers are getting a mean volume equal to 32 ounces of beer in their quart bottles. The Bureau of Weights and Measures randomly selects 36 bottles and obtains the following measures in ounces:

32.09	31.89	31.06	32.03	31.42	31.39	31.75	31.53	32.42	31.56
31.95	32.00	31.39	32.09	31.67	31.47	32.45	32.14	31.86	32.09
32.34	32.00	30.95	33.53	32.17	31.81	31.78	32.64	31.06	32.64
32.20	32.11	31.42	32.09	33.00	32.06				

Using the sample standard deviation as an estimate for σ, test the claim of the brewery at the 0.05 significance level.

6-25. Since α is the probability of making a type I error, why not let α = 0.001, or 0.00001, or even 0? That is, why do we not make α small in all hypothesis tests?

6-3 t TEST

In Section 6-2 we introduce the general method for testing hypotheses, but all of the examples and exercises involve situations in which the normal distribution can be used. The population standard deviation σ is given, the samples are large, and each hypothesis tested relates to a population mean. In those cases, we can apply the Central Limit Theorem and use the normal distribution as an approximation to the distribution of sample means. A very unrealistic feature of those examples and exercises is the assumption that σ is known. If σ is unknown and the sample is large, we can use the sample standard deviation s as an estimate for the population standard deviation σ and proceed as in Section 6-2. This estimation of σ by s is reasonable because large random samples tend to be representative of the population. But small random samples may exhibit unusual behavior and they cannot be so trusted. Here we consider tests of hypotheses about a population mean, where the samples are small and σ is unknown. We begin by referring to Figure 6-8 which outlines the theory we are describing.

Starting at the top of Figure 6-8, we see that our immediate concerns lie only with hypotheses made about one population mean. (In following sections we will accommodate hypotheses made about population parameters other than the mean.) Figure 6-8 proceeds to summarize the following observations:

1. In *any* population, the distribution of sample means can be approximated by the normal distribution as long as the random samples are large. This is justified by the Central Limit Theorem.

2. In populations with distributions that are essentially normal, samples of *any* size will yield means having a distribution that is approximately normal. The value of μ would correspond to the null hypothesis, and the value of σ must be known. If σ is unknown and the samples are large, we can estimate σ by the sample standard deviation s since large random samples tend to be representative of the populations from which they come.

3. In populations with distributions that are essentially normal, assume that we randomly select small samples and we do not know the value of σ. The distribution of these sample means is approximately a *student t distribution*, which is described in this section.

4. If our random samples are small, σ is unknown, and the population is grossly nonnormal, then we can use nonparametric methods, some of which are discussed in Chapter 11.

We refer to large and small samples, but the number that separates large and small samples is not objectively derived with absolute exactness. It is somewhat arbitrary. However, there is widespread agreement that samples of size $n > 30$ are large enough so that the distribution of their means can be approximated by a normal distribution. Accordingly, samples of size $n \leq 30$ will be considered small so that the normal distribution is not always a suitable approximation to the distribution of sample means.

Around 1908, William S. Gosset (1876–1937) developed specific distributions for small samples in response to certain applied problems he encountered while working for a brewery. The Irish brewery where he worked had a policy which did not allow the publication of research results, so Gosset published under the pseudonym "Student." The small sample results are

FIGURE 6-8

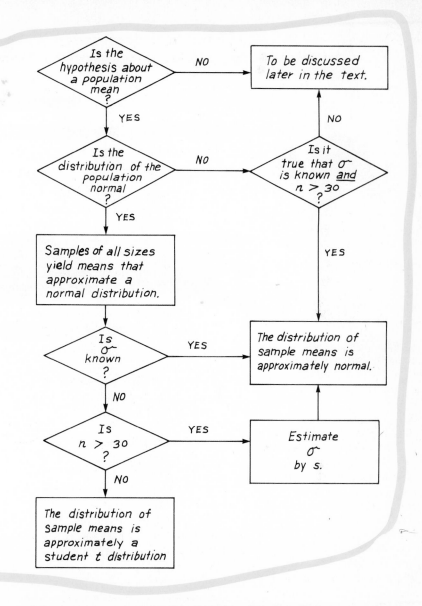

extremely valuable because there are many real and practical cases for which large samples are impossible or impractical and for which σ is unknown. Gosset, in his original research that led to small sample techniques, found that temperature and ingredient changes associated with brewing often allowed experimentation that provided only small samples. Factors such as cost and time often severely limited the size of a sample, so the normal distribution could not be an appropriate approximation to the distribution of these small sample means. As a result of those earlier experiments and studies of small samples, we can now use the student t distributions instead.

STUDENT *t* DISTRIBUTION

> *If a population is essentially normal, then the distribution of*
>
> $$t = \frac{\bar{x} - \mu}{s/\sqrt{n}}$$
>
> *is essentially a **student t distribution** for all samples of size* n.

DOES COLLEGE REALLY PAY?

*In **The Case Against College**, author Caroline Bird argues that a high school graduate would accrue greater lifetime earnings by not going to college and by putting the cost of a college education in a savings bank. A serious flaw in that argument is that high school graduates rarely have $30,000 to invest in a savings account. However, Ms. Bird does refer to a more valid study by Christopher Jencks. Professor Jencks concludes that the degree of lifetime financial success is more dependent upon luck and social class than on the number of years spent in college.*

We will not discuss the complicated mathematical equations which correspond to this student *t* distribution. Instead, we list critical values for this distribution in Table A-4. (Note that the student *t* distribution of Table A-4 involves only the critical values of *t* corresponding to common choices of α.) The critical *t* value is obtained by locating the proper value for degrees of freedom* in the left column and then proceeding across that corresponding row until reaching the number directly below the applicable value of α. In tests on a mean, the number of degrees of freedom is simply the sample size minus 1.

$$\text{degrees of freedom} = n - 1$$

For example, suppose we are testing at the $\alpha = 0.05$ significance level the hypothesis that $\mu = 100$ for some population, and we have only a sample of 20 for which $\bar{x} = 102$ and $s = 5$. (We assume that the sample is random and that the population is essentially normal.) The critical *t* value is extracted from Table A-4 by noting the following:

1. The test involves two tails since H_0 is $\mu = 100$ and H_1 becomes $\mu \neq 100$.
2. $\alpha = 0.05$ for the two-tailed case.
3. The sample size is $n = 20$, so there are $20 - 1$ or 19 degrees of freedom.

Locating 19 at the left column and 0.05 (two tails) at the top row, we determine that the critical *t* value is 2.093. Actually, $t = \pm 2.093$ since the test is two-tailed. The test statistic is obtained by computing

$$t = \frac{\bar{x} - \mu}{s/\sqrt{n}} = \frac{102 - 100}{5/\sqrt{20}} = 1.789$$

We would therefore fail to reject H_0 since the test statistic would not be in the critical region. (We will soon present another example in much more detail and with the critical region, critical values, and test statistic depicted in an appropriate graph.)

*Roughly stated, degrees of freedom correspond to the number of values that may vary after certain restrictions have been imposed on all values. For example, if 10 scores must total 50, we can freely assign values to the first 9 scores, but the tenth score would then be determined so that there would be 9 degrees of freedom.

In order to use the student t distribution, we require that the parent population be essentially normal. The population may not be exactly normal, but if it has only one mode and is basically symmetric, we will generally get good results if we use the student t distribution. If there is strong evidence that the population has a very nonnormal distribution, then nonparametric methods (see Chapter 11) may apply.

The student t distribution has the same general shape and symmetry of the normal distribution, but it reflects the greater variability that is expected with small samples. The normal and student t distributions both have a mean of zero, but the standard deviation for the normal distribution is one while the student t distribution has a standard deviation greater than one. For a visual comparison, see Figure 6-9. In Figure 6-9 we can also see that, as n gets larger, the student t distribution gets closer to the normal distribution. For values of $n > 30$, the differences between the student t and normal distributions are so small that we can use the critical z values instead of developing a much larger table of critical t values. Consequently, the values in the bottom row of Table A-4 are equal to the corresponding critical z values found from the normal distribution.

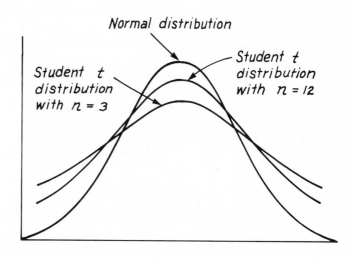

FIGURE 6-9

The student t distribution has the same general shape and symmetry as the normal distribution, but it reflects the greater variability that is expected with small samples.

Let's use a few examples to illustrate the use of the student t distribution. Remember that the student t distribution applies when we test a claim about a population mean and the following conditions are met:

1. The sample is small ($n \leq 30$).

2. σ is unknown.

3. The parent population is essentially normal.

Example   A pilot training program usually takes an average of 57.2 hours, but new teaching methods were used on the last class of 25 students. Computations reveal that for this experimental class, the completion times had a mean of 54.8 hours and a standard deviation of 4.3 hours. At the $\alpha = 0.05$ significance level, test the claim that the new teaching techniques reduce the instruction time.

Solution   Let μ represent the mean completion time for the new teaching method. The claim that it reduces instruction time is equivalent to the claim that $\mu < 57.2$ hours. We use the format of Figure 6-2 to outline the test of this claim, and the results are shown in Figures 6-10 and 6-11. We compute the test statistic as follows:

$$t = \frac{\bar{x} - \mu}{s/\sqrt{n}} = \frac{54.8 - 57.2}{4.3/\sqrt{25}} \stackrel{\circ}{=} -2.791$$

We find the critical t value from Table A-4 where we locate $25 - 1$ or 24 degrees of freedom at the left column and $\alpha = 0.05$ (one tail) across the top. The critical t value of 1.711 is obtained, but since small values of $\bar{x}$ will cause the rejection of H_0, we recognize that $t = -1.711$ is the actual t value that is the boundary for the critical region.

It is easy to lose sight of the underlying rationale as we go through this hypothesis testing procedure, so let's review the essence of the test. We set out to determine whether the sample mean of 54.8 hours is *significantly* below the value of 57.2 hours. Knowing the distribution of sample means (of which 54.8 is one) and choosing a level of significance (5% or $\alpha = 0.05$), we are able to determine the cutoff for what is a significant difference and what is not. Any sample mean equivalent to a t score below -1.711 represents a significant difference. The mean of 54.8 hours is significantly below 57.2 hours, so it appears as though the new teaching method does reduce the instruction time.

Example   A tobacco company claims that its best selling cigarettes contain 40 milligrams of nicotine. Test this claim at the 1% significance level by using the results of 15 randomly selected cigarettes for which $\bar{x} = 42.6$ milligrams and $s = 3.7$ milligrams.

Solution   In this example we will list only certain key elements of the solution. The null and alternate hypotheses are as follows.

$$H_0: \mu = 40 \text{ milligrams}$$
$$H_1: \mu \neq 40 \text{ milligrams}$$

The significance level of 1% corresponds to $\alpha = 0.01$. The sample mean

FIGURE 6-10

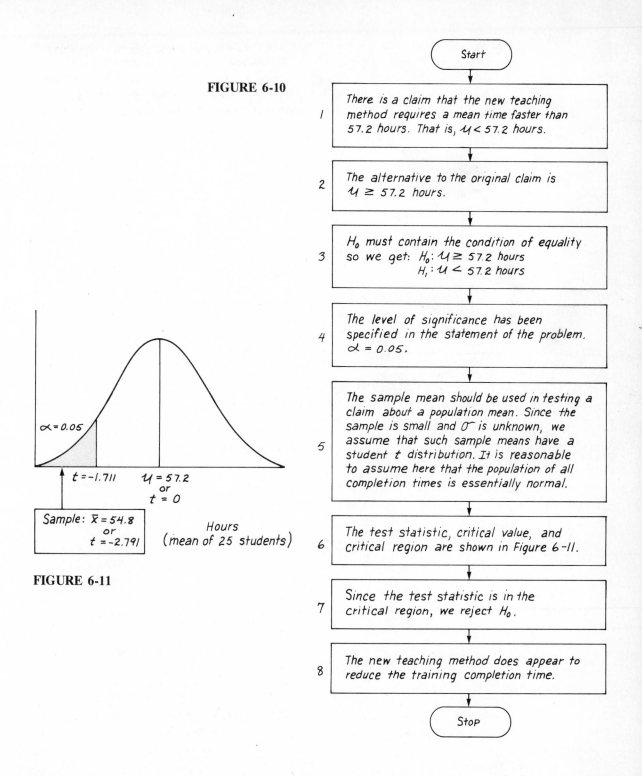

FIGURE 6-11

Start

1 | There is a claim that the new teaching method requires a mean time faster than 57.2 hours. That is, $\mu < 57.2$ hours.

2 | The alternative to the original claim is $\mu \geq 57.2$ hours.

3 | H_0 must contain the condition of equality so we get: $H_0: \mu \geq 57.2$ hours
$H_1: \mu < 57.2$ hours

4 | The level of significance has been specified in the statement of the problem. $\alpha = 0.05$.

5 | The sample mean should be used in testing a claim about a population mean. Since the sample is small and σ is unknown, we assume that such sample means have a student t distribution. It is reasonable to assume here that the population of all completion times is essentially normal.

6 | The test statistic, critical value, and critical region are shown in Figure 6-11.

7 | Since the test statistic is in the critical region, we reject H_0.

8 | The new teaching method does appear to reduce the training completion time.

Stop

$\alpha = 0.05$

$t = -1.711$ $\mu = 57.2$
or
$t = 0$

Sample: $\bar{x} = 54.8$
or
$t = -2.791$

Hours
(mean of 25 students)

should be used in testing a claim about the population mean, and we assume that such sample means have a student t distribution because:

1. It is reasonable to expect that the nicotine contents of all cigarettes of a certain brand are essentially a normal distribution.

2. σ is unknown.

3. The sample size ($n = 15$) is small. It is less than or equal to 30.

The test statistic is found as follows:

$$t = \frac{\bar{x} - \mu}{s/\sqrt{n}} = \frac{42.6 - 40}{3.7/\sqrt{15}} \doteq 2.722$$

In this two-tailed test, the critical t values of ± 2.977 are found from Table A-4 by noting that $\alpha = 0.01$ (two tails) and degrees of freedom $= 15 - 1$ or 14. Since the test statistic of $t = 2.722$ does not fall in the critical regions, we do not reject H_0 and we conclude that there is insufficient evidence to warrant rejection of the tobacco company's claim.

In the next section we test hypotheses relating to claims made about population proportions or percentages.

EXERCISES A: t TEST

6-26. Find the critical t value suggested by the given data:

(a) H_0: $\mu = 12$
$n = 27$
$\alpha = 0.05$

(b) H_0: $\mu \leq 50$
$n = 17$
$\alpha = 0.10$

(c) H_0: $\mu \geq 1.36$
$n = 6$
$\alpha = 0.01$

(d) H_0: $\mu = 1.36$
$n = 6$
$\alpha = 0.01$

(e) H_0: $\mu \geq 10.75$
$n = 29$
$\alpha = 0.01$

(f) H_0: $\mu \leq 100$
$n = 27$
$\alpha = 0.10$

(g) H_1: $\mu \neq 500$
$n = 16$
$\alpha = 0.05$

(h) H_1: $\mu < 67.5$
$n = 12$
$\alpha = 0.05$

(i) H_1: $\mu > 98.4$
$n = 7$
$\alpha = 0.05$

(j) H_1: $\mu \neq 75$
$n = 24$
$\alpha = 0.05$

In Exercises 6-27 through 6-32, assume that the population is normally distributed.

6-27. Test the claim that $\mu \leq 10$ given a sample of 9 for which $\bar{x} = 11$ and s = 2. Use a significance level of $\alpha = 0.05$.

6-28. Test the claim that $\mu \leq 32$ given a sample of 27 for which $\bar{x} = 33.5$ and $s = 3$. Use a significance level of $\alpha = 0.01$.

6-29. Test the claim that $\mu \leq 75$ given a sample of 15 for which $\bar{x} = 77.6$ and $s = 5$. Use a significance level of $\alpha = 0.05$.

6-30. Test the claim that $\mu \leq 500$ given a sample of 20 for which $\bar{x} = 541$ and $s = 115$. Use a significance level of $\alpha = 0.10$.

6-31. Test the claim that $\mu \geq 100$ given a sample of 22 for which $\bar{x} = 95$ and $s = 18$. Use a 5% level of significance.

6-32. Test the claim that $\mu \geq 98.6$ given a sample of 18 for which $\bar{x} = 98.2$ and $s = 0.8$. Use a significance level of $\alpha = 0.025$.

In Exercises 6-33 through 6-45, test the given hypothesis by following the procedure suggested by Figure 6-2. Draw the appropriate graph.

6-33. A manufacturer of chains claims that her product has a mean breaking level of at least 3000 pounds. Test this claim at the $\alpha = 0.005$ significance level if a random sample of 10 chains produces a mean and standard deviation of 2892 pounds and 480 pounds, respectively.

6-34. A battery indicates that it supplies 1.50 volts, but a sample of 19 such batteries is tested and yields a mean of 1.54 volts and a standard deviation of 0.10 volt. At the $\alpha = 0.02$ significance level, can we conclude that these batteries have a mean voltage level of 1.50 volts?

6-35. A pill is supposed to contain 20.0 grams of phenobarbitol. A random sample of 30 pills yields a mean and standard deviation of 20.5 grams and 1.5 grams, respectively. Are these sample pills acceptable at the $\alpha = 0.02$ significance level?

6-36. The Federal Aviation Administration randomly selects 5 light aircraft of the same type and tests the left wings for their loading capacities. The sample mean and standard deviation are 16,735 pounds and 978 pounds, respectively. At the 5% level of significance, test the claim that the mean loading capacity for all such aircraft is equal to 17,850.

6-37. A standard test for braking reaction times has produced an average of 0.75 second for young males. A driving instructor claims that her class of young males exhibits an overall reaction time that is below the average. Test her claim if it is known that 13 of her students (randomly selected) produced a mean of 0.71 second and a standard deviation of 0.06 second. Test at the 1% level of significance.

6-38. The mean down time for a computer has been 6.34 hours per week. A new supervisor initiates new procedures; the down times (based on a sample of 23 different weeks) now have a mean of 5.77 hours and a standard deviation of 1.82 hours. At the 5% level of significance, test the new supervisor's claim that the mean down time has been reduced.

6-39. A sociologist designs a test to measure prejudicial attitudes and claims that the mean population score is 60. The test is then administered to 28 randomly selected subjects and the results produced a mean and standard deviation of 69 and 12, respectively. At the 5% level of significance, test the sociologist's claim.

6-40. Teaching method A produces a mean score of 77 on a standard test. Teaching method B has been introduced and the first sample of 19 students achieved a mean score of 79 with a standard deviation of 8. At the 5% level of significance, test the claim that method B is better.

6-41. The skidding properties of a snow tire have been tested and the mean skid distance of 154 feet has been established for standardized conditions. A new, more expensive tire is developed, but tests on a sample of 20 new tires yield a mean skid distance of 141 feet with a standard deviation of 12 feet. Because of the cost involved, the new tires will be purchased only if they skid less at the $\alpha = 0.005$ significance level. Based on the sample, will the new tires be purchased?

6-42. A long-range missile misses its target by an average of 0.88 mile. A new steering device is supposed to increase accuracy, and a random sample of eight missiles is equipped with this new mechanism and tested. These eight missiles miss by distances with a mean of 0.76 mile and a standard deviation of 0.04 mile. At $\alpha = 0.01$, does the new steering mechanism lower the miss distance?

6-43. A biofeedback experiment involves measurement of muscle tension by using an instrument attached to a person's forehead. Sixteen randomly selected subjects are tested. The mean and standard deviation for this sample group are found to be 6.3 microvolts and 1.8 microvolts, respectively. At the 5% level of significance, test the claim that the mean of all subjects equals 5.4 microvolts.

6-44. Ten randomly selected power lawnmowers are tested for noise. The mean and standard deviation are found to be 95 decibels and 5 decibels, respectively. At the $\alpha = 0.05$ significance level, test the claim that the mean level for all such lawnmowers is below 100 decibels.

6-45. An aircraft manufacturer randomly selects 12 planes of the same model and tests them to determine the distance they require for takeoff. The sample mean and standard deviation are found to be 524 meters and 23 meters, respectively. At the 5% level of significance, test the claim that the mean for all such planes is more than 500 meters.

EXERCISES B: *t* TEST

6-46. A standardized final examination in an elementary statistics course produces a mean score of 75. At the 5% level of significance, test the claim that the following sample scores reflect an above-average class:

79	79	78	74	82	89	74	75	78	73
74	84	82	66	84	82	82	71	72	83

6-47. A sample of beer cans labeled 16 ounces is randomly selected and the actual contents accurately measured. The results (in ounces) are as follows. Is the consumer being cheated?

15.8	16.2	16.3	15.9	15.5
15.9	16.0	15.6	15.8	

6-48. A battery is sold with the claim that it lasts at least 26 months. A random sample of these batteries is tested with the following results (given in months that the battery functioned). Is the seller's claim valid?

34.0	37.2	35.5	30.7	33.1	32.2	34.0	32.3
35.7	32.6	36.1	35.6	29.3	33.4	37.6	

6-49. Obtain a random sample of ten heights of adult males and test the claim that all adult males have a mean height of 6 feet.

6-50. Obtain a random sample of ten heights of adult females and test the claim that all adult females have a mean height of more than 63 inches.

6-4 TESTS OF PROPORTIONS

The data that we collect are basically either quantitative or qualitative. Quantitative data are also called **variable data** and can be counted or measured. Qualitative data are also called **attribute data** and can be classified or described, but they cannot be counted or measured. Actual incomes of workers in various occupations would be an example of variable data since they consist of specific numbers or quantities. A list of occupations, however, would be an example of attribute data since they can only be described qualitatively. We are able to apply standard statistical methods to attribute data by representing that data in the form of a proportion or percentage.

As an example, suppose that you have just surveyed 720 randomly selected registered voters in your county and 390 have indicated that they favor the Democratic candidate in the upcoming presidential election. Each of the 720 voter preferences is a qualitative piece of information, but we can quantify these collective results by stating that, for this sample of 720 registered voters, 390/720 or 54.2% favored the Democratic candidate. In this section we show how hypothesis testing is used to decide, on the basis of sample data, how a true population proportion (or percentage) compares to some given value.

Throughout this section we assume that individual trials are independent so that the relevant probability remains constant for each trial. Recall that with a fixed number of independent trials having constant probabilities, as long as each trial has two outcomes, we have a binomial experiment (see Section 4-4). When we deal with proportions or percentages, we can usually make the assumptions that enable us to use a binomial distribution.

In Section 5-5 we saw that under suitable circumstances (namely, $np \geq 5$ and $nq \geq 5$)* the binomial distribution can be approximated by a normal distribution with the mean and standard deviation given by $\mu = np$ and $\sigma = \sqrt{npq}$. Replacing μ and σ by their binomial counterparts, we get

$$z = \frac{x - \mu}{\sigma} = \frac{x - np}{\sqrt{npq}}$$

where x is the number of successes in n trials.

To test a hypothesis made about a population proportion or percentage, we will therefore follow the standard procedure described in Section 6-2, but the value of the test statistic will be found by computing

$$z = \frac{x - np}{\sqrt{npq}}$$

where

n = number of trials
p = population proportion (given in the null hypothesis)
$q = 1 - p$

Let's assume that your county Republican party leader has claimed that the Democratic presidential candidate will receive no more than 48% of all votes cast in the county. Let us also assume that this claim is to be tested on the basis of the survey of 720 randomly selected registered voters of which 390 have indicated a preference for the Democratic candidate. The sample proportion of 390/720 (or 54.2%) is more than the claimed value of 48%, but we must now determine whether it is *significantly* more. We will assume a significance level of 5%. The sample data can now be summarized:

$$n = 720$$

$$x = 390$$

*If $np \geq 5$ and $nq \geq 5$ are not both true, we may be able to use Table A-2 or the binomial probability formula described in Section 4-4, but this section deals only with situations in which the normal distribution is a suitable approximation for the distribution of sample proportions.

MAGAZINE ACKNOWLEDGES BIAS IN READER SURVEY

Better Homes and Gardens conducted a survey by including questionnaires in two of its issues. The number of respondents was an impressively large 302,602. One result showed that 76% of the respondents felt that "family life in America is in trouble." Along with an analysis of the survey results, the magazine presented data which indicated that the poll results did not really represent a cross section of our nation's population. They pointed out, for example, that a disproportionately high number of respondents had attended college. The median income level of respondents was also well above the national median. It was also noted that only readers with "a particularly strong interest in the subject" would be likely to complete the questionnaire. It was finally concluded that the survey results represent the views of only the 302,602 respondents. These survey results cannot be treated as sample data from which broader inferences can be drawn. Unlike Better Homes and Gardens, many people often make the mistake of concluding that sample data can be used for inferences simply because the sample is large. Biased sample data cannot be used for inferences despite the large sample size.

The population parameters p and q are suggested by the claim that we are testing:

$$p = 0.48$$
$$q = 1 - 0.48 = 0.52$$

The significance level of 5% indicates that $\alpha = 0.05$. Following the pattern developed in Section 6-2 (see Figure 6-2), we get the following null and alternative hypotheses:

$$H_0: p \leq 0.48$$
$$H_1: p > 0.48$$

The statistic relevant to this test is

$$z = \frac{x - np}{\sqrt{npq}}$$

and it is normally distributed. Specifically, we get

$$z = \frac{x - np}{\sqrt{npq}} = \frac{390 - (720)(0.48)}{\sqrt{(720)(0.48)(0.52)}} \doteq 3.31$$

This test statistic, the critical region, and critical value are shown in Figure 6-12. The test statistic does fall in the critical region, and we therefore reject the null hypothesis H_0. This indicates that the sample result of a 54.2% Democratic preference among the 720 voters does represent a proportion of voters *significantly* greater than the 48% value claimed by the county Republican leader. If the 48% value is correct, there is less than a 5% chance of getting the sample results used in this example. Rather than concluding that an unusual sample has been obtained, we conclude that the true population proportion is more likely to be greater than 0.48 or 48%. This conclusion may in reality be wrong, even though it follows from accepted standard statistical techniques. But that is the nature of statistics. We are led to likely, but not indubitable, conclusions.

FIGURE 6-12

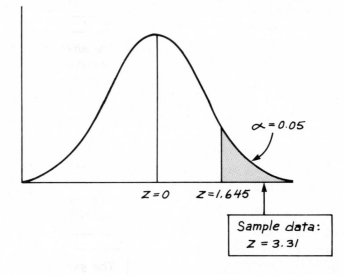

If the original problem is stated with a percentage, convert the percentage to the equivalent decimal form. For example, the claim that "at least 45% of all married people are women" would suggest the null hypothesis $H_0: p \geq 0.45$. In our computations, we should assume that $p = 0.45$ and $q = 0.55$. In addition to accommodating proportions and percentages, the methods described here also apply to tests of hypotheses made about probabilities. Whether we have a proportion, a percentage, or probability, the value of p must be between 0 and 1, and the sum of p and q must be exactly 1.

Example　A senator claims that 60% or more of his constituents favor a gun control bill. An independent pollster contacts 500 constituents selected at random and finds 273 that favor the bill. What can we conclude about the senator's claim? Assume a 5% level of significance.

Solution　We summarize the solution to this problem by following the pattern suggested by Figure 6-2. The results are shown in Figures 6-13 and 6-14 where the test statistic is computed as follows:

$$z = \frac{x - np}{\sqrt{npq}} = \frac{273 - (500)(0.6)}{\sqrt{(500)(0.6)(0.4)}} \doteq -2.465$$

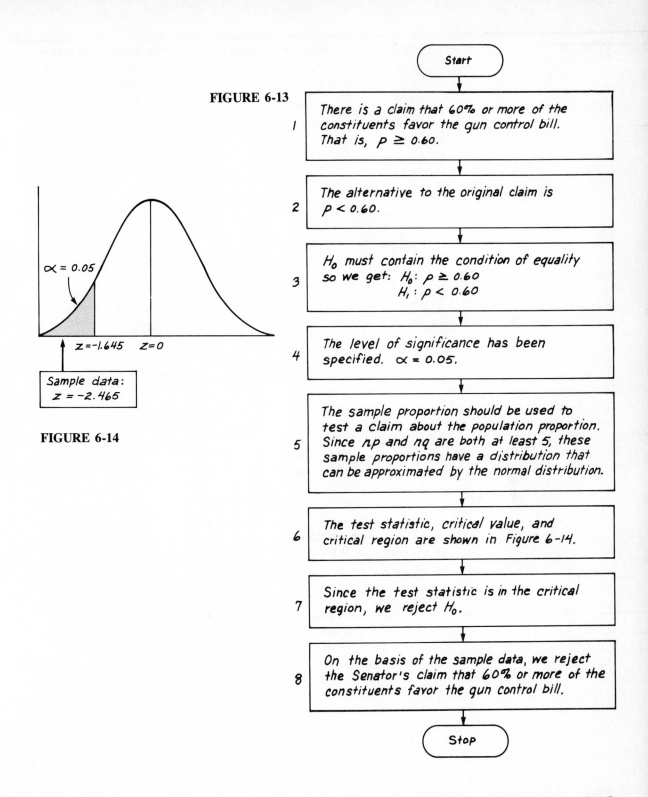

FIGURE 6-13

$\alpha = 0.05$

$z = -1.645$ $z = 0$

Sample data:
$z = -2.465$

FIGURE 6-14

Start

1   There is a claim that 60% or more of the constituents favor the gun control bill. That is, $p \geq 0.60$.

2   The alternative to the original claim is $p < 0.60$.

3   H_0 must contain the condition of equality so we get: $H_0: p \geq 0.60$
$H_1: p < 0.60$

4   The level of significance has been specified. $\alpha = 0.05$.

5   The sample proportion should be used to test a claim about the population proportion. Since np and nq are both at least 5, these sample proportions have a distribution that can be approximated by the normal distribution.

6   The test statistic, critical value, and critical region are shown in Figure 6-14.

7   Since the test statistic is in the critical region, we reject H_0.

8   On the basis of the sample data, we reject the Senator's claim that 60% or more of the constituents favor the gun control bill.

Stop

POLITICAL CANDIDATES ARE GUIDED BY POLLSTERS AND STATISTICIANS

Statisticians have become key advisors to political candidates who seek the best campaign strategies and the best allocation of time and money. Campaign themes can be "market tested" before they are presented on a large scale. Patrick Caddell is a pollster who helped Jimmy Carter win the Presidency. Caddell says that "we serve as a focal point, not only for strategy, but also for tactics on where you spend your money, where you put your media advertising, where you put your candidate's time, which issues you can use most effectively."

As the last example of this section, we will consider the close 1960 presidential election and, for the sake of simplicity, we will exclude consideration of electoral votes and popular votes cast for minor candidates. We will assume that the votes cast represent a random sample taken from the population of all eligible voters. The results are summarized as follows:

Popular votes cast for Kennedy: 34,227,000

Popular votes cast for Nixon: 34,108,000

Total number of eligible voters: 108,297,000

Since Kennedy won by about 119,000 votes and about 40 million eligible voters did not vote, it is reasonable to wonder whether the sample of 68,335,000 votes cast reflected the true proportion of preferences. Let's test the claim that the true population proportion for Kennedy, denoted by p, exceeded 50% or 0.5 despite the small margin of victory. This claim suggests the following null and alternative hypotheses:

$$H_0: p \leq 0.5$$
$$H_1: p > 0.5$$

Let's be conservative and choose a significance level of 0.01. We compute the test statistic based on the election results:

$$z = \frac{x - np}{\sqrt{npq}} = \frac{34,227,000 - (68,335,000)(0.5)}{\sqrt{(68,335,000)(0.5)(0.5)}} \overset{\circ}{=} 14.40$$

The critical value of $z = 2.33$ determines a critical region which certainly contains the test statistic of $z = 14.40$. As a result, we reject the null hypothesis; we reject the claim that at most 50% of the voters actually favored Kennedy. That the apparently slim margin of victory is so significant might seem surprising when we consider the margin of 119,000 votes in comparison to the 40 million votes that were not cast. It's results like this that make the subject of statistics fascinating and useful.

The tests of proportions are very useful in a variety of applications including surveys, polls, and quality control considerations involving the proportions of defective parts.

EXERCISES A: TESTS OF PROPORTIONS

6-51. In each case, determine whether the information is attribute data or variable data.
(a) The salaries of 50 farm workers.
(b) The occupations of 50 neighborhood residents.
(c) The colors of 50 randomly selected cars.
(d) The weights of 50 randomly selected fashion models.
(e) The answers on a true-false test.
(f) The votes of 50 citizens.
(g) The answers on a multiple-choice test.
(h) The I.Q. scores of 50 subjects.
(i) The shelf-lives (in hours) of 50 batteries.
(j) The responses to a survey question about capital punishment.

In Exercises 6-52 through 6-72, test the given hypotheses. Include the steps listed in Figure 6-2 and draw the appropriate graph.

6-52. At the 0.05 significance level, test the claim that the proportion of defects p for a certain product equals 0.3. Sample data consist of $n = 100$ randomly selected products of which 45 are defective.

6-53. At the 0.05 significance level, test the claim that the proportion of males p at a given college equals 0.6. Sample data consist of 80 randomly selected students of which 54 are males.

6-54. At the 0.01 level of significance, test the claim that the proportion of voters p who favor nuclear disarmament is equal to 0.7. Sample data consist of 50 randomly selected voters of whom 43 favor nuclear disarmament.

6-55. At the 0.01 significance level, test the claim that the percentage of Catholics in the United States is more than 20%. Sample data consist of 150 randomly selected Americans of whom 35 are Catholics.

6-56. At the 0.05 significance level, test the claim that fewer than 10% of U.S. colleges use a four-semester calendar. Sample data consist of a survey of 60 randomly selected colleges of which 10 use a four-semester calendar.

6-57. At the 0.05 significance level, test the claim that at least 80% of the students who take a statistics course receive a passing grade. Sample data consist of 90 randomly selected statistics students of whom 64 received a passing grade.

6-58. A lawyer claims in court that exactly 50% of all accidental deaths are related to motor vehicles. A random sample of 650 accidental deaths included exactly 342 that were related to motor vehicles. At the 0.05 significance level, test the lawyer's claim.

6-59. A sociologist claims that at least 22% of black males between the ages of 18 and 21 are enrolled in college. A random sample of 927 blacks in that age bracket produces 188 who are enrolled in college. At the 0.05 significance level, test the sociologist's claim.

6-60. A manufacturer considers his production process to be out of control when defects exceed 3%. A random sample of 500 items includes exactly 22 defects, but the manager claims that this represents a chance fluctuation and that production is not really out of control. Test the manager's claim at the 5% level of significance.

6-61. One company's birth control pill has a known failure rate of 1.2%. A second pharmaceutical company produces a new pill which is tested on a sample of 1000 randomly selected subjects. Of these 1000 trials, the pill is ineffective in eight cases. The second company claims that this new pill has a lower failure rate. Test the claim at the 0.01 significance level.

6-62. The unemployment rate was 8.2% in February. In March the Bureau of the Census polled 50,000 randomly selected employable subjects. Of these, 3785 subjects were classified as unemployed. At the 5% level of significance, test the assertion of the labor leader who claimed that the March unemployment rate was not below the February rate.

6-63. A senator claims that at least 60% of all Americans oppose abortions. A poll of 1500 randomly selected Americans produces 869 who register opposition. Test the senator's claim at the 5% level of significance.

6-64. A Gallup poll of 1553 adult Americans reveals that 47% of these subjects are able to correctly identify the name Freud. At the 0.01 significance level, test the claim that "most adult Americans do not know who Freud is."

6-65. In a poll of 1553 randomly selected adult Americans, 92% recognize the name Christopher Columbus. At the 0.01 significance level, test the claim that exactly 95% of all adult Americans recognize the name Columbus.

6-66. An obstetrician develops a drug that is claimed to increase the probability of a baby being male. To test the effectiveness of the drug, it is given to 150 volunteers before conception, and the 150 births result in 84 boys and 66 girls. Is the claim supported by the experimental data? Use a 1% level of significance.

6-67. The mathematics department has had a consistent failure rate of 12%. In one experimental semester, more active counseling plus rigorous enforcement of prerequisites results in 338 failures among 3217 mathematics students. At the 0.05 significance level, test the department's claim that the failure rate has been lowered.

6-68. A door-to-door vacuum cleaner salesperson has had a success rate of 8% over the past several years. After using a new mouthwash, the salesperson enjoyed success in 22 of 212 attempts. Upon learning of this, the mouthwash producer claimed in its advertisements that greater sales success is achieved by users of its product. Test that claim at the 0.02 significance level.

6-69. In a genetics experiment, the Mendelian law is followed as expected if one-eighth of the offspring exhibit a certain recessive trait. Analysis of 500 randomly selected offspring indicates that 83 exhibited the necessary recessive trait. Is the Mendelian law being followed as expected? Use a 2% level of significance.

6-70. In testing depressed patients, a doctor observes that a certain drug causes an unfavorable reaction in 15% of the cases. The doctor then alters the drug to determine whether there is any significant change in the unfavorable reaction rate. The new drug is used on 100 depressed patients and only 7 unfavorable reactions occur. At the 2% level of significance, test the claim that there is no change in the 15% rate.

6-71. A poll is conducted to test the claim of the Republican presidential candidate that she is favored by 56% of the voters. Fifteen hundred randomly selected subjects were polled and 54% of them indicated a preference for the Republican. Test this candidate's claim at the 0.02 significance level.

6-72. An airline reservations system suffers from a 7% rate of no shows. A new procedure is instituted whereby reservations are confirmed on the day preceding the actual flight, and a study is then made of 5218 randomly selected reservations made under the new system. If 333 no shows are recorded, test the claim that the no show rate is lower with the new system. Use a 5% level of significance.

EXERCISES B: TESTS OF PROPORTIONS

6-73. Show that $z = (x - np)/\sqrt{npq}$ is equivalent to

$$z = \frac{\bar{p} - p}{\sqrt{p(1 - p)/n}}$$

where $\bar{p} = x/n$.

6-74. At the 5% level of significance, test the claim that a certain coin favors heads when it is flipped. Sample data consist of seven heads of eight flips. (The normal distribution is not a suitable approximation to the distribution of the number of heads that will occur in eight flips.)

6-75. A reporter claims that 10% of the residents of his city feel that the mayor is doing a good job. Test his claim if it is known that, in a random sample of 15 residents, there are none who feel that the mayor is doing a good job. Use a 5% level of significance. (The normal distribution is not a suitable approximation to the distribution of sample proportions in this case.)

6-5 TESTS OF VARIANCES

The preceding sections of this chapter deal with tests of hypotheses made about means and proportions. In testing these hypotheses, we use the normal and student t distributions in ways that are very similar. They have the same basic bell shapes, they are both symmetric about zero, and the test statistics involve comparable computations. Here, we encounter a very different distribution as we test hypotheses made about a population variance or standard deviation. Recall that the standard deviation is simply the square root of the variance, so if we know the value of one then we really know the value of the other.

σ:	Population standard deviation
σ^2:	Population variance
s:	Sample standard deviation
s^2:	Sample variance

For this reason, the comments we make about variances will also apply to standard deviations.

Many real and practical situations demand decisions or inferences about variances. In manufacturing, quality control engineers want to ensure that a product is, on the average, acceptable. But the engineers also want to produce items of *consistent* quality so there are only a few defective products. This consistency is measured by the variance.

HOW VALID ARE CRIME STATISTICS?

Police departments are often judged by the number of arrests since that statistic is readily available and clearly understood. However, this encourages police to concentrate on easily solved minor crimes (such as marijuana smoking) at the expense of more serious crimes which are difficult to solve. A recent study of Washington, D.C. police records revealed that police can manipulate statistics another way. The study showed that more than 1000 thefts in excess of $50 were intentionally valued at less than $50 so that they would be classified as petty larceny and not major crime.

As a specific example, let us consider aircraft altimeters. Due to mass-production techniques and a variety of other factors, these altimeters do not all give exact readings. (Some errors are built in.) It would be easy to change the overall average reading by simply shifting the scale on the altimeters, but it would be very difficult to change the variance of the readings. Even if the overall average were perfect, a very large variance would indicate that some altimeters give excessively high or low readings and seriously jeopardize safety. In this case, quality control engineers want to keep the variance below some tolerable level. When the variance exceeds that level, production is considered to be out of control and corrective action must be taken.

In this section, **we assume that the population in question has normally distributed values.** We made this same assumption earlier in this chapter, but it is a more critical assumption here. In using the student t distribution of Section 6-3, for example, we require that the population of values be approximately normal, and we can accept deviations away from normality that are not too severe. However, when we deal with variances by using the distribution to be introduced shortly, departures away from normality will lead to gross errors. Consequently the assumption of a normally distributed population must be followed much more strictly.

In a normally distributed population with variance σ^2, we randomly select independent samples of size n and compute the variance s^2 for each sample. The quantity $(n - 1)s^2/\sigma^2$ has a distribution called the **chi-square distribution**. We denote chi-square by χ^2. (The specific mathematical equations used to define this distribution are not given here since they are confusing at this stage. Instead, you can refer to Table A-5 for the critical values of the chi-square distribution.)

The test statistic used in tests of hypotheses about variances is χ^2 (chi-square):

$$\chi^2 = \frac{(n - 1)s^2}{\sigma^2}$$

where

n = sample size
s^2 = sample variance
σ^2 = population variance (given in the null hypothesis)

The chi-square distribution resembles the student t distribution in that there is actually a different distribution for each sample size n. In Figure 6-15 we see that the chi-square distribution does not have the same symmetric bell shape of the normal and student t distributions. The chi-square distribution has a longer right tail. Unlike the normal and student t distributions, the chi-square distribution does not include negative numbers.

In using Table A-5 to determine critical values of the chi-square distribution, we must first determine the degrees of freedom which are $n - 1$ as in the student t distribution.

$$\text{degrees of freedom} = n - 1$$

We can use this expression for Section 6-3 and this section, but in later chapters we will encounter situations in which the degrees of freedom are not $n - 1$. For that reason, you should not universally equate degrees of freedom with $n - 1$.

Once we have determined degrees of freedom, the significance level α, and the type of test (left-tailed, right-tailed, or two-tailed), we can use Table A-5 to find the critical chi-square values. An important feature of this table is that each critical value separates an area to the *right* which corresponds to the value given in the top row.

Example

Find the critical values of χ^2 which determine critical regions containing areas of 0.025 in each tail. Assume that the relevant sample size is 10 so that the degrees of freedom are $10 - 1$ or 9.

Solution

See Figure 6-16 and refer to Table A-5. The critical value to the right (19.023) is obtained in a straightforward manner by locating 9 in the degrees-of-freedom column at the left and 0.025 across the top. The left critical value of 2.700 again corresponds to 9 in the degrees-of freedom column, but we must locate 0.975 across the top since the values in the top row are always *areas to the right* of the critical value.

In a right-tailed test, the value of α will correspond exactly to the areas given in the top row of Table A-5. In a left-tailed test, the value of $1 - \alpha$ will correspond exactly to the areas given in the top row of Table A-5. In a two-tailed test, the values of $\alpha/2$ and $1 - \alpha/2$ will correspond exactly to the areas given in the top row of Table A-5.

Having discussed the underlying theory and the mechanics of using Table A-5, we give an example of a hypothesis test.

Example

Test the claim that scores on a standard I.Q. test have a variance equal to 225 if a sample of 41 randomly selected subjects achieve scores with a variance of 258. Use a significance level of $\alpha = 0.05$.

Solution

We again use the basic method for testing hypotheses outlined in Figure 6-2. The claim that ''the variance equals 225'' is $\sigma^2 = 225$ in symbolic form. The alternative is $\sigma^2 \neq 225$ and we get the following null and alternate hypotheses:

$$H_0: \sigma^2 = 225$$
$$H_1: \sigma^2 \neq 225$$

We can see that this is a two-tailed test since a sample variance significantly above or below 225 will be a basis for rejecting H_0. The significance level of

FIGURE 6-15
Chi-square distribution

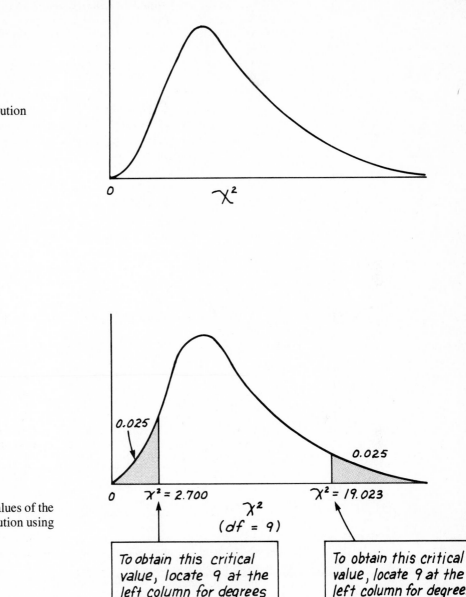

FIGURE 6-16
Finding critical values of the
chi-square distribution using
Table A-5.

0.025

0.025

$\chi^2 = 2.700$ $\chi^2 = 19.023$

χ^2
$(df = 9)$

To obtain this critical
value, locate 9 at the
left column for degrees
of freedom and then
locate 0.975 across the
top. The total area to
the right of this critical
value is 0.975 which
we get by subtracting
0.025 from 1.

To obtain this critical
value, locate 9 at the
left column for degrees
of freedom and then
locate 0.025 across
the top.

$\alpha = 0.05$ has already been stipulated. The sample variance should obviously be used in testing a claim about the population variance, and the χ^2 distribution is therefore appropriate. We compute the test statistic as follows:

$$\chi^2 = \frac{(n-1)s^2}{\sigma^2} = \frac{(41-1) \cdot 258}{225} \stackrel{\circ}{=} 45.867$$

We find the right critical χ^2 value of 59.342 in Table A-5 by locating 40 degrees of freedom and an area of 0.025. We find the left critical χ^2 value of 24.433 in Table A-5 by locating 40 degrees of freedom and an area of 0.975. (Since $\alpha = 0.05$, there is an area of 0.025 in each tail, but the left critical value is found by locating the area to its right. See Figure 6-17.) The test statistic based on the sample data is not in the critical region so we fail to reject H_0. On the basis of the available evidence, we cannot refute the claim that the variance equals 225.

FIGURE 6-17

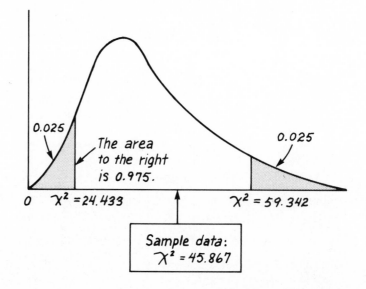

One example of variance occurs in the waiting lines of banks. In the past, customers traditionally entered a bank and selected one of several lines formed at different windows. A different system growing in popularity involves one main waiting line which feeds the various windows as vacancies occur. The mean waiting time isn't reduced, but the variability among waiting times is decreased and the irritation of being caught in a slow line is also diminished.

Example   With individual lines at its various windows, a bank finds that the standard deviation for waiting times on Friday afternoons is 6.2 minutes. The bank experiments with a single main waiting line and finds that for a random sample of 25 customers, the waiting times have a standard deviation of 3.8 minutes. At the $\alpha = 0.05$ significance level, test the claim that a single line causes lower variation among the waiting times.

Solution We wish to test $\sigma < 6.2$ based on a sample of $n = 25$ for which $s = 3.8$. We begin by identifying the null and alternative hypotheses. (We will use σ^2 instead of σ since the χ^2 distribution involves variances directly.)

$$H_0: \sigma^2 \geq (6.2)^2$$
$$H_1: \sigma^2 < (6.2)^2$$

The significance level of $\alpha = 0.05$ has already been selected so we proceed to compute the value of χ^2 based on the given data:

$$\chi^2 = \frac{(n-1)s^2}{\sigma^2} = \frac{(25-1)(3.8)^2}{(6.2)^2} \overset{\circ}{=} 9.016$$

This test is left-tailed since H_0 will be rejected only for small values of χ^2; with $\alpha = 0.05$ and $n = 25$, we go to Table A-5 and align 24 degrees of freedom with an area of 0.95 to obtain the critical χ^2 value of 13.848 (see Figure 6-18 on the following page). Since the test statistic falls within the critical region, we reject H_0 and conclude that the 3.8-minute standard deviation is significantly less than the 6.2-minute standard deviation that corresponds to multiple waiting lines. That is, the single main line does appear to lower the variation among waiting times.

FIGURE 6-18

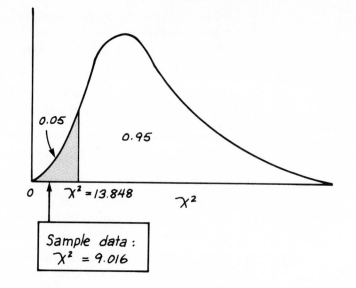

We have already illustrated some of the important uses of the chi-square distribution. Other valuable uses, such as situations in which we wish to analyze the significance of differences between expected frequencies and the frequencies that actually occur, are considered in Chapter 10.

EXERCISES A: TESTS OF VARIANCES

6-76. Use Table A-5 to find the critical values of χ^2 based on the given data.

(a) $\alpha = 0.05$
$n = 20$
$H_0: \sigma^2 = 256$

(b) $\alpha = 0.05$
$n = 20$
$H_0: \sigma^2 \geq 256$

(c) $\alpha = 0.01$
$n = 23$
$H_0: \sigma^2 = 10$

(d) $\alpha = 0.01$
$n = 23$
$H_0: \sigma^2 \leq 10$

(e) $\alpha = 0.005$
$n = 15$
$H_1: \sigma^2 < 23.4$

(f) $\alpha = 0.10$
$n = 6$
$H_1: \sigma^2 < 100$

(g) $\alpha = 0.05$
$n = 40$
$H_1: \sigma^2 > 500$

(h) $\alpha = 0.025$
$n = 81$
$H_0: \sigma^2 \geq 144$

(i) $\alpha = 0.01$
$n = 50$
$H_0: \sigma^2 = 225$

(j) $\alpha = 0.05$
$n = 75$
$H_1: \sigma^2 \neq 31.5$

In Exercises 6-77 through 6-95, test the given hypotheses. Follow the pattern outlined in Figure 6-2 and draw the appropriate graph. In all cases, assume that the population is normally distributed.

6-77. At the $\alpha = 0.05$ significance level, test the claim that $\sigma^2 > 100$ if a random sample of 27 yields a variance of 194.

6-78. At the $\alpha = 0.05$ significance level, test the claim that $\sigma^2 > 225$ if a random sample of 30 yields a variance of 380.

6-79. At the $\alpha = 0.05$ significance level, test the claim that $\sigma^2 \leq 9.00$ if a random sample of 81 yields a variance of 12.25.

6-80. At the $\alpha = 0.05$ significance level, test the claim that $\sigma^2 \geq 416$ if a random sample of 17 yields a variance of 247.

6-81. At the $\alpha = 0.10$ significance level, test the claim that $\sigma^2 \geq 90$ if a random sample of 21 yields a variance of 53.

6-82. At the $\alpha = 0.01$ significance level, test the claim that $\sigma^2 = 2.38$ if a random sample of 12 yields a variance of 5.00.

6-83. At the $\alpha = 0.05$ significance level, test the claim that $\sigma^2 = 100$ if a random sample of 27 yields a variance of 57.

6-84. At the $\alpha = 0.05$ significance level, test the claim that $\sigma = 10.0$ if a random sample of 18 yields a standard deviation of 12.3.

6-85. At the $\alpha = 0.05$ significance level, test the claim that $\sigma = 52.0$ if a random sample of 18 yields a standard deviation of 71.2.

6-86. Test the claim that scores on a standard I.Q. test have a variance equal to 225 if a sample of 41 randomly selected subjects achieve scores with a variance of 135. Use a significance level of $\alpha = 0.05$.

6-87. Test the claim that scores on a standard I.Q. test have a standard deviation equal to 15 if a sample of 24 randomly selected subjects yields a standard deviation of 10. Use a significance level of $\alpha = 0.01$.

6-88. Repeat Exercise 6-87 after changing the significance level to $\alpha = 0.05$.

6-89. A manufacturer produces batteries whose lives have a standard deviation of 52 hours. A sample of 30 new experimental batteries are produced and their lives have a standard deviation of 60 hours. At the 0.05 significance level, test the claim that the standard deviation for this new type of battery is equal to 52 hours.

6-90. The variance for scores on a standard I.Q. test is known to be 225. An experimental I.Q. test is administered to 50 randomly selected subjects, and the standard deviation for this sample group is 18. At the 0.01 significance level, test the claim that the variance for the experimental test equals 225.

6-91. The reaction times for adults on a standard test have a standard deviation of 0.21 second. Twenty-four subjects are randomly selected and given two glasses of beer before the test. Their scores produce a standard deviation of 0.29 second. At the 0.05 significance level, test the claim that the standard deviation for the reaction times of all adults who drink two glasses of beer is greater than 0.21 second.

6-92. If the standard deviation for the weekly down times of a computer is low, then availability of the computer is predictable and planning is facilitated. Twelve weekly down times for a computer are randomly selected and the standard deviation is computed to be 2.85 hours. At the 0.025 significance level, test the claim that $\sigma > 2.00$ hours.

6-93. A machine pours soda into a cup in such a way that the standard deviation of the weights is 0.15 ounce. A new machine is tested on 71 cups, and the standard deviation for this group is 0.12 ounce. At the 0.05 significance level, test the claim that the new machine produces less variance.

6-94. The weights of 20 newborn girls are randomly selected and the standard deviation for this group is computed to be 1.18 pounds. Test the claim that the weights of newborn girls have a greater standard deviation than the weights of all newborn boys. (Assume that $\sigma = 0.99$ pound for boys and let $\alpha = 0.05$.)

6-95. A course taught the traditional way produces final grades with a standard deviation of eight. An experimental self-paced course is taken by 22 randomly selected students and the final grades for this group yield a standard deviation of ten. At the 0.05 significance level, test the claim that the scores for all self-paced students in this course have a standard deviation equal to eight.

EXERCISES B: TESTS OF VARIANCE

6-96. A computer or calculator capable of finding variances is required for this project. Using a telephone book, obtain random digits from 0 through 9 by selecting the last digit of telephone numbers. Obtain 50 groups of ten digits in each group and then compute the 50 sample variances. Compute:

$$C = \frac{(n-1)s^2}{\sigma^2} = \frac{9s^2}{8.25}$$

for the 50 samples and then make a histogram of the 50 values of C. Compare the result to Figure 6-15. (Note that the original population is not normally distributed.)

REVIEW

In this chapter, we studied a standard method used in statistical tests of hypotheses made about the values of population means, proportions, percentages, variances, and standard deviations. (A hypothesis is simply a statement that something is true.) When sample data conflict with the given hypothesis, we want to decide whether the differences are due to chance fluctuations or whether the differences are so significant that they are not likely to occur by chance. We are able to select exact **levels of significance**; 0.05 and 0.01 are common values. Sample results are said to reflect significant differences when their occurrences have probabilities less than the chosen level of significance.

Section 6-2 presented in detail the procedure for testing hypotheses. The essential steps are summarized in Figure 6-2. We defined **null hypothesis**, **alternative hypothesis**, **type I error**, **type II error**, **test statistic**, **critical region**, and **critical value**. All of these standard terms are commonly used in discussing tests of hypotheses. We also identified the three basic types of tests: **right-tailed**, **left-tailed**, and **two-tailed** (see Figure 6-3).

We introduced the method of testing hypotheses in Section 6-2 by using examples in which only the normal distribution applies, and we introduced other distributions in subsequent sections. The brief table that follows serves as a rough guide since it does not include some of the necessary assumptions, such as the normality of the population.

Parameter referred to in hypothesis	Applicable distribution	Test statistic	Table of critical values
μ (population mean)	Normal (if σ is known or if $n > 30$)	$z = \dfrac{\bar{x} - \mu_{\bar{x}}}{\sigma/\sqrt{n}}$	Table A-3
	Student t (if σ is unknown and $n \leq 30$)	$t = \dfrac{\bar{x} - \mu}{s/\sqrt{n}}$	Table A-4
p (population proportion)	Normal	$z = \dfrac{x - np}{\sqrt{npq}}$	Table A-3
σ^2 (population variance)	Chi-square	$\chi^2 = \dfrac{(n-1)s^2}{\sigma^2}$	Table A-5

In addition to presenting the method for testing hypotheses, Chapter 6 also introduced the student t and chi-square distributions. In Chapter 7 we again refer to these distributions as we investigate estimations of parameters and ways to determine how large certain samples should be.

REVIEW EXERCISES

In all tests of hypotheses, follow the procedure outlined in Figure 6-2 and draw the appropriate graphs. Both of the following samples are randomly selected from very large populations of normally distributed values.

Sample A	Sample B
$n = 100$	$n = 20$
$\bar{x} = 8.2$ seconds	$\bar{x} = 15.0$ grams
$s = 6.0$ seconds	$s = 2.8$ grams

6-97. At the 0.05 significance level, test the claim that $\mu = 10.0$ seconds for the population from which the Sample A results were obtained.

6-98. At the 0.05 significance level, test the claim that $\mu = 12.5$ grams for the population from which the Sample B results were obtained.

6-99. At the 0.01 significance level, test the claim that $\sigma = 4.0$ grams for the population from which the Sample B results were obtained.

6-100. Of 80 workers randomly selected and interviewed, 55 were opposed to an increase in social security taxes. At the 5% level of significance, test the claim that the majority (more than 50%) of all such workers are opposed to the increase in taxes.

6-101. The following sample scores have been randomly selected from a normally distributed population. At the 0.01 significance level, test the claim that the population has a mean below 100.

101	106	98	92	97	80	89	88
110	112	100	100	103	97	97	

6-102. Using the sample data in Exercise 6-101, test the claim that $\sigma = 12$. Use a 1% level of significance.

6-103. Test the claim that the mean female reaction time to a highway signal is less than 0.700 second. Eighteen females are randomly selected and tested; their mean is 0.668 second. Assume that s = 0.100 second and use a 5% level of significance.

6-104. Of 200 females randomly selected and interviewed in the midwest, 27% believed (incorrectly) that birth control pills prevent venereal disease. Use these sample results to test (at the 5% level of significance) the claim that "more than one-fifth of all midwestern women believe that birth control pills prevent venereal disease."

6-105. Find the appropriate critical values.

(a) $\alpha = 0.10$
$n = 15$
$H_0: \mu = 1.23$

(b) $\alpha = 0.10$
$n = 15$
$H_0: \sigma^2 = 123$

(c) $\alpha = 0.06$
$n = 100$
$H_0: \mu = 72.3$

(d) $\alpha = 0.05$
$n = 10$
$H_0: \sigma = 15$

(e) $\alpha = 0.01$
$n = 30$
$H_1: \sigma < 5.8$

6-106. At the 5% level of significance, we want to test the claim that the mean reading in a biofeedback experiment is 6.2 microvolts, and we are waiting for sample data.

(a) Give the null hypothesis in symbolic form.

(b) Give the value of α.

(c) Is this test left-tailed, right-tailed, or two-tailed?

(d) In simple terms devoid of symbolism and technical language, describe the type I error for this problem.

(e) In simple terms devoid of symbolism and technical language, describe the type II error for this problem.

(f) What is the probability of making a type I error?

ESTIMATES AND SAMPLE SIZES

7-1 OVERVIEW

Chapter 6 introduced one aspect of inferential statistics, hypothesis testing. We used sample data to make **decisions** about claims or hypotheses. In this chapter we use sample data to make **estimates** of the values of population parameters. Section 7-2 begins by using the statistic $\bar{x}$ in estimating the value of the parameter μ. In subsequent sections we use sample proportions and sample variances to estimate the values of population proportions and population variances. We also identify some ways of determining how large samples should be.

We apply these methods of estimating and determining sample size to population means, proportions, percentages, variances, and standard deviations. The same distributions whose parameters you studied in Chapter 6 are included in the methods we develop in this chapter. Like hypothesis testing, the fundamental concepts of this chapter are very important and basic to the subject of inferential statistics. Practical applications will become apparent through the examples and exercises.

CHILDREN SEE MUCH TELEVISION VIOLENCE

The National Association for Better Broadcasting has estimated that, between the ages of 5 and 15, a child will see about 13,000 violent deaths on television. This estimate was based on a 1-week monitoring of Los Angeles television stations. Results were projected according to the viewing habits of children in the 5-year to 15-year age bracket. This statistic has caused concern because many people believe that television violence has an adverse effect on child behavior.

7-2 ESTIMATES AND SAMPLE SIZES OF MEANS

Let's assume that the 50 following I.Q. scores represent a randomly selected sample of the seniors in a large suburban high school.

110	122	119	95	98	100	100	105	112	111
111	116	110	109	108	112	115	123	145	102
85	90	126	127	135	112	113	127	129	99
144	140	97	98	88	117	77	83	119	121
103	100	133	130	109	112	105	106	107	115

$$\bar{x} = 111.4$$
$$s = 15.2$$

Using only these sample results, we want to estimate the value of the mean I.Q. of *all* the seniors in that high school.

We could use statistics such as the sample median, midrange, or mode as estimates of μ, but the sample mean $\bar{x}$ usually provides the best estimate of μ. This is not an intuitive conclusion. It is based on careful study and an analysis of the distributions of various estimators. For many populations, the distribution of sample means $\bar{x}$ has a smaller variance than the distribution of the other possible estimators, so $\bar{x}$ tends to be more consistent. For all populations we say that $\bar{x}$ is an **unbiased** estimator. This means that the distribution of $\bar{x}$ values tends to center about the value of μ. For these reasons, we will use $\bar{x}$ as the best estimate of μ. Because $\bar{x}$ is a single number that corresponds to a point on the number scale, we call it a point estimate.

DEFINITION

*The sample mean $\bar{x}$ is the best **point estimate of the population mean** μ.*

Computing the sample mean of the preceding 50 I.Q. scores, we get $\bar{x} =$ 111.4, which becomes our point estimate of the mean I.Q. for all the seniors in the high school. Even though 111.4 is our *best* estimate of μ, we really have no indication of just how good that estimate is. (Sometimes even the best is very poor.) Suppose that we have only the first two sample I.Q. scores of 110 and 122. Their mean of 116 is the best point estimate of μ, but we cannot expect this best estimate to be very good since it is based on such a small sample. Mathematicians have developed a better estimator which does reveal how good it is. We will first present this estimator, illustrate its use, and then explain the underlying rationale.

NOTATION

$z(\alpha/2)$ is the positive standard z value that separates an area of $\alpha/2$ in the right tail of the standard normal distribution (see Figure 7-1). Note that $z(\alpha/2)$ is a single number and is not the product of z and $\alpha/2$.

FIGURE 7-1

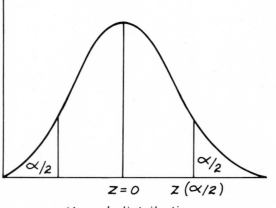

Normal distribution

Example

If $\alpha = 0.05$, then $z(\alpha/2) = 1.96$. 1.96 is the standard z value that separates a right-tail region with an area of 0.05/2 or 0.025. We find $z(\alpha/2)$ by noting that the region to its left must be $0.5 - 0.025$ or 0.475. That is, the area bounded by the centerline and $z(\alpha/2)$ is 0.475. In Table A-3, an area of 0.4750 corresponds exactly to a z score of 1.96.

DEFINITION

*The **confidence interval** (or **interval estimate**) for the population mean is given by*

$$\bar{x} - z(\alpha/2)\frac{\sigma}{\sqrt{n}} < \mu < \bar{x} + z(\alpha/2)\frac{\sigma}{\sqrt{n}}$$

We will use the above form of the confidence interval, but another equivalent form is given below.

$$\mu = \bar{x} \pm z(\alpha/2)\frac{\sigma}{\sqrt{n}}$$

DEFINITION

*The **degree of confidence** is the probability $1 - \alpha$ that the parameter μ is contained in the confidence interval. (The probability is often expressed as the equivalent percentage value.)*

With $\alpha = 0.05$ we get a 0.95 degree of confidence. This means that for 95% of all samples, the resulting interval will ''cover'' or contain the true population mean.

Example

Using the sample data from the beginning of this section, find the 95% confidence interval for the mean I.Q. of all the seniors in the high school.

Solution

To find the confidence interval for μ (the mean I.Q. of all the seniors), we must first determine the values of $\bar{x}$, n, $z(\alpha/2)$, and σ. We have already established the fact that $\bar{x} = 111.4$. The sample size is 50, so $n = 50$. $z(\alpha/2) = 1.96$ since 95% or a probability of 0.95 suggests that $\alpha = 0.05$. (In the previous example we concluded that $z(\alpha/2) = 1.96$ when $\alpha = 0.05$.) For the standard I.Q. test, we will assume that σ is 15, and the sample standard deviation of 15.2 lends credence to this assumption. Using the known values of $\bar{x}$, n, $z(\alpha/2)$, and σ,

$$\bar{x} - z(\alpha/2)\frac{\sigma}{\sqrt{n}} < \mu < \bar{x} + z(\alpha/2)\frac{\sigma}{\sqrt{n}}$$

becomes

$$111.4 - 1.96 \cdot \frac{15}{\sqrt{50}} < \mu < 111.4 + 1.96 \cdot \frac{15}{\sqrt{50}}$$

or

$$111.4 - 4.16 < \mu < 111.4 + 4.16$$

or

$$107.24 < \mu < 115.56$$

Based on the sample results, we are 95% confident that μ is between 107.24 and 115.56.

EXCERPTS FROM A DEPARTMENT OF TRANSPORTATION CIRCULAR

The following excerpts from a Department of Transportation circular concern some of the accuracy requirements for navigation equipment used in aircraft. Note the use of the confidence interval.

"The total of the error contributions of the airborne equipment, when combined with the appropriate flight technical errors listed, should not exceed the following with a 95% confidence (2-sigma) over a period of time equal to the update cycle."

"The system of airways and routes in the United States has widths of route protection used on a VOR system use accuracy of ±4.5 degrees on a 95% probability basis."

FIGURE 7-2
There is a $1 - \alpha$ probability that a sample mean will be in error by less than $z(\alpha/2)\sigma/\sqrt{n}$.

Thus far we have defined the confidence interval for the mean and illustrated its use. We now proceed to explain why the confidence interval has the form given in the definition.

The basic underlying idea relates to the Central Limit Theorem, which indicates that the distribution of sample means is approximately normal as long as the samples are large ($n \geq 30$). The Central Limit Theorem was also used to determine that sample means have a mean of μ while the standard deviation of means from samples of size n is $\sigma/\sqrt{n}$. That is,

$$\mu_{\bar{x}} = \mu$$

$$\sigma_{\bar{x}} = \frac{\sigma}{\sqrt{n}}$$

Recall that $\sigma_{\bar{x}}$ is called the standard error of the mean. It is the standard deviation of means computed from samples of size n. Since a z score is the number of standard deviations a value is away from the mean, we conclude that $z(\alpha/2)\dfrac{\sigma}{\sqrt{n}}$ represents a number of standard deviations away from μ. **There is a probability of $1 - \alpha$ that a sample mean will differ from μ by less than $z(\alpha/2)\dfrac{\sigma}{\sqrt{n}}$.** See Figure 7-2 and note that the unshaded inner regions total $1 - \alpha$ and correspond to sample means which differ from μ by less than $z(\alpha/2)\dfrac{\sigma}{\sqrt{n}}$. In other words, a sample mean error of $\mu - \bar{x}$ will be between $- z(\alpha/2)\dfrac{\sigma}{\sqrt{n}}$ and $z(\alpha/2)\dfrac{\sigma}{\sqrt{n}}$. This can be expressed as one inequality:

$$-z(\alpha/2)\frac{\sigma}{\sqrt{n}} < \mu - \bar{x} < z(\alpha/2)\frac{\sigma}{\sqrt{n}}$$

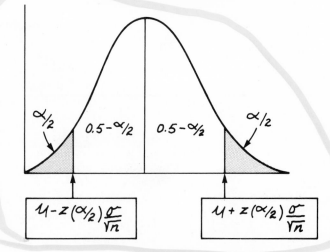

Adding $\bar{x}$ to each component of the inequality, we get the confidence interval given in the definition.

Applying these general concepts to the specific I.Q. data given at the beginning of this section, we can see that there is a 95% chance that the sample mean of 111.4 will be in error by less than

$$z(\alpha/2) \frac{\sigma}{\sqrt{n}} = 1.96 \cdot \frac{15}{\sqrt{50}} = 4.16$$

That is, there is a 95% chance that μ is between $111.4 - 4.16$ and $111.4 + 4.16$. This corresponds to our confidence interval of $107.24 < \mu < 115.56$.

In the following definition we formalize the fact that a sample mean has a probability of $1 - \alpha$ of being in error by less than $z(\alpha/2) \, (\sigma/\sqrt{n})$.

DEFINITION

> At the $1 - \alpha$ degree of confidence, the **maximum error of the estimate of μ is given by**
>
> $$E = z(\alpha/2) \frac{\sigma}{\sqrt{n}}$$

Example At the 0.95 degree of confidence, find the maximum error E of the estimate of μ for a random sample of 50 scores taken from a normally distributed population with $\sigma = 15$.

Solution We find the maximum error of the estimate of μ as follows:

$$E = z(\alpha/2) \frac{\sigma}{\sqrt{n}} = 1.96 \cdot \frac{15}{\sqrt{50}} = 4.16$$

If we begin with the expression for E and proceed to solve for n we get

Formula 7-1
$$n = \left[\frac{z(\alpha/2)\sigma}{E} \right]^2$$

which may be used in determing the sample size necessary to produce results accurate to a desired degree of confidence. This equation should be used when we know the value of σ and we want to determine the sample size necessary to establish, with a probability of $1 - \alpha$, the value of μ to within $\pm E$. The existence of such an equation is somewhat remarkable, since it implies that the sample size does not depend upon the size of the population.

Example We wish to be 99% sure that a random sample of I.Q. scores yields a mean that is within two units of the true mean. How large should the sample be? Assume that σ is 15.

Solution We seek n given that $\alpha = 0.01$ (from 99% confidence) and the maximum allowable error is two. Applying the equation for sample size n, we get

$$n = \left[\frac{z(\alpha/2)\sigma}{E} \right]^2 = \left[\frac{2.575 \cdot 15}{2} \right]^2 = 19.3125^2 \overset{\circ}{=} 373 \qquad (7\text{-}1)$$

Therefore we should obtain at least 373 randomly selected I.Q. scores if we require 99% confidence that our sample mean is within two units of the true mean.

If we can settle for less accurate results and accept a maximum error of four instead of two, we can see that the required sample size is reduced from 373 to 94. Direct application of the equation for n produces a value of 93.24, which is *rounded up* to 94. (We always round up in sample size computations so that the required number is at least adequate instead of being slightly inadequate.)

Doubling the maximum error from two to four caused the required sample size to decrease to one-fourth of its original value. Conversely, if we want to halve the maximum error, we must quadruple the sample size. In the equation, n is inversely proportional to the square of E and directly proportional to the square of $z(\alpha/2)$. All of this implies that we can obtain more accurate results with greater confidence, but the sample size will be substantially increased.

Unfortunately, the application of the equation for sample size, (equation (7-1)), requires prior knowledge of σ. Realistically, σ is usually unknown unless previous research results are available. When we do sample from a population in which σ is unknown, a preliminary study must be conducted so that σ can be reasonably estimated. Only then can we determine the sample size required to meet our error tolerance and confidence demands. However, if we intend to construct a confidence interval, do not know σ, and do not plan a preliminary study, we can use the student t distribution as long as the population is essentially normal. In a normally distributed population with unknown σ, we can estimate σ by s if $n > 30$. Otherwise we can replace $z(\alpha/2)$ by $t(\alpha/2)$ to get

$$\bar{x} - t(\alpha/2) \frac{s}{\sqrt{n}} < \mu < \bar{x} + t(\alpha/2) \frac{s}{\sqrt{n}}$$

Example Twenty randomly selected high school seniors are given I.Q. tests with the following sample results:

$$\bar{x} = 111.4$$

$$s = 15.2$$

Assuming that σ is unknown, construct the 95% confidence interval for the mean I.Q. of all seniors.

Solution It is reasonable to assume here that the population is essentially normal. With σ unknown and a sample of 20 scores, we know from Section 6-3 that the distribution of such sample means is a student t distribution.

$$\bar{x} - t(\alpha/2)\frac{s}{\sqrt{n}} < \mu < \bar{x} + t(\alpha/2)\frac{s}{\sqrt{n}}$$

becomes

$$111.4 - 2.093 \cdot \frac{15.2}{\sqrt{20}} < \mu < 111.4 + 2.093 \cdot \frac{15.2}{\sqrt{20}}$$

or

$$111.4 - 7.11 < \mu < 111.4 + 7.11$$

or

$$104.29 < \mu < 118.51$$

In this case, we are 95% confident that the population mean is between 104.29 and 118.51.

Note that in choosing between the normal and student t distributions, we use the same criteria that were used in Sections 6-2 and 6-3.

AIRCRAFT ENGINE FAILURES

An author of an article in a national magazine on flying pointed out that he checked some of the busier flight schools and found that the average flying time between engine failures in flight ranged from 20,000 hours to 90,000 hours. He then wrote that "using a guess or two and some numbers from another survey on this subject, the time between actual failures appears to be about 25,000 hours." He then goes on to conclude that, if that last figure is correct and if he flies 2000 hours as a pilot, his chances of having an engine failure are "1 in 12.5."

EXERCISES A: ESTIMATES AND SAMPLE SIZES OF MEANS

7-1. (a) If $\alpha = 0.05$, find $z(\alpha/2)$.
      (b) If $\alpha = 0.02$, find $z(\alpha/2)$.
      (c) If $\alpha = 0.01$, find $z(\alpha/2)$.
      (d) If $\alpha = 0.10$, find $z(\alpha/2)$.
      (e) If $\alpha = 0.25$, find $z(\alpha/2)$.
      (f) Find $z(\alpha/2)$ for the value of α corresponding to a confidence level of 96%.
      (g) Find $z(\alpha/2)$ for the value of α corresponding to a confidence level of 95%.
      (h) Find $z(\alpha/2)$ for the value of α corresponding to a confidence level of 80%.
      (i) If $\alpha = 0.05$, find $t(\alpha/2)$ for a sample of 20 scores.
      (j) If $\alpha = 0.01$, find $t(\alpha/2)$ for a sample of 15 scores.
      (k) If $\alpha = 0.10$, find $t(\alpha/2)$ for a sample of 10 scores.
      (l) If $\alpha = 0.02$, find $t(\alpha/2)$ for a sample of 25 scores.

7-2. Find the 95% confidence interval for μ if $\sigma = 5$, $\bar{x} = 70.4$, and $n = 36$.

7-3. Find the 99% confidence interval for μ if $\sigma = 2$, $\bar{x} = 98.6$, and $n = 100$.

7-4. Find the 90% confidence interval for μ if $\sigma = 5.5$, $\bar{x} = 123.6$, and $n = 75$.

7-5. A sample of 40 randomly selected adults is tested for pulse rates. The resulting mean is 75.8 beats per minute. Assuming that $\sigma = 10$ beats per minute, find the 95% confidence interval for the mean pulse rate of all adults.

7-6. A sample of 40 randomly selected I.Q. scores produces a mean of 96.8. Assuming that $\sigma = 15$, find the 99% confidence interval of μ based on these sample results.

7-7. A psychologist administers a test to determine the braking reaction time of drivers. Two hundred subjects produce a mean time of 0.61 second. Assuming that $\sigma = 0.12$ second, find the 95% confidence interval for the mean reaction time of all drivers.

7-8. A random sample of 40 newborn boys produces a mean weight at birth of 7.61 pounds. Assuming that $\sigma = 1$ pound, find the 99% confidence interval for the mean weight of all newborn boys.

7-9. An electronics company manufactures radios and conducts extensive tests on 30 randomly selected models of a type used in airplanes. The mean time between failures is found to be 322.4 hours for the sample group. Assuming that $s = 45$ hours, find the 99% confidence interval for the mean time between failures for all such radios.

7-10. A magazine reporter is conducting independent tests to determine the distance a certain car will travel while consuming only 1 gallon of gas. A sample of five cars is tested and a mean of 28.2 miles is obtained. Assuming that $\sigma = 2.7$ miles, find the 99% confidence interval for the mean distance traveled by all such cars using 1 gallon of gas.

7-11. An assembly line foreman observes 50 employees assigned to a given task, and in each case he records the time required for the employee to learn the operation. A mean of 36.2 minutes results from these times. Assuming that $\sigma = 5$ minutes, find the 95% confidence interval for the mean learning time for all such employees.

7-12. Find the sample size necessary to estimate a population mean to within three units if $\sigma = 16$ and we want 95% confidence in our results.

7-13. Find the sample size necessary to estimate a population mean. Assume that $\sigma = 20$, the maximum allowable error is 1.5, and we want 95% confidence in our results.

7-14. On a standard I.Q. test, σ is 15. How many random I.Q. scores must be obtained if we want to find the true population mean (with an allowable error of 0.5) and we want 99% confidence in the results?

7-15. Do Exercise 7-14 assuming that the maximum error is one unit instead of 0.5 unit.

7-16. Do Exercise 7-14 assuming that the maximum error is 0.25 unit instead of 0.5 unit.

7-17. Assume that $\sigma = 2.64$ inches for heights of adult males. Find the sample size necessary to estimate the mean height of all adult males to within 0.5 inch if we want 99% confidence in our results.

7-18. The effectiveness of a certain drug is measured in the resulting decrease in blood pressure (measured in millimeters of mercury). If we want to estimate the mean decrease to within 1.5 millimeters of mercury, how many people should we include in the sample group? Assume that $\sigma = 4$ and that we want 95% confidence in our results.

7-19. A sample of 25 I.Q. scores is selected from a special group of students meeting specific criteria. This sample group produces a mean and standard deviation of 123.7 and 9.5, respectively. Construct the 95% confidence interval for the mean I.Q. score of all students in this special category.

7-20. A sample of 15 experimental batteries is analyzed and the sample mean and standard deviation of their voltages are 12.8 and 0.6, respectively. Find the 95% confidence interval for the mean voltage of all such batteries.

7-21. A sample of seven monkeys is studied. Among the results we find that the times required for the monkeys to learn a task had a mean of 14.7 minutes and a standard deviation of 2.5 minutes. If previous testing has shown that $\sigma = 2.5$ minutes, find the 95% confidence interval for the mean time required to learn the task.

7-22. The birth process of a newly discovered mammal is being studied, and the lengths of 18 observed pregnancies have been recorded. The mean and standard deviation for these 18 times is 97.3 days and 2.2 days, respectively. Find the 95% confidence interval for the mean time of pregnancy.

7-23. We want to determine the mean weight of all boxes of cereal labeled 400 grams. We need to be 98% confident that our sample mean is within 3 grams of the population mean, and a pilot study suggests that $\sigma = 10$ grams. How large must our sample be?

7-24. We want to determine the mean weight of a certain type of bird, and we want to be 95% confident that our sample mean is within 15 grams of the true population mean. A pilot study suggests that $\sigma = 120$ grams. How large must our sample be?

7-25. We want to estimate the mean voltage of batteries produced by a certain manufacturer. We want to be 98% confident that our sample mean is within 0.6 volt of the true population mean. Our quality control engineer suggests that $\sigma = 2.8$ volts. How large must our sample be?

EXERCISES B: ESTIMATES AND SAMPLE SIZES OF MEANS

7-26. Solve for n and show all work:

$$E = \frac{z(\alpha/2)\sigma}{\sqrt{n}}$$

7-27. Why *don't* we take

$$E = \frac{t(\alpha/2)s}{\sqrt{n}}$$

and solve for n to get an equation that can be used to determine the sample size required for cases in which the student t distribution applies?

7-28. The standard error of the mean is $\sigma/\sqrt{n}$ provided that the population size is infinite. If the population size is finite and is denoted by N, then the correction factor

$$\sqrt{\frac{N - n}{N - 1}}$$

should be used whenever $n > 0.05N$. This correction factor multiplies the standard error of the mean. Find the 95% confidence interval for the mean of 100 I.Q. scores if a sample of 30 scores produces a mean and standard deviation of 132 and 10, respectively.

7-3 ESTIMATES AND SAMPLE SIZES OF PROPORTIONS

In this section we consider the same concepts of estimating and sample size determination that were in Section 7-2, but we apply the concepts to proportions instead of means. We already studied proportions in Section 6-4 where we presented a method of conducting tests of hypotheses about population proportions. As in Section 6-4, we again assume that the conditions for the binomial distribution are essentially satisfied. We consider binomial experiments for which $np \geq 5$ and $nq \geq 5$. These assumptions enable us to use the normal distribution as an approximation to the binomial distribution.

Although we make repeated references to proportions, you should remember that the theory and procedures also apply to probabilities and percentages. Proportions and probabilities are both expressed in decimal or fraction form. If we intend to deal with percentages, we can easily convert them to proportions by deleting the percentage sign and dividing by 100. The symbol p may therefore represent a proportion, a probability, or the decimal equivalent of a percentage. We continue to use p as the population proportion in the same way that we use μ to represent the population mean. In Section 6-4, we represented a sample proportion by x/n where x was the number of successes in n trials. Continued use of x/n here would lead to some awkward expressions, so we introduce the following new notation.

$$p_s = \frac{x}{n}$$

In this way, p represents the population proportion while p_s represents the sample proportion. In previous chapters we stipulated that $q = 1 - p$, so it now becomes natural to stipulate that $q_s = 1 - p_s$.

The term p_s denotes the sample proportion that is analogous to the relative frequency definition of a probability. As an example, suppose that a pollster is hired to determine the proportion of adult Americans who favor socialized medicine. Let's assume that 2000 adult Americans are surveyed with 1347 favorable reactions. The pollster seeks the value of p, the true proportion of all adult Americans favoring socialized medicine. Sample results indicate that $x = 1347$ and $n = 2000$ so that

$$p_s = \frac{x}{n} = \frac{1347}{2000} = 0.6735$$

Just as $\bar{x}$ was selected as the point estimate of μ, we now select p_s as the best point estimate of p.

DEFINITION

> *The sample proportion* p_s *is the best* **point estimate of the population proportion** p.

Of the various estimators that could be used for p, p_s is deemed best because it is unbiased and the most consistent. It is unbiased in the sense that the distribution of sample proportions tends to center about the value of p. It is most consistent in the sense that the variance of sample proportions tends to be smaller than the variance of the other unbiased estimators.

We assume in this section that the binomial conditions are essentially satisfied and that the normal distribution can be used as an approximation to the distribution of sample proportions. This allows us to draw from results established in Section 5-5 and to conclude that the mean number of successes

μ and the standard deviation of the number of successes σ are given by

$$\mu = np$$

$$\sigma = \sqrt{npq}$$

where p is the probability of a success. Both of these parameters pertain to n trials, and we now convert them to a "per trial" basis simply by dividing by n.

$$\text{mean of sample proportions} = \frac{np}{n} = p$$

$$\text{standard deviation of sample proportions} = \frac{\sqrt{npq}}{n} = \sqrt{\frac{npq}{n^2}} = \sqrt{\frac{pq}{n}}$$

The first result may seem trivial since we have already stipulated that the true population proportion is p. The second result is nontrivial and very useful. In the last section, we saw that the sample mean $\bar{x}$ has a probability of $1 - \alpha$ of being within $z(\alpha/2) \dfrac{\sigma}{\sqrt{n}}$ of μ. Similar reasoning leads us to conclude that p_s has a probability of $1 - \alpha$ of being within $z(\alpha/2) \sqrt{pq/n}$ of p. But if we already know the value of p or q, we have no need for estimates or sample size determinations. Consequently, we must replace p and q by their point estimates of p_s and q_s so that an error factor can be computed in real situations. This leads to the following definition of a confidence interval.

DEFINITION

> *The confidence interval (or interval estimate) for the population proportion* p *is given by*
>
> $$p_s - z(\alpha/2) \sqrt{\frac{p_s q_s}{n}} < p < p_s + z(\alpha/2) \sqrt{\frac{p_s q_s}{n}}$$

The preceding confidence interval indicates that p is probably (with a probability of $1 - \alpha$) within a certain error factor of the sample proportion p_s. The error factor E, which is given by

$$E = z(\alpha/2) \sqrt{\frac{p_s q_s}{n}}$$

results from our use of the normal distribution as an approximation to the distribution of sample proportions, and the use of $\sqrt{p_s q_s / n}$ as the standard deviation of sample proportions. For example, we can be 95% confident that p is within $1.96 \sqrt{p_s q_s / n}$ of p_s. The following example illustrates the construction of a confidence interval for a proportion.

Example   Of 856 Americans polled, 360 smoked at least one cigarette in the last week. Find the 95% confidence interval for the true population proportion of Americans who smoked at least one cigarette in the last week.

Solution   The sample results show that $n = 856$ and $x = 360$, so $p_s = 360/856 = 0.421$ and $q_s = 1 - p_s = 1 - 0.421 = 0.579$. A confidence level of 95% requires that $\alpha = 0.05$ so that $z(\alpha/2) = 1.96$. Substituting these values into the general expression for a confidence interval for a proportion, we get

$$0.421 - 1.96 \sqrt{\frac{(0.421)(0.579)}{856}} < p < 0.421 + 1.96 \sqrt{\frac{(0.421)(0.579)}{856}}$$

or

$$0.421 - 0.033 < p < 0.421 + 0.033$$

or

$$0.388 < p < 0.454$$

Based upon the sample data, we are 95% confident that the true proportion of smokers is between 0.388 and 0.454.

Having discussed point estimates and confidence intervals for p, we now consider the problem of determining how large a sample should be when we want to find the approximate value of a population proportion. In the previous section we started with the expression for the error E and solved for n. Following that reasonable precedent, we begin with

$$E = z(\alpha/2) \sqrt{\frac{p_s q_s}{n}}$$

and we solve for n to get

$$n = \frac{[z(\alpha/2)]^2 \, p_s q_s}{E^2}$$

But if we are going to determine the necessary sample size, we can assume that the sampling has not yet taken place, so p_s and q_s are not known. Mathematicians have cleverly circumvented this problem by showing that, in the absence of p_s and q_s, we can assign the value of 0.5 to each of those statistics and the resulting sample size will be at least sufficient. The underlying reason for the assignment of 0.5 is found in the conclusion that the product $p_s \cdot q_s$ achieves a maximum possible value of 0.25 when $p_s = 0.5$ and $q_s = 0.5$. In practice, this means that no knowledge of p_s or q_s requires that the preceding expression for n evolves into the following:

$$n = \frac{[z(\alpha/2)]^2 \cdot 0.25}{E^2}$$

where the occurrence of 0.25 reflects the substitution of 0.5 for both p_s and q_s. If we have evidence supporting specific known values of p_s or q_s, we can substitute those values and thereby reduce the sample size accordingly.

The next two examples are intended to illustrate these points. Again we are faced with the useful and fascinating result that the size of a sample depends, not on the total population size, but on the levels of accuracy and confidence we desire.

Example

We want to estimate, with a maximum error of 0.01, the true proportion of all people who believe in life on other planets, and we want 99% confidence in our results. Also, we assume we have no prior information suggesting a possible value of p.

Solution

With a confidence level of 99%, we have $\alpha = 0.01$, so $z(\alpha/2) = 2.575$. We are given $E = 0.01$, but in the absence of p_s or q_s we use the last expression for n.

$$n = \frac{[z(\alpha/2)]^2 \cdot 0.25}{E^2} = \frac{[2.575]^2 \cdot 0.25}{(0.01)^2} = \frac{1.65765}{0.0001} \stackrel{\circ}{=} 16,577$$

To be 99% confident that we come within 0.01 of the true proportion of believers, we should poll 16,577 randomly selected people.

Example

We want to estimate the proportion of those who believe that the extended use of marijuana weakens self discipline. We want an error of no more than 0.02 and a confidence level of 96%. A previous survey indicates that p should be close to 0.85. How large should our sample be?

Solution

With a 96% confidence level, we have $\alpha = 0.04$ and $z(\alpha/2) = 2.05$. We are given $E = 0.02$, and since $p_s \stackrel{\circ}{=} 0.85$, we conclude that $q_s = 1 - p_s = 1 - 0.85 = 0.15$. We can now use our original expression for n to get

$$n = \frac{[z(\alpha/2)]^2 \, p_s q_s}{E^2} = \frac{[2.05]^2 \, (0.85)(0.15)}{(0.02)^2} \stackrel{\circ}{=} 1339.55 \stackrel{\circ}{=} 1340$$

Rounding *up*, we find that the sample size should be 1340. If we had no prior knowledge of the value of p, we would have used 0.25 for $p_s q_s$ and our required sample size would have been 2627, almost twice as large!

Newspaper, television, and radio reports often give poll results. A good reporter supplies information that will reveal the quality of the results. One statement from a national news magazine follows:

> The study is drawn from personal interviews conducted during the month of June with a representative sample of 1016 men and women of voting age. The figures have an error factor of plus or minus 3%.

**WHO IS
SHAKESPEARE?**

*A poll of 1553 randomly
selected adult Americans
revealed that:
89% could identify
Shakespeare.
58% could identify
Napoleon.
47% could identify Freud.
92% could identify
Columbus.
71% know what happened
in 1776.*

*Ronald Berman, a past
director of the National
Endowment for the
Humanities, said "I don't
worry about them (those
who don't know what
happened in 1776). I worry
about the people who take
polls." He goes on to say
that the poll should ask
substantive questions, such
as: What is the difference
between democracy and
totalitarianism?*

If we let $E = 0.03$ and choose a confidence level of 95%, we will find that $n = 1068$, so we can see that the data is approximately correct. We could verify the statement another way by letting $E = 0.03$ and $n = 1016$. Solving for $z(\alpha/2)$ we get 1.91, indicating a confidence level of 94.4% which is reasonably close to the claimed value of 95%.

Polling is an important and common practice in the United States. The concepts you have just studied should help to remove much of the mystery and misunderstanding often created by polls.

EXERCISES A: ESTIMATES AND SAMPLE SIZES OF PROPORTIONS

7-29. In a binomial experiment, a trial is repeated n times with x successes. For the given data, find p_s, q_s, and the best point estimate for the value of p.
(a) $n = 1000$, $x = 450$
(b) $n = 283$, $x = 172$
(c) $n = 17519$, $x = 873$
(d) $n = 1879$, $x = 1653$
(e) $n = 2366$, $x = 1103$

In Exercises 7-30 through 7-34, use the given data to find the appropriate confidence interval for the population proportion p.

7-30. $n = 400$, $x = 100$, 95% confidence

7-31. $n = 900$, $x = 400$, 95% confidence

7-32. $n = 512$, $x = 309$, 98% confidence

7-33. $n = 12485$, $x = 3456$, 99% confidence

7-34. $n = 1500$, $x = 607$, 99% confidence

7-35. A pollster is hired to determine the percentage of voters favoring the Republican presidential nominee. If we require 99% confidence that the estimated value is within two percentage points of the true value, how large should the random sample be?

7-36. A quality control engineer wants to determine the proportion of defective flashbulbs that her employer produces. How many bulbs should be tested if she needs to have 95% confidence that the estimated proportion is within 0.03 of the true proportion? Previous experience suggests that p is close to 0.02.

7-37. You know that several years ago 82% of those interviewed were opposed to busing students from one school to another to achieve racial balance. If you want to determine whether that attitude has changed, how many random subjects would you survey? Assume that you want 95% confidence in your results with a maximum error of 4 percentage points.

7-38. An educator intends to verify the claim that the illiteracy rate in the United States is about 1%. How many randomly selected subjects should be tested if we want 95% confidence that our sample is off by no more than one-half of one percentage point?

7-39. A pollster surveys 1500 randomly selected eligible voters and finds that 855 favored the Republican presidential nominee. Construct the 95% confidence interval for the true proportion of all voters favoring the Republican presidential nominee.

7-40. A quality control analyst tests 500 flashbulbs and finds 13 that are defective. Construct the 98% confidence interval for the true proportion of defective flashbulbs.

7-41. Of 150 interviews, 119 registered opposition to busing students to achieve racial balance. Construct the 95% confidence interval for the true proportion of all who are opposed to busing students.

7-42. Of 3000 random subjects tested, 27 were illiterate. Find the 99% confidence interval for the true proportion of illiterates.

7-43. A multiple-choice test question is considered easy if at least 80% of the responses are correct. A sample of 6503 responses to one question indicates that 5463 of those responses were correct. Construct the 99% confidence interval for the true proportion of correct responses. Is it likely that this question is really easy?

7-44. Of 281 aviation accidents, 95 resulted in fatalities. Based on this sample, find the 98% confidence interval for the true proportion of all aviation accidents that result in fatalities.

7-45. A survey of 777 audited tax returns shows that 536 resulted in additional tax payments. Construct the 99% confidence interval for the true proportion of all audited returns that result in additional payments.

7-46. A political candidate is planning his campaign strategy and needs to know the extent to which his name is recognized. How many voters should be polled if the candidate wants to be 98% sure that the sample percentage of those who recognize his name is within three percentage points of the true percentage?

7-47. A newspaper article has indicated that, of 1533 adults interviewed in a Gallup poll, 47% correctly identified Freud. To verify that percentage in a new poll, how many adults should be interviewed if we can tolerate an error no larger than one percentage point, and we insist on 99% confidence?

7-48. A poll conducted for *Time* reported that its results were based on a representative sample of 1016 people of voting age. An error factor of plus or minus 3% was also reported. Verify that the 3% error is approximately correct.

EXERCISES B: ESTIMATES AND SAMPLE SIZES OF PROPORTIONS

7-49. In Exercise 7-47 we see that 47% of 1533 adults were able to identify Freud correctly. Assuming that the 47% sample estimate is within two percentage points of the true percentage, find the corresponding confidence level.

7-50. A newspaper article indicates that an estimate of the unemployment rate involves a sample of 47,000 people. If the reported unemployment rate must have an error no larger than 0.2 percentage point and the rate is known to be about 8%, find the corresponding confidence level.

7-51. A rectangle has sides of length p and q, and the perimeter is 2 units. What values of p and q will cause the rectangle to have the largest possible area? To what concept of Section 7-3 does this problem relate?

7-52. A sample of 1200 scores produces a sample proportion that has an error of no more than 0.03. What is the corresponding level of confidence?

7-53. A sample of 500 scores produces a sample proportion that has an error of no more than 0.02. What is the corresponding level of confidence?

7-54. Solve for n showing all steps:

$$E = z(\alpha/2) \sqrt{\frac{p_s q_s}{n}}$$

7-4 ESTIMATES AND SAMPLE SIZES OF VARIANCES

In Section 6-5 we considered tests of hypotheses made about population variances or standard deviations. We noted that many real and practical situations, such as quality control in a manufacturing process, require inferences about variances or standard deviations. In addition to making products with good average quality, the manufacturer must make products of *consistent* quality that do not run the gamut from extremely poor to extremely good. This consistency can be measured by variance and standard deviation, so these statistics become important in maintaining the quality of products. There are many other situations in which variance and standard deviation are critically important.

As in Section 6-5 we assume here that the population in question has normally distributed values. This assumption is again a strict requirement since the chi-square distribution is so sensitive to departures from normality

In order to earn a coveted reputation, a good wine must be of a high and consistent quality. The consistency of the quality refers to the variation among different samples of the same type of wine, and that consistency can be measured by the variance or standard deviation.

that gross errors can easily arise. We describe this sensitivity by saying that inferences about σ^2 (or σ) made on the basis of the chi-square distribution are not *robust* against departures from normality. In contrast, inferences made about μ based on the student t distribution are reasonably robust since departures from normality which are not too extreme will not lead to gross errors.

In this section we extend the concepts of the previous sections to variances. These concepts are point estimates, confidence intervals, and sample size determinations. Since the sample mean $\bar{x}$ was given as the best point estimate of μ and the sample proportion $p_s = x/n$ was given as the best estimate of p, you should not be surprised by the following definition.

The sample variance s^2 *is the best* **point estimate of the population variance** σ^2. *The sample standard deviation* s *is the best* **point estimate of the population standard deviation** σ.

Since sample variances tend to center about the value of the population variance, we say that s^2 is an unbiased estimator of σ^2. Also, the variance of s^2 values tends to be smaller than the variance of the other unbiased estimators. For these reasons we decree that, among the various possible statistics we could use to estimate σ^2, the best is s^2. But like all point estimates, s^2 does not reveal just how good it is as an estimate. To compensate for that deficiency, we need to develop a more informative interval estimate (or confidence interval).

FIGURE 7-3

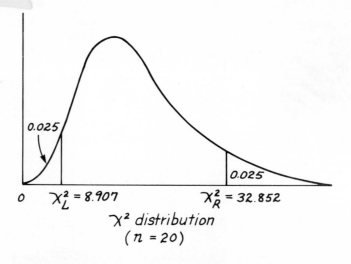

χ^2 distribution
($n = 20$)

Recall that in Section 6-5 we tested hypotheses about σ^2 by using the test statistic $(n - 1)s^2/\sigma^2$, which has a chi-square distribution. In Figure 7-3 we illustrate the chi-square distribution of $(n - 1)s^2/\sigma^2$ for samples of size $n = 20$. We also show the two critical values of χ^2 corresponding to a 95% level of significance. In a normal or t distribution, the left and right critical values are the same numbers with opposite signs, but from Figure 7-3 we can see that this is not the case with the chi-square distribution. Because we will use those values in developing confidence intervals for standard deviations and variances, we introduce the following notation.

With a total area of α divided equally between the two tails of a chi-square distribution, χ^2_L denotes the left-tailed critical value and χ^2_R denotes the right-tailed critical value.

A sample of 20 scores corresponds to 19 degrees of freedom, and if we refer to Table A-5 we find that the critical values shown in Figure 7-3 separate areas to the right of 0.025 and 0.975.

Figure 7-3 shows that, for a sample of 20 scores taken from a normally distributed population, the statistic $(n-1)s^2/\sigma^2$ has a 0.95 probability of falling between 8.907 and 32.852. In general, there is a probability of $1 - \alpha$ that the statistic $(n-1)s^2/\sigma^2$ will fall between χ_L^2 and χ_R^2. In other words (and symbols), there is a $1 - \alpha$ probability that both of the following are true:

$$\frac{(n-1)s^2}{\sigma^2} < \chi_R^2$$

and

$$\frac{(n-1)s^2}{\sigma^2} > \chi_L^2$$

If we multiply both of the preceding inequalities by σ^2 and divide each inequality by the appropriate critical value of χ^2, we see that the two inequalities can be expressed in the following equivalent forms:

$$\frac{(n-1)s^2}{\chi_R^2} < \sigma^2$$

and

$$\frac{(n-1)s^2}{\chi_L^2} > \sigma^2$$

These last two inequalities can be combined into one inequality which is given below:

$$\frac{(n-1)s^2}{\chi_R^2} < \sigma^2 < \frac{(n-1)s^2}{\chi_L^2}$$

There is a probability of $1 - \alpha$ that the population variance σ^2 is contained in the above interval.

These results provide the foundation for the following definition.

DEFINITION

> *The **confidence interval** (or **interval estimate**) for the population variance σ^2 is given by*
>
> $$\frac{(n-1)s^2}{\chi_R^2} < \sigma^2 < \frac{(n-1)s^2}{\chi_L^2}$$

The confidence interval (or interval estimate) for σ can be found by simply taking the square root of each component of the preceding inequality:

$$\sqrt{\frac{(n-1)s^2}{\chi_R^2}} < \sigma < \sqrt{\frac{(n-1)s^2}{\chi_L^2}}$$

TABLE 7-1

To be 95% confident that s^2 is within:	of the value of σ^2, the sample size n should be at least:
5%	4200
10%	750
20%	250
30%	100

To be 99% confident that s^2 is within:	of the value of σ^2, the sample size n should be at least:
5%	5150
10%	1800
20%	300
30%	125

To be 95% confident that s is within:	of the value of σ, the sample size n should be at least:
2.5%	4200
5%	750
10%	250
15%	100

To be 99% confident that s is within:	of the value of σ, the sample size n should be at least:
2.5%	5150
5%	1800
10%	300
15%	125

Example

A bank experiments with a single waiting line that feeds all windows as openings occur. An employee observes 30 randomly selected customers, and their waiting times produce a standard deviation of 3.8 minutes. Construct the 95% confidence interval for the true value of σ^2.

Solution

With a sample of size $n = 30$, we get $n - 1$ or 29 degrees of freedom. Since we seek 95% confidence, we divide the 5% chance of error between the two tails so that each tail contains a proportion of 0.025. In the 29th row of Table A-5, we find that 0.025 in the left and right tails indicates that $\chi_L^2 = 16.047$ and $\chi_R^2 = 45.722$. With these values of χ_L^2 and χ_R^2, with $n = 30$, and with $s = 3.8$, we use the preceding definition to obtain

$$\frac{(30 - 1)(3.8)^2}{45.722} < \sigma^2 < \frac{(30 - 1)(3.8)^2}{16.047}$$

which becomes $9.159 < \sigma^2 < 26.096$. That is, we are 95% confident that the actual variance of the waiting times of all customers is between 9.159 and 26.096.

If the previous problem had required the 95% confidence interval for σ instead of σ^2, we could have taken the square root of each component to get $\sqrt{9.159} < \sigma < \sqrt{26.096}$, which becomes $3.026 < \sigma < 5.108$.

The problem of determining the sample size necessary to estimate σ^2 to within given tolerances and confidence levels becomes much more complex than it was in similar problems that dealt with means and proportions. Instead of developing very complicated procedures, we supply Table 7-1 which lists approximate sample sizes.

Example   You wish to estimate σ^2 to within 10% and you need 99% confidence in your results. How large should your sample be? Assume that the population is normally distributed.

Solution   From Table 7-1, 99% confidence and an error of 10% for σ^2 correspond to a sample of size 1800. You should randomly select 1800 values from the population.

EXERCISES A: ESTIMATES AND SAMPLE SIZES OF VARIANCES

7-55.   Use the given data to find the best point estimate of σ^2.

     (a) $n = 100$, $\bar{x} = 103$, $s^2 = 12.5$
     (b) $n = 500$, $\bar{x} = 320$, $s = 15.3$
     (c) $n = 10$, $\bar{x} = 98.6$, $s = 1.2$
     (d) $n = 3$, $\bar{x} = 74$, $s = 1.9$

7-56.   Find the χ_L^2 and χ_R^2 values for a sample of 25 scores and a confidence level of 99%.

7-57.   Find the χ_L^2 and χ_R^2 values for a sample of 15 scores and a confidence level of 95%.

7-58.   Find the χ_L^2 and χ_R^2 values for a sample of 11 scores and a confidence level of 95%.

7-59.   Find the χ_L^2 and χ_R^2 values for a sample of 27 scores and a confidence level of 90%.

7-60.   Construct a 95% confidence interval about σ^2 if a random sample of 21 scores is selected from a normally distributed population and the sample variance is 100.

7-61. Construct a 95% confidence interval about σ^2 if a random sample of 10 scores is selected from a normally distributed population and the sample variance is 225.

7-62. A standard test for motor skills is being developed by a manufacturer, and a sample of 27 randomly selected adults produces scores with a variance of 22. Construct the 95% confidence interval for the true variance of all scores on this test.

7-63. The weekly down times of a computer are being studied for planning and replacement purposes. If the down times of 15 randomly selected weeks produces a standard deviation of 2.85 hours, find the 90% confidence interval for the true value of the variance of all weekly down times.

7-64. The weights of 20 randomly selected newborn girls have a standard deviation of 535 grams. Find the 95% confidence interval for the true standard deviation of the weights of all newborn girls.

7-65. A vending machine pours soup into cups. The standard deviation of 20 randomly selected cups is 4.4 milliliters. Find the 95% confidence interval for the true standard deviation of all cups filled by this machine.

7-66. Twenty-two normal adults are randomly selected and their pulse rates measured. The standard deviation for these pulse rates is 9.25 beats per minute. Find the 99% confidence interval for the true variance of the pulse rates of all normal adults.

7-67. A fuse manufacturer randomly selects 30 blown fuses for detailed analysis. The standard deviation for the 30 amperage values that caused the fuses to blow is 0.874 ampere. Find the 95% confidence interval for the variance of the critical amperage levels.

7-68. A drug is intended to lower blood pressure by an amount equivalent to the pressure of 10 millimeters of mercury. To test the success of this drug, a clinic administers it to ten randomly selected normal adults. The resulting decreases produce a standard deviation of 3.1 millimeters of mercury. Find the 90% confidence interval for the true variance of all such decreases.

7-69. Use Table 7-1 to find the approximate minimum sample size necessary to estimate the parameter with the given tolerances and levels of confidence.
   (a) σ^2 with a 30% maximum error and 95% confidence.
   (b) σ^2 with a 5% maximum error and 95% confidence.
   (c) σ with a 2.5% maximum error and 95% confidence.
   (d) σ with a 2.5% maximum error and 99% confidence.
   (e) σ with a 10% maximum error and 99% confidence.
   (f) σ with a 15% maximum error and 99% confidence.
   (g) σ with a 5% maximum error and 95% confidence.

7-70. The following scores are randomly selected from a normally distributed population: 16, 18, 21, 20, 19, 19, 23, 16, 15, 23. Find the 95% interval estimate of σ^2.

7-71. The following scores are randomly selected from a normally distributed population: 103, 103, 101, 102, 107, 106, 108, 102. Find the 99% interval estimate of σ^2.

7-72. The following reaction times (in seconds) are randomly obtained from a normally distributed population: 0.60, 0.61, 0.63, 0.72, 0.91, 0.72. Find the 99% confidence interval for σ.

7-73. The following weights (in grams) are randomly obtained from a normally distributed population: 201, 203, 212, 222, 213, 215, 217, 230, 205, 208, 217, 225. Find the 95% confidence interval for σ.

EXERCISES B: ESTIMATES AND SAMPLE SIZES OF VARIANCES

7-74. Under what conditions, if any, can the numerical value of σ exceed that of σ^2?

REVIEW

In this chapter we continued our study of inferential statistics by introducing the concepts of **point estimate, confidence interval** (or **interval estimate**), and ways of determining the **sample size** necessary to estimate parameters to within given error factors. In Chapter 6 we used sample data to make **decisions** about hypotheses, but the central concern of this chapter is the **estimate** of parameter values. The parameters are population means, proportions, and variances.

Table 7-2 summarizes some of the key results of this chapter. It serves as a rough guide since it does not include underlying assumptions (such as the assumption of normality) and other relevant information.

TABLE 7-2

Parameter	Point estimate	Confidence interval	Sample size
μ	$\bar{x}$	$\bar{x} - z(\alpha/2)\dfrac{\sigma}{\sqrt{n}} < \mu < \bar{x} + z(\alpha/2)\dfrac{\sigma}{\sqrt{n}}$ or $\bar{x} - t(\alpha/2)\dfrac{s}{\sqrt{n}} < \mu < \bar{x} + t(\alpha/2)\dfrac{s}{\sqrt{n}}$	$n = \left[\dfrac{z(\alpha/2)\sigma}{E}\right]^2$
p	$p_s = \dfrac{x}{n}$	$p_s - z(\alpha/2)\sqrt{\dfrac{p_s q_s}{n}} < p < p_s + z(\alpha/2)\sqrt{\dfrac{p_s q_s}{n}}$	$n = \dfrac{[z(\alpha/2)]^2 \, p_s q_s}{E^2}$ or $n = \dfrac{[z(\alpha/2)]^2 \cdot 0.25}{E^2}$
σ^2	s^2	$\dfrac{(n-1)s^2}{\chi_R^2} < \sigma^2 < \dfrac{(n-1)s^2}{\chi_L^2}$	See Table 7-1

REVIEW EXERCISES

7-75. (a) Evaluate $z(\alpha/2)$ for $\alpha = 0.10$.
    (b) Evaluate χ_L^2 and χ_R^2 for $\alpha = 0.05$ and a sample of 10 scores.
    (c) Evaluate $t(\alpha/2)$ for $\alpha = 0.05$ and a sample of 10.
    (d) What is the maximum possible value of $p \cdot q$?

7-76. $n = 60$, $\bar{x} = 83.2$ kilograms, $s = 4.1$ kilograms. Assume that the given statistics represent sample data randomly selected from a normally distributed population.
    (a) What is the best point estimate of μ?
    (b) Construct the 95% confidence interval about μ.

7-77. Using the sample data given in Exercise 7-76:
    (a) What is the best point estimate of σ?
    (b) Construct the 95% confidence interval about σ.

7-78. $n = 16$, $\bar{x} = 83.2$ kilograms, $s = 4.1$ kilograms. Assume that the given statistics represent sample data randomly selected from a normally distributed population.
    (a) What is the best point estimate of μ?
    (b) Construct the 95% confidence interval about μ.

7-79. Using the sample data given in Exercise 7-78:
    (a) What is the best point estimate of σ?
    (b) Construct the 95% confidence interval about σ.

7-80. A pollster is expected to determine the percentage of voters favoring some form of capital punishment. If there must be 95% confidence that the estimated value is within two percentage points of the true value, how large should the random sample be?

7-81. A medical researcher wishes to estimate the cholesterol level of all 6-year-old boys, and there is strong evidence suggesting that $\sigma = 50$. If the researcher wants to be 95% confident of obtaining a sample mean that is off by no more than four units, how large must the sample be?

7-82. An educator wishes to estimate the standard deviation of all I.Q. scores for 6-year-old girls. If the educator wants to be 95% confident that the sample standard deviation is off by no more than 10% of σ, how large should the sample be?

7-83. An intrauterine device is used by 300 women for the purpose of preventing pregnancy. Eighteen women from this group become pregnant. Construct the 95% confidence interval for the proportion of failures for this birth control device.

7-84. A psychologist is collecting data on the time it takes to learn a certain task. For 50 randomly selected adult male subjects, the sample mean and standard deviation are computed to be 16.40 minutes and 4.00 minutes, respectively. Construct the 98% interval estimate for the mean time required by all adult males.

TESTS COMPARING TWO PARAMETERS

8-1 OVERVIEW

Why would we want to compare two parameters, such as μ_1 and μ_2, that come from two populations? In reality, there are many cases where the main objective is a comparison of *two* groups of data instead of a comparison of *one* group to some known value. For example, a manufacturer may want to know which of two different production techniques yields better results. A college dean may be interested in the differences between the grades given at two different schools. A psychologist may need to know whether two different I.Q. tests produce similar or different results. A doctor may be interested in the comparative effectiveness of two different cold medicines. A sociologist may want to compare last year's college entrance examination scores to those of 20 years ago. A psychologist may wish to compare variability in reaction times between a group of nondrinkers and a group of people, each of whom has consumed a double martini. In all of these cases, we want to compare two population parameters. In this chapter we extend the methods of hypothesis testing (see Chapter 6) to cases involving the comparison of two variances, two means, or two proportions.

Chapters 6 and 7 started with means and then moved on to proportions, followed by variances. Here we depart from this order because one of the results of tests on two variances serves as a *prerequisite* for a test involving two means. Consequently, we begin this chapter with a discussion of the method for testing hypotheses involving a comparison of two variances.

8-2 TESTS COMPARING TWO VARIANCES

In many cases where we want to compare parameters from each of two separate populations, the **variances** are the most relevant parameters. For example, a manufacturer of automobile batteries may want to compare two different production methods which result in batteries that last, on the average, approximately 4 years. Consider the hypothetical case in Table 8-1.

In general, we assume throughout this section that we have two independent populations that are approximately normally distributed. A sample of size n_1 is drawn from the first population, while a sample of size n_2 is drawn from the second. Since we want to compare variances, we compute the respective sample variances s_1^2 and s_2^2 and use those statistics in testing for equality of σ_1^2 and σ_2^2.

From Table 8-1 we see that both production methods seem to result in batteries that last the same length of time, but the batteries produced by method A exhibit much less consistency. The two production methods appear to differ radically in the variability of the battery lives, and this is reflected in the large difference between the two sample variances. Thus, it is through a comparison of the two *variances* that we find the crucial difference between the two production methods. Since not all comparisons of variances involve

TABLE 8-1 Life (in years) of car batteries

Production method A	Production method B
2.0	3.7
2.1	3.9
2.5	3.9
3.0	3.9
3.3	4.0
4.2	4.0
4.2	4.0
4.3	4.1
6.8	4.2
7.6	4.3
$\bar{x}_1 = 4.00$	$\bar{x}_2 = 4.00$
$s_1^2 = 3.59$	$s_2^2 = 0.03$

such obvious differences, we need more standardized and objective procedures. Even for the data in Table 8-1, the difference between the variances of 3.59 and 0.03 must be weighed against the sample sizes so that we can determine whether this "obvious" difference is statistically significant.

For simplicity of notation, let's stipulate that s_1^2 **always represents the larger of the two sample variances** and s_2^2 represents the smaller of the two sample variances. We can do this since identification of the samples through subscript notation is arbitrary. Extensive analyses have shown that **for two normally distributed populations with equal variances (that is, $\sigma_1^2 = \sigma_2^2$), the sampling distribution of**

$$F = \frac{s_1^2}{s_2^2}$$

is the F distribution shown in Figure 8-1 and described in Table A-6.

FIGURE 8-1
Properties of the F distribution.
(1) All values of F are nonnegative ($F \geq 0$).
(2) Instead of being symmetric, the F distribution is skewed to the right.
(3) There is a different F distribution for each different pair of degrees of freedom for numerator and denominator.

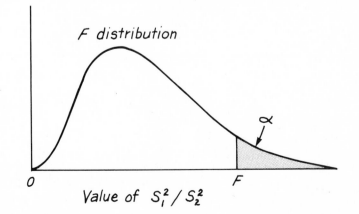

LOWER FERTILITY RATE CLOSES SCHOOLS

In a recent 5-year period, about 500 schools were closed in the United States. Martin Frankel, a spokesman for the National Center for Educational Statistics, says that a declining fertility rate is causing many schools to close and is seriously affecting most others. The fertility rate, now below 2.1, is the number of births per 1000 women between the ages of 15 and 48. This is different from the birth rate, which is the number of births per 1000 people. Mr. Frankel says that the fertility rate is a more meaningful figure than the birth rate.

If the two populations really do have equal variances, then $F = s_1^2/s_2^2$ tends to be close to 1 since s_1^2 and s_2^2 tend to be close in value. But if the two populations have radically different variances, s_1^2 and s_2^2 tend to be very different numbers. Denoting the larger of the sample variances by s_1^2, we see that the ratio s_1^2/s_2^2 will be a large number whenever s_1^2 and s_2^2 are far apart in value. Consequently, a value of F near 1 will be evidence in favor of the conclusion that $\sigma_1^2 = \sigma_2^2$. A large value of F will be evidence against the conclusion of equality of the population variances.

Ideally we would describe the mathematics underlying the F distribution, but the complexity of that distribution makes such a discussion beyond the scope of this text. For now, we must accept the rigorous proof on faith. Believe us when we say that repeated samplings from two normally distributed populations for which $\sigma_1^2 = \sigma_2^2$ yield sample variances whose ratios follow the F distribution. The critical F values are summarized in Table A-6.

When we use Table A-6 we obtain critical F values that are determined by the following three values:

1. The significance level α.

2. The degrees of freedom for the numerator $(n_1 - 1)$.

3. The degrees of freedom for the denominator $(n_2 - 1)$.

We should be careful to ensure that n_1 corresponds to the sample yielding a variance of s_1^2 while n_2 corresponds to the sample having variance s_2^2. In Table A-6, we choose the appropriate value of α and then intersect the column representing the degrees of freedom for s_1^2 with the row representing the degrees of freedom for s_2^2. The appropriateness of α is determined by the level of significance desired and whether the test is one-tailed or two-tailed. In a one-tailed test we use the F distribution corresponding to the significance level α, but in a two-tailed test we divide α equally between the two tails and refer to the section of Table A-6 identified by $\alpha/2$. The following examples illustrate the method of testing hypotheses about two population variances.

Example

Two different machines are used to produce the same type of resistor. Samples are randomly selected and the relevant data are as follows.

Machine A	Machine B
$n = 21$	$n = 30$
$s^2 = 40$ ohms^2	$s^2 = 25$ ohms^2

At the 0.05 significance level, test the claim that the variance of all resistors from Machine A exceeds that of Machine B.

Solution

The larger sample variance of 40 ohms² will be identified with sample 1, so $n_1 = 21$, $s_1^2 = 40$, $n_2 = 30$, and $s_2^2 = 25$.

The null and alternate hypotheses are as follows:

$$H_0: \sigma_1^2 \leq \sigma_2^2$$

$$H_1: \sigma_1^2 > \sigma_2^2$$

With α already selected as 0.05, we compute the F statistic based on the sample data:

$$F = \frac{s_1^2}{s_2^2} = \frac{40}{25} = 1.6$$

In Table A-6 we select the section for which $\alpha = 0.05$. We locate the column representing $21 - 1$ or 20 degrees of freedom and the row representing $30 - 1$ or 29 degrees of freedom. The critical F score of 1.9446 indicates that the computed F score of 1.6 is not large enough to warrant rejection of the null hypothesis H_0. We lack sufficient evidence to support the claim that the variance of resistors from Machine A exceeds that of Machine B.

Example

At the 0.02 significance level, test the claim that $\sigma_1^2 = \sigma_2^2$ given the following sample data which came from normally distributed populations:

Sample 1	Sample 2
$n_1 = 16$	$n_2 = 10$
$s_1^2 = 225$	$s_2^2 = 100$

Solution

The solution is summarized in Figures 8-2 and 8-3. With the claim of $\sigma_1^2 = \sigma_2^2$ and the alternative of $\sigma_1^2 \neq \sigma_2^2$, we have a two-tailed test, so the significance level of $\alpha = 0.02$ leads to an area of 0.01 in the left-tailed critical region and an area of 0.01 in the right-tailed critical region. But as long as we stipulate that the larger of the two sample variances be placed in the numerator, we need concern ourselves only with the right-tailed critical value. With an area of 0.01 in the right tail, with $n_1 = 16$ and $n_2 = 10$, we locate the critical F value of 4.9621 which corresponds to 15 degrees of freedom in the numerator and 9 degrees of freedom in the denominator. Refer to Figures 8-2 and 8-3 for the completed hypothesis test.

So far in this section we have discussed only tests comparing two population variances, but the same methods and theory can be used to compare two population standard deviations. Any claim about two population standard deviations can be easily restated in terms of the corresponding

FIGURE 8-2

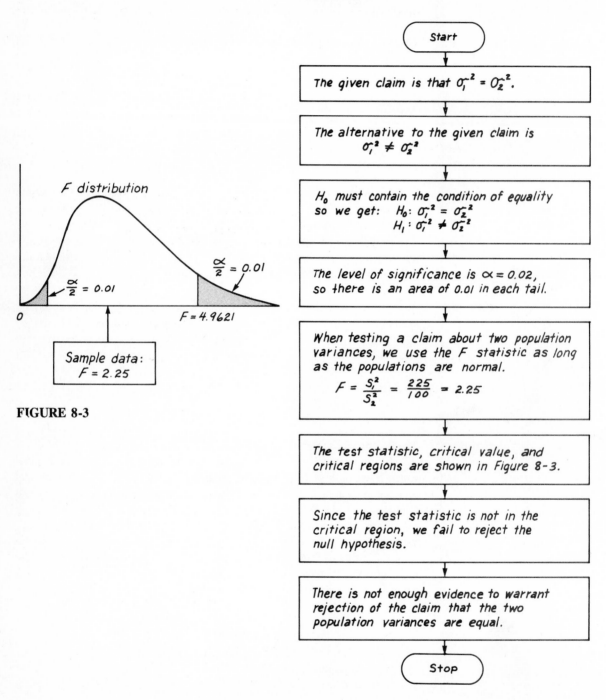

F distribution

$\frac{\alpha}{2} = 0.01$

$\frac{\alpha}{2} = 0.01$

0

$F = 4.9621$

Sample data:
$F = 2.25$

FIGURE 8-3

Start

The given claim is that $\sigma_1^2 = \sigma_2^2$.

The alternative to the given claim is
$\sigma_1^2 \neq \sigma_2^2$

H_0 must contain the condition of equality
so we get: $H_0: \sigma_1^2 = \sigma_2^2$
            $H_1: \sigma_1^2 \neq \sigma_2^2$

The level of significance is $\alpha = 0.02$,
so there is an area of 0.01 in each tail.

When testing a claim about two population
variances, we use the F statistic as long
as the populations are normal.
$$F = \frac{S_1^2}{S_2^2} = \frac{225}{100} = 2.25$$

The test statistic, critical value, and
critical regions are shown in Figure 8-3.

Since the test statistic is not in the
critical region, we fail to reject the
null hypothesis.

There is not enough evidence to warrant
rejection of the claim that the two
population variances are equal.

Stop

population variances. For example, suppose we want to test the claim that $\sigma_1 = \sigma_2$ and we are given the following sample data taken from normal populations.

	Sample 1	Sample 2
	$n_1 = 25$	$n_2 = 17$
	$s_1 = 12.72$	$s_2 = 8.43$

Restating the claim as $\sigma_1^2 = \sigma_2^2$, we get $s_1^2 = (12.72)^2$ and $s_2^2 = (8.43)^2$, and we can proceed with the F test in the usual way. Rejection of $\sigma_1^2 = \sigma_2^2$ is equivalent to rejection of $\sigma_1 = \sigma_2$. Failure to reject $\sigma_1^2 = \sigma_2^2$ implies failure to reject $\sigma_1 = \sigma_2$.

Note that in all tests of hypotheses made about population variances and standard deviations, the values of the means are irrelevant. In the next section we consider tests comparing two population means, and we see that the tests sometimes require and incorporate hypothesis tests of the type just discussed.

EXERCISES A: TESTS COMPARING TWO VARIANCES

In Exercises 8-1 through 8-6, test the claim that the two samples come from populations having equal variances. Use a significance level of $\alpha = 0.05$ and assume that all populations are normally distributed. Follow the pattern suggested by Figure 6-2 and draw the appropriate graphs.

8-1. Sample A: $n_1 = 10$, $s_1^2 = 50$
      Sample B: $n_2 = 10$, $s_2^2 = 25$

8-2. Sample A: $n = 10$, $s^2 = 50$
      Sample B: $n = 15$, $s^2 = 25$

8-3. Sample A: $n = 20$, $s^2 = 110$
      Sample B: $n = 25$, $s_1^2 = 265$

8-4. Sample A: $n = 41$, $s^2 = 15.1$
      Sample B: $n = 16$, $s_1^2 = 33.2$

8-5. Sample A: $n = 5$, $s_1 = 2.1$
      Sample B: $n = 15$, $s = 1.1$

8-6. Sample A: $n = 25$, $s = 3.9$
      Sample B: $n = 10$, $s_1 = 6.3$

In Exercises 8-7 through 8-10, test the claim that the variance of population A exceeds that of population B. Use a 5% level of significance and assume that all populations are normally distributed. Follow the pattern outlined in Figure 6-2 and draw the appropriate graphs.

8-7. Sample A: $n = 10$, $s^2 = 48$
     Sample B: $n = 10$, $s^2 = 12$

8-8. Sample A: $n = 50$, $s^2 = 18.2$
     Sample B: $n = 20$, $s^2 = 8.7$

8-9. Sample A: $n = 16$, $s^2 = 225$
     Sample B: $n = 200$, $s^2 = 160$

8-10. Sample A: $n = 35$, $s^2 = 42.3$
      Sample B: $n = 25$, $s^2 = 16.2$

8-11. On a test of manual strength, a sample of 25 randomly selected males earn scores that have a variance of 130 points. Twenty-seven randomly selected girls take the same test, and the variance of their scores is 75 points. At the 0.02 significance level, test the claim that the population variances are not equal.

8-12. Fifteen randomly selected adult males are given a test on reaction times, and their scores produce a variance of 1.04. Seventeen other randomly selected adult males are given a double martini before taking the same test, and their scores produce a variance of 3.26. At the 0.05 significance level, test the claim that the population of all drinkers will have a variance larger than the population of all non-drinkers.

8-13. A scientist wants to compare two delicate weighing instruments by repeated weighings of the same object. The first scale is used 30 times and the weights have a standard deviation of 72 milligrams. The second scale is used 41 times and the weights have a standard deviation of 98 milligrams. At the 0.05 significance level, test the claim that the second scale produces greater variance.

8-14. An experiment is devised to study the variability of grading procedures among college professors. Two different professors are asked to grade the same set of 25 exam solutions and their grades have variances of 103.4 and 39.7, respectively. At the 0.05 significance level, test the claim that the first professor's grading exhibits greater variance.

EXERCISES B: TESTS COMPARING TWO VARIANCES

8-15. Two samples of equal size produce variances of 37 and 57. At the 0.05 significance level, we test the claim that the variance of the second population exceeds that of the first, and that claim is upheld by the data. What is the approximate minimum size of each sample?

8-16. Do Exercise 8-15 after changing the significance level to 0.01.

8-17. A sample of 21 scores produces a variance of 67.2 and another sample of 25 produces a variance that causes rejection of the claim that the two populations have equal variances. If this test is conducted at the 0.02 level of significance, find the maximum variance of the second sample if you know that it is smaller than that of the first sample.

8-18. Given the following samples, test the claim that they come from populations with equal variances. Assume that both populations are normally distributed and use a 5% level of significance.

Sample A: 47 52 54 49 38 68 52 47 48 51
Sample B: 101 86 105 91 93 99 96 112 94 102 115 86 117 108

8-19. Given the following sample I.Q. scores, test the claim that they come from populations with equal standard deviations. Assume that both populations are normally distributed and use a 5% level of significance.

Sample A: 100 101 110 115 95 88 120 109 116
102 107 93 99 95
Sample B: 112 113 123 127 107 100 132 118 121
128 114 119 105 111 120 135

8-20. Given the following sample I.Q. scores, test the claim that they come from populations with equal standard deviations. Assume that both populations are normally distributed and use a 5% level of significance.

Sample A: 95 98 90 99 100 103 102 85 95 97 94 94
Sample B: 115 118 110 124 124 102 115 105 117 101

8-3 TESTS COMPARING TWO MEANS

In this section we consider tests of hypotheses made about two population means. Many real and practical situations use such tests successfully. For example, an educator may want to compare mean test scores produced by two teaching methods. A manager may want to test for a difference in the mean weight of cereal loaded into boxes by two machines. A car manufacturer may want to test for a difference in the mean longevity of batteries produced by two suppliers. A psychologist may want to test for a difference in mean reaction times between men and women. A farmer may want to test for a difference in mean crop production for two irrigation methods. A medical researcher may want to test for a difference between a new drug and one currently in use.

The way in which we compare means using sample data taken from two populations is affected by the presence or absence of a relationship between those samples.

> Two samples are **dependent** if one is related to the other in some way.

AIR IS HEALTHIER THAN TOBACCO

A consumer testing group studied the tar and nicotine levels of various brands of cigarettes. ''Now'' was advertised as being lowest in tar and nicotine, but it seemed to burn more quickly than other brands. Samples of Now and Winston were randomly selected and weighed. Both brands were products of the R. J. Reynolds Tobacco Company and Winston was their best seller at the time of the comparison. Results showed that the average weight of the tobacco in Now was about two-thirds the average weight of the tobacco in a Winston. With one-third less tobacco, it isn't difficult to get lower tar and nicotine levels. The study also noted that smokers tend to consume more cigarettes when they are of the low tar-nicotine variety. The net effect for Now was that more cigarettes could be sold at lower production costs.

A sample of wives' ages and another sample of their husbands' ages would exhibit dependency since marriages tend to occur between members of the same age group. In the first two examples, each x value is related to the corresponding y value. In the first example, each x and y pair represents the ages of a husband and wife. In the second example, each x and y pair represents the before-and-after blood pressures of the same person. In the third example, we test only subjects who are twenty years old, and we record their reaction times as measured by a standard instrument. In this last example, we do not expect to find any relationship between an x value and the corresponding y value. Consequently, while the first two examples involve dependent samples, the third example involves independent samples.

	Smith	Jones	Brown	Carter	Mayer	
x (age of wife)	22	48	27	29	32	
y (age of husband)	23	51	25	29	32	

	Ann	Bob	Carol	Don	Eve	
x (blood pressure before treatment)	120	150	136	166	192	
y (blood pressure after treatment)	118	130	136	152	170	

x (reaction times of females)	0.70	0.68	0.59	0.72	0.74		
y (reaction times of males)	0.63	0.55	0.58	0.59	0.57	0.60	0.59

When dealing with two dependent samples, it is very wasteful to reduce the sample data to $\bar{x}_1$, s_1, n_1, $\bar{x}_2$, s_2, and n_2 since the relationship between pairs of data would be completely lost. Instead, we compute the *differences* between the pairs of data as follows:

x	22	48	27	29	32
y	23	51	25	29	32
$x - y$	-1	-3	2	0	0

NOTATION | Let $\bar{d}$ denote the mean value of $x - y$ for the paired sample data.

Example For the $x - y$ values of -1, -3, 2, 0, 0 taken from the preceding table, we get

$$\bar{d} = \frac{(-1) + (-3) + 2 + 0 + 0}{5} = \frac{-2}{5} = -0.40$$

NOTATION | Let s_d denote the standard deviation of the $x - y$ values for the paired sample data.

Example For the $x - y$ values of -1, -3, 2, 0, 0 we get $s_d = 1.82$.

NOTATION | n denotes the number of *pairs* of data.

Example For the data of the last table, $n = 5$.

In repeated random sampling from two normal and dependent populations in which the difference between the means is μ_d, the values of $\bar{d}$ possess a student t distribution with mean μ_d and standard deviation $s_d/\sqrt{n}$. This means that our test statistic is obtained by computing

$$t = \frac{\bar{d} - \mu_d}{s_d/\sqrt{n}}$$

If we claim that there is no difference between the two population means, then we are claiming that $\mu_d = 0$. This becomes reasonable when we recognize that $\bar{d}$ should be around zero if there is no difference between the two population means. In conducting these tests, there will be $n - 1$ degrees of freedom.

In the following example we illustrate a complete hypothesis test for a situation involving dependent samples.

Example

A private educational service offers a course designed to increase scores on I.Q. tests. To test the effectiveness of this course, 10 randomly selected subjects agreed to take I.Q. tests before and after the course. The results follow. At the 5% level of significance, test the claim that the course does increase I.Q. scores.

Subjects

	A	B	C	D	E	F	G	H	I	J
I.Q. test scores before course	96	110	98	113	88	92	106	119	100	97
I.Q. test scores after course	99	112	107	110	88	101	107	123	91	99
difference	-3	-2	-9	3	0	-9	-1	-4	9	-2

Solution

Since each pair of scores is for the same person, we can conclude that the values are dependent. Each difference represents the "before score − after score." If the course is effective, we would expect the before scores to be less than the after scores, so the differences would tend to be negative and $\bar{d}$ would be negative and *significantly* below zero. Thus, the claim of improved scores is equivalent to the claim that $\mu_d < 0$. Figures 8-4 and 8-5 summarize the key features of this hypothesis test.

The mean of the differences found in the sample data is $\bar{d}$. It is computed as follows:

$$\bar{d} = \frac{(-3) + (-2) + (-9) + 3 + 0 + (-9) + (-1) + (-4) + 9 + (-2)}{10}$$

$$= -1.8$$

The standard deviation of the differences s_d is computed to be 5.3. There are 10 pairs of data, so $n = 10$. These components are used in finding the value of the test statistic.

$$t = \frac{\bar{d} - \mu_d}{s_d/\sqrt{n}} = \frac{-1.8 - 0}{5.3/\sqrt{10}} = -1.074$$

The critical t value of -1.833 can be found in Table A-4 after you note that this is a left-tailed test, $\alpha = 0.05$, and there are $10 - 1$ or 9 degrees of freedom. (The test is left-tailed since only negative values of $\bar{d}$ significantly less than zero will cause rejection of the null hypothesis.)

FIGURE 8-4

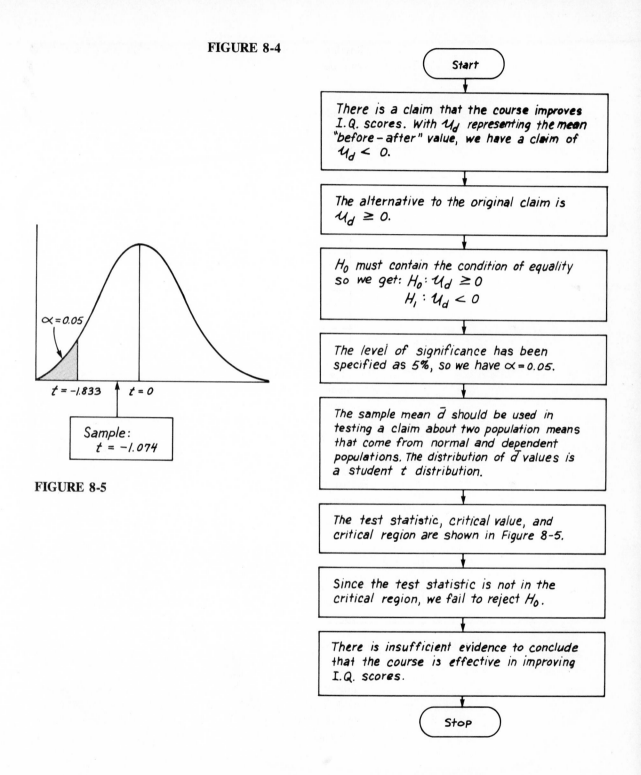

α = 0.05

$t = -1.833$ $t = 0$

Sample:
$t = -1.074$

FIGURE 8-5

Start

There is a claim that the course improves I.Q. scores. With u_d representing the mean "before – after" value, we have a claim of $u_d < 0$.

The alternative to the original claim is $u_d \geq 0$.

H_0 must contain the condition of equality so we get: $H_0 : u_d \geq 0$
$H_1 : u_d < 0$

The level of significance has been specified as 5%, so we have α = 0.05.

The sample mean $\bar{d}$ should be used in testing a claim about two population means that come from normal and dependent populations. The distribution of $\bar{d}$ values is a student t distribution.

The test statistic, critical value, and critical region are shown in Figure 8-5.

Since the test statistic is not in the critical region, we fail to reject H_0.

There is insufficient evidence to conclude that the course is effective in improving I.Q. scores.

Stop

As we consider other tests of hypotheses made about two population means, we begin to encounter a maze which can easily lead to confusion. Most of the confusion can be avoided by referring to Figure 8-6, which summarizes the procedures discussed in this section. We illustrate the use of Figure 8-6 through specific examples and then present the underlying theory

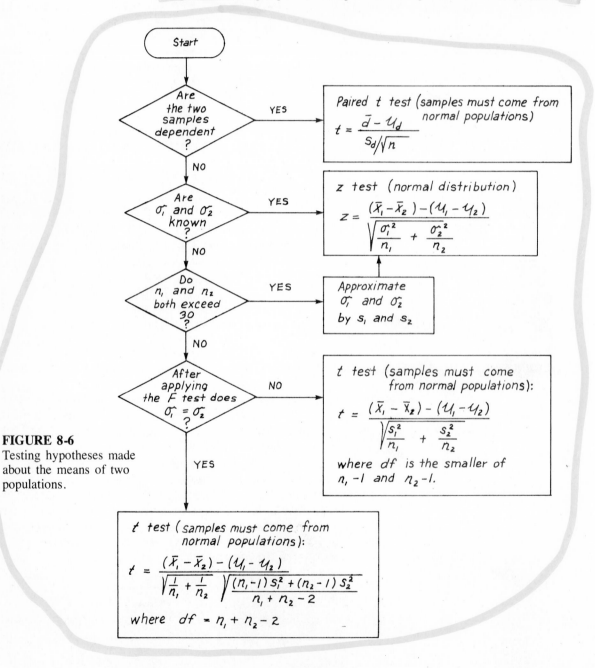

FIGURE 8-6
Testing hypotheses made about the means of two populations.

which led to its development. Since we have already presented an example involving dependent populations, our next examples will involve independent populations.

Example

Two machines fill packages, and samples selected from each machine produce the following results:

Machine A	Machine B
$n_1 = 50$	$n_2 = 100$
$\bar{x}_1 = 4.53$ kilograms	$\bar{x}_2 = 4.01$ kilograms

If the standard deviations of the contents filled by Machine A and Machine B are 0.80 kilogram and 0.60 kilogram, respectively, test the claim that the mean contents produced by Machine A equal the mean for Machine B. Assume a 0.05 significance level.

Solution

The two means are independent and σ_1 and σ_2 are known. Referring to Figure 8-6, we see that we should use a normal distribution test with

$$z = \frac{(\bar{x}_1 - \bar{x}_2) - (\mu_1 - \mu_2)}{\sqrt{\dfrac{\sigma_1^2}{n_1} + \dfrac{\sigma_2^2}{n_2}}} = \frac{(4.53 - 4.01) - 0}{\sqrt{\dfrac{0.80^2}{50} + \dfrac{0.60^2}{100}}} \cong 4.061$$

With the null and alternate hypotheses described as

$$H_0: \mu_1 = \mu_2$$

$$H_1: \mu_1 \neq \mu_2$$

and $\alpha = 0.05$, we conclude that the test involves two tails. From Table A-3 we extract the critical z values of 1.96 and -1.96. The test statistic of 4.061 is well into the critical region, and we therefore reject H_0 and conclude that the population means corresponding to the two machines are not equal.

The preceding example is somewhat contrived because we seldom know the values of σ_1 and σ_2. It is rare to sample from two populations with unknown means but known standard deviations. To cover more realistic cases involving independent samples with unknown standard deviations or variances, we next examine the sizes of the two samples as suggested by Figure 8-6.

If both samples are large (greater than 30), we can estimate σ_1 and σ_2 by s_1 and s_2. We can then proceed as in the last example. However, if either sample is small, we must apply the F test to determine whether the two sample variances are indicative of equal population variances. The next example illustrates these points since the populations are independent, both population standard deviations are unknown, and both samples are small (less than or equal to 30). We see that the F test suggests that the two population standard deviations are not equal and, from Figure 8-6, we see that the circumstances of this next example cause us to turn right at the last diamond.

Example Random samples of similar calculators are obtained from two competing manufacturers. Analyses of the length of their lives are as follows:

Brand X	Brand Y
$n_1 = 11$	$n_2 = 15$
$\bar{x}_1 = 4.3$ years	$\bar{x}_2 = 4.6$ years
$s_1 = 1.1$ years	$s_2 = 0.4$ year

At the 0.05 significance level, test the claim that the mean lives of both brands are equal.

Solution Since the samples are independent, neither standard deviation is known, and both samples are small, Figure 8-6 indicates that we should begin by applying the F test which was discussed in Section 8-2. We want to decide whether $\sigma_1 = \sigma_2$, so we formulate the following null and alternative hypotheses:

$$H_0: \sigma_1 = \sigma_2$$

$$H_1: \sigma_1 \neq \sigma_2$$

With $\alpha = 0.05$, we do a complete hypothesis test to decide whether or not $\sigma_1 = \sigma_2$. We then do another complete hypothesis test of the claim that $\mu_1 = \mu_2$ using the appropriate student t distribution. For the preliminary test we get

$$F = \frac{s_1^2}{s_2^2} = \frac{1.1^2}{0.4^2} = 7.5625$$

The critical F value obtained from Table A-6 is 3.1469. (The test involves two tails with $\alpha = 0.05$, and the degrees of freedom for the numerator and the denominator are 10 and 14, respectively.) These results cause us to reject the null hypothesis of equal variances and we decide that $\sigma_1 \neq \sigma_2$. In Figure 8-6, we answer no to the question contained in the fourth diamond, and we test the claim that $\mu_1 = \mu_2$ by using the student t distribution and the test statistic given in the box to the right of the fourth diamond. With

$$H_0: \mu_1 = \mu_2$$

$$H_1: \mu_1 \neq \mu_2$$

$$\alpha = 0.05$$

we compute the test statistic based on the sample data.

$$t = \frac{(\bar{x}_1 - \bar{x}_2) - (\mu_1 - \mu_2)}{\sqrt{\dfrac{s_1^2}{n_1} + \dfrac{s_2^2}{n_2}}} = \frac{(4.3 - 4.6) - 0}{\sqrt{\dfrac{1.1^2}{11} + \dfrac{0.4^2}{15}}} = -0.8636$$

This is a two-tailed test with $\alpha = 0.05$ and 10 degrees of freedom, so the critical t values obtained from Table A-4 are $t = 2.228$ and $t = -2.228$. The computed t value of -0.8636 does not fall within the critical region and we fail to reject the null hypothesis of equal means. Based on the available sample data, it appears that the two brands of calculators have equal mean lives.

The next example illustrates a hypothesis test comparing two means for a situation featuring the following characteristics:

- The two samples come from independent and normal populations.
- σ_1 and σ_2 are unknown.
- Both sample sizes are small (≤ 30).
- The sample variances suggest, through the F test, that $\sigma_1 = \sigma_2$.

The conditions inherent in this next example cause us to follow the path leading to the bottom of the flow chart in Figure 8-6.

Example

A psychologist experimenting with two methods of teaching obtains the following sample results from two separate groups using the two methods.

Method A	Method B
$n_1 = 16$	$n_2 = 12$
$\bar{x}_1 = 73.2$	$\bar{x}_2 = 80.7$
$s_1 = 12.8$	$s_2 = 8.5$

At the 0.05 significance level, test the claim that both methods produce the same mean score.

Solution

Referring to Figure 8-6 we begin by questioning the independence of the two populations and conclude that they are independent because separate groups of subjects are used. We continue with the flow chart by noting that σ_1 and σ_2 are not known and that neither n_1 nor n_2 exceeds 30. At this stage, the flow chart brings us to the last diamond, which requires application of the F test. With H_0: $\sigma_1^2 = \sigma_2^2$, H_1: $\sigma_1^2 \neq \sigma_2^2$, and $\alpha = 0.05$, we compute

$$F = \frac{s_1^2}{s_2^2} = \frac{12.8^2}{8.5^2} \overset{\circ}{=} 2.268$$

With $\alpha = 0.05$ in a two-tailed F test and with 15 and 11 as the degrees of freedom for the numerator and the denominator, respectively, we use Table A-6 to obtain the critical F value of 3.3299. Since the computed test statistic of 2.268 is not within the critical region, we fail to reject the null hypothesis of equal variances (or standard deviations). Leaving the last diamond in Figure 8-6, we follow the yes path and apply the required t test as follows:

$$H_0: \mu_1 = \mu_2$$

$$H_1: \mu_1 \neq \mu_2$$

$$\alpha = 0.05$$

$$t = \frac{(\bar{x}_1 - \bar{x}_2) - (\mu_1 - \mu_2)}{\sqrt{\frac{1}{n_1} + \frac{1}{n_2}} \sqrt{\frac{(n_1 - 1)s_1^2 + (n_2 - 1)s_2^2}{n_1 + n_2 - 2}}}$$

$$= \frac{(73.2 - 80.7) - 0}{\sqrt{\frac{1}{16} + \frac{1}{12}} \sqrt{\frac{(16 - 1)(12.8)^2 + (12 - 1)(8.5)^2}{16 + 12 - 2}}} = -1.756$$

With $\alpha = 0.05$ in this two-tailed t test, and with $n_1 + n_2 - 2$ or $16 + 12 - 2 = 26$ degrees of freedom, we obtain critical t values of 2.056 and -2.056. The computed test statistic of -1.756 does not fall in the critical region, so we fail to reject the null hypothesis of equal means. The populations appear to have equal means.

One of the most difficult aspects of tests comparing two means is the determination of the correct test that should be used. Careful and consistent use of Figure 8-6 should help you avoid the common error of using the wrong procedures for a situation involving a hypothesis test. There is a danger of being overwhelmed by the overall complexity of the work when you deal with the five different cases considered here. However, you can use Figure 8-6 to decompose a complex problem into simpler components which can be treated individually.

It is not practical to outline in full detail the derivations leading to the general test statistics given in Figure 8-6, but we can give some reasons for their existence. We have already discussed the case for dependent populations having normal distributions. Actual experiments and mathematical derivations show that, in repeated random samplings from two normal and dependent populations, the values of $\bar{d}$ possess a student t distribution with mean μ_d and standard deviation s_d/n.

Two cases of Figure 8-6 lead to the normal distribution with the test statistic given by

$$z = \frac{(\bar{x}_1 - \bar{x}_2) - (\mu_1 - \mu_2)}{\sqrt{\sigma_1^2/n_1 + \sigma_2^2/n_2}}$$

This expression is essentially an application of the Central Limit Theorem, which tells us that sample means $\bar{x}$ will, as the sample size increases, tend to approach a normal distribution with mean μ and standard deviation $\sigma/\sqrt{n}$.

In Section 5-6 we saw that, when samples are size 30 or larger, the normal distribution serves as a reasonable approximation to the distribution of sample means. By similar reasoning, the values of $\bar{x}_1 - \bar{x}_2$ also tend to approach a normal distribution with mean $\mu_1 - \mu_2$. When both samples are large, we conclude that the values of $\bar{x}_1 - \bar{x}_2$ will have a standard deviation of

$$\sqrt{\frac{\sigma_1^2}{n_1} + \frac{\sigma_2^2}{n_2}}$$

by using a property of variances: **the variance of the differences between two random variables tends to equal the variance of the first random variable plus the variance of the second random variable.** That is, the variance of values of $\bar{x}_1 - \bar{x}_2$ will tend to equal $\sigma_{\bar{x}_1}^2 + \sigma_{\bar{x}_2}^2$. This is a difficult concept, and we therefore illustrate it by the specific data given in Table 8-2. The x and y scores were randomly selected as the last digits of numbers in a telephone book. The variance of the x values is 7.68, the variance of the y values is 8.48, and the variance of the $x - y$ values is 17.22.

$$s_x^2 = 7.68$$
$$s_y^2 = 8.48 \longrightarrow s_{x-y}^2 \stackrel{\circ}{=} s_x^2 + s_y^2$$
$$s_{x-y}^2 = 17.22$$

We can see that $s_x^2 + s_y^2$ is roughly equal to s_{x-y}^2. By comparing the x, y and $x - y$ values, we see that the variance is largest for the $x - y$ values. The x values range from 0 to 9, the y values range from 0 to 9, but the $x - y$ values exhibit greater variation by ranging from -8 to 9. If our x, y, and $x - y$ sample sizes were much larger than the sample sizes of 36 for Table 8-2, we would approach these theoretical values: $s_x^2 = 8.25$, $s_y^2 = 8.25$, and $s_{x-y}^2 = 16.50$.

These theoretical values illustrate that $\sigma_{x-y}^2 = \sigma_x^2 + \sigma_y^2$. When we deal with means of large random sample sizes, the Central Limit Theorem indicates that the standard deviation of those sample means approach $\sigma/\sqrt{n}$, so the variance will approach σ^2/n.

We can now combine our additive property of variances with the Central Limit Theorem's expression for variance of sample means to obtain the following result:

$$\sigma_{\bar{x}_1 - \bar{x}_2}^2 = \sigma_{\bar{x}_1}^2 + \sigma_{\bar{x}_2}^2 = \frac{\sigma_1^2}{n_1} + \frac{\sigma_2^2}{n_2}$$

In the preceding expression, we assume that we have Population 1 with variance σ_1^2 and Population 2 with variance σ_2^2. Samples of size n_1 are ran-

TABLE 8-2

x	y	$x - y$
1	7	-6
3	0	3
1	0	1
3	4	-1
2	6	-4
8	3	5
3	9	-6
3	7	-4
9	0	9
1	1	0
4	2	2
9	4	5
4	9	-5
9	6	3
5	2	3
6	7	-1
5	6	-1
5	8	-3
3	3	0
3	6	-3
1	3	-2
0	2	-2
4	4	0
3	1	2
3	9	-6
6	4	2
9	5	4
5	1	4
2	2	0
9	6	3
2	2	0
6	1	5
1	9	-8
7	0	7
9	1	8
6	5	1

domly drawn from Population 1 and the means $\bar{x}_1$ computed. The same is done for Population 2. $\sigma^2_{\bar{x}_1 - \bar{x}_2}$ denotes the variance of $\bar{x}_1 - \bar{x}_2$ values. This result shows that the standard deviation of $\bar{x}_1 - \bar{x}_2$ values is

$$\sqrt{\frac{\sigma^2_1}{n_1} + \frac{\sigma^2_2}{n_2}}$$

Since z is a standard score that corresponds in general to

$$z = \frac{\text{(sample statistic)} - \text{(population mean)}}{\text{(population standard deviation)}}$$

we get

$$z = \frac{(\bar{x}_1 - \bar{x}_2) - (\mu_1 - \mu_2)}{\sqrt{\sigma^2_1/n_1 + \sigma^2_2/n_2}}$$

by noting that the sample values of $\bar{x}_1 - \bar{x}_2$ will have a mean of $\mu_1 - \mu_2$ and the standard deviation just given. The same reasons that lead to the preceding expression can be used to justify the test statistic

$$t = \frac{(\bar{x}_1 - \bar{x}_2) - (\mu_1 - \mu_2)}{\sqrt{\frac{s^2_1}{n_1} + \frac{s^2_2}{n_2}}}$$

which is used in tests of hypotheses about two means coming from independent and normal populations when the samples are small and the two population variances appear to be different. In such cases, the distribution of $\bar{x}_1 - \bar{x}_2$ values is a student t distribution with mean $\mu_1 - \mu_2$ and standard deviation

$$\sqrt{\frac{s^2_1}{n_1} + \frac{s^2_2}{n_2}}$$

which reflects the additive property of variances already described.

The test statistic located at the bottom of Figure 8-6 is appropriate for tests of hypotheses about two means from independent and normal populations when the samples are small and the population variances appear to be equal. The numerator of $(\bar{x}_1 - \bar{x}_2) - (\mu_1 - \mu_2)$ again describes the difference between the sample statistic $(\bar{x}_1 - \bar{x}_2)$ and the mean of all such values $(\mu_1 - \mu_2)$. The denominator represents the standard deviation of $(\bar{x}_1 - \bar{x}_2)$ values, which come from repeated sampling when the stated assumptions are satisfied. Since these assumptions include equality of population variances, the denominator should be an estimate of

$$\sqrt{\frac{\sigma^2}{n_1} + \frac{\sigma^2}{n_2}} = \sigma \sqrt{\frac{1}{n_1} + \frac{1}{n_2}}$$

Both n_1 and n_2 will be known, so we need to estimate only σ and we pool the sample variances to get the best possible estimate. The pooled estimate of σ is

IF YOU'RE A RAT, DON'T RIDE SUBWAYS

Scientists have conducted an experiment in which rats prone to high blood pressure were put in an environment designed to simulate a ride on a New York City subway. Each morning and afternoon these hapless rats were put in plastic containers that shook and jerked to the tape-recorded sounds of subway noises. In a control group, luckier rats were put in similar containers, but they were not shaken or exposed to the noise. After 4 months, 4 of the 25 subway rats died, while none of the 24 rats in the control group died. Researchers concluded that the difference in the number of deaths is highly unlikely to have occurred by chance. It was more likely caused by aggravated high blood pressure or hypertension.

$$\sqrt{\frac{(n_1 - 1)s_1^2 + (n_2 - 1)s_2^2}{n_1 + n_2 - 2}}$$

where the expression under the square root sign is essentially a weighted average of s_1^2 and s_2^2. (Weights of $n_1 - 1$ and $n_2 - 1$ are used.) See Exercise 8-50.

EXERCISES A: TESTS COMPARING TWO MEANS

8-21. A pill designed to lower systolic blood pressure is administered to 10 randomly selected volunteers. The results follow. At the $\alpha = 0.05$ significance level, test the claim that systolic blood pressure is not affected by the pill.

Before pill	120	136	160	98	115	110	180	190	138	128
After pill	118	122	143	105	98	98	180	175	105	112

8-22. A test of driving ability is given to a random sample of 10 student drivers before and after they completed a formal driver education course. The results follow. At the $\alpha = 0.05$ significance level, test the claim that the mean score is not affected by the course.

Score before course	100	121	93	146	101	109	149	130	127	120
Score after course	136	129	125	150	110	138	136	130	125	129

$-26 \quad -8 \quad -32 \quad -4 \quad -9 \quad -29 \quad +13 \quad 0 \quad +2 \quad -9$

$d = -11.2$

8-23. Fifty students from two separate colleges are randomly selected and their grade point averages computed. College *A* produces a sample mean of 2.78 and a sample standard deviation of 0.40, while College *B* yields a sample mean and standard deviation of 3.21 and 0.22, respectively. At the $\alpha = 0.05$ significance level, test the claim that the mean grade point averages of both schools are equal.

8-24. A sample of 75 randomly selected newborn babies produce the following results. At the $\alpha = 0.05$ significance level, test the claim that there is no difference between the mean weights of newborn boys and girls.

Boys	Girls
$n_1 = 40$	$n_2 = 35$
$\bar{x}_1 = 3.39$ kilograms	$\bar{x}_2 = 3.18$ kilograms
$s_1 = 0.44$ kilogram	$s_2 = 0.53$ kilogram

8-25. Two separate I.Q. tests are designed so that the first produces a standard deviation of 15 while the second produces a standard deviation of 16.5. Twenty-five randomly selected subjects are given the first test and the resulting mean is 103.5. Thirty different subjects are randomly selected and given the second test, and their mean score is 111.1. At the 0.05 significance level, test the claim that both tests produce the same mean score.

8-26. Two machines pour beer into liter containers in such a way that the standard deviations of the contents are 0.10 liter for the first machine and 0.15 liter for the second machine. A sample of 20 containers from the first machine is found to have a mean volume of 1.00 liter, while a sample of 15 containers from the second machine has a mean volume of 0.95 liter. At the 0.05 significance level, test the claim that the mean volumes put out by both machines are equal.

8-27. Samples of two different car models are tested for fuel economy by determining the miles traveled using 1 gallon of gas. The results follow. At the 0.05 significance level, test the claim that the means of the number of miles traveled by the two models are equal.

Car A	Car B
$n_1 = 10$	$n_2 = 12$
$\bar{x}_1 = 27.2$ miles	$\bar{x}_2 = 31.6$ miles
$s_1 = 4.1$ miles	$s_2 = 2.1$ miles

8-28. Do Exercise 8-27 after changing s_1 to 3.7 miles.

8-29. Cigarettes randomly selected from two brands are analyzed to determine their nicotine content. The results follow. At the $\alpha = 0.05$ significance level, test the claim that both brands have equal nicotine content.

Brand X	Brand Y
$n_1 = 20$	$n_2 = 40$
$\bar{x}_1 = 35.1$ milligrams	$\bar{x}_2 = 36.5$ milligrams
$s_1 = 4.2$ milligrams	$s_2 = 3.3$ milligrams

8-30. Two procedures are used for controlling the aircraft traffic at an airport. Sample results based on each of the two procedures follow. At the 0.05 significance level, test the claim that the use of System 2 results in a mean number of operations per hour exceeding the mean for System 1.

System 1	System 2
$n_1 = 24$ (hours)	$n_2 = 24$ (hours)
$\bar{x}_1 = 63.0$ (operations per hour)	$\bar{x}_2 = 60.1$ (operations per hour)
$s_1 = 5.2$ (operations per hour)	$s_2 = 3.2$ (operations per hour)

8-31. A student hears that fish is a "brain food" that helps to make people more intelligent. He participates in an experiment involving 12 randomly selected volunteers who take an I.Q. test and then begin a diet consisting solely of fish. A second I.Q. test is given at the end of the experiment and the results are listed in the following table. At the 0.05 significance level, test the claim that the fish diet has no effect on I.Q. scores.

I.Q. score before diet	98 110 105 121 100 88 112 92 99 109 103 104
I.Q. score after diet	98 112 106 118 102 97 115 90 99 110 105 109

8-32. Samples of two competing cold medicines are tested for the amount of acetaminophelan, and the results follow. At the 0.05 significance level, test the claim that the mean amount of acetaminophelan is the same in each brand.

Brand X	Brand Y
$n = 25$	$n = 35$
$\bar{x} = 503$ milligrams	$\bar{x} = 520$ milligrams
$s = 14$ milligrams	$s = 18$ milligrams

8-33. A small commuter airline owns two jets and has subcontracted its maintenance operations to two separate firms. Samples of the monthly down times for these subcontractors follow. At the 0.05 significance level, test the claim that the mean monthly down times of both firms are equal.

Firm A	Firm B
$n = 12$	$n = 14$
$\bar{x} = 14.4$ hours	$\bar{x} = 16.3$ hours
$s = 3.1$ hours	$s = 1.6$ hours

8-34. In writing an anti-union article, a management consultant presents statistics that purportedly show nonunion masons are more productive than their union counterparts. At the 0.05 significance level, test the consultant's claim that the mean number of bricks set in 1 hour by nonunion workers exceeds the corresponding mean for union masons. The following data are based on random samples consisting of bricks set in 1 hour.

Nonunion	Union
$n = 15$	$n = 10$
$\bar{x} = 24.3$	$\bar{x} = 23.3$
$s = 3.6$	$s = 1.8$

8-35. Fifteen randomly selected adult females are given I.Q. tests, and 15 randomly selected adult males are given the same tests with the following results. At the 0.05 significance level, test the claim that the mean score for females is higher than that for males.

Males	Females
$\bar{x} = 96.8$	$\bar{x} = 103.5$
$s = 14.2$	$s = 16.3$

8-36. A course is designed to increase readers' speed and comprehension. To evaluate the effectiveness of this course, a test is given both before and after the course, and sample results follow. At the 0.05 significance level, test the claim that the scores are higher after the course.

Before	100	110	135	167	200	118	127	95	112	116
After	136	160	120	169	200	140	163	101	138	129

8-37. A college entrance examination is given to seniors from two different high schools and the sample results follow. At the 0.05 significance level, test the claim that the mean scores of the two groups are equal.

School 1	School 2
$n = 35$	$n = 50$
$\bar{x} = 421$	$\bar{x} = 347$
$s = 122$	$s = 85$

8-38. The following chart lists a random sampling of the ages of married couples. The age of the husband is listed above the age of his wife. At the 0.05 significance level, test the claim that there is no difference between the mean ages of husbands and wives.

Husband's age	28.1	33.0	29.8	53.1	56.7	41.6	50.6	21.4	62.0	19.7
Wife's age	28.4	27.6	32.7	52.0	58.1	41.2	50.7	20.6	61.1	18.1

8-39. A manufacturer produces two models of car batteries. Sample results of each type follow. Assuming that Model A and Model B have standard deviations of 1 month and 2 months, respectively, test the claim that the mean lives of each type are equal. Use a 0.05 level of significance.

Model A	Model B
$n = 15$	$n = 18$
$\bar{x} = 38$ months	$\bar{x} = 35$ months

8-40. A hearing sensitivity test is given to two groups of employees. Group A consists of clerical personnel working in a quiet environment, while Group B consists of assembly workers constantly exposed to loud noises. The sample results follow. At the 0.05 significance level, test the claim that there is no difference between the two population means.

Group A	Group B
$n = 50$	$n = 40$
$\bar{x} = 76$	$\bar{x} = 64$
$s = 12$	$s = 16$

8-41. Two types of string are tested for strength and the sample data follow. At the 0.05 significance level, test the claim that there is no difference between the two types of string.

Breaking load in kilograms

String A	23	25	25	28	19	31	35	30	26			
String B	18	17	16	24	20	21	25	15	15	16	18	21

8-42. Researchers study commercial air filtering systems for noise pollution with sample results as follows. At the 5% level of significance,
test the claim that there is no difference in the mean noise levels.

Unit A	Unit B
$n = 8$	$n = 6$
$\bar{x} = 87.5$ decibels	$\bar{x} = 91.3$ decibels
$s = 0.8$ decibel	$s = 1.1$ decibels

8-43. Random samples of two brands of snow tires are tested for stopping
distances under standardized ice conditions and the results follow. At
the 5% level of significance, test the claim that there is no difference
between the two brands.

Badmonth	Brimstone
$n = 36$	$n = 20$
$\bar{x} = 42.7$ meters	$\bar{x} = 46.3$ meters
$s = 6.1$ meters	$s = 6.2$ meters

8-44. At the 5% level of significance, test the claim that the sample x and y
values come from populations having equal means. Assume that the
populations are independent.

x	1	2	2	3	5	6	8	4	6	7	2	5	3	2
y	2	0	1	4	4	4	9	3	5	7	2	6	2	1

8-45. Do Exercise 8-44 assuming that the two populations are dependent
and normal.

8-46. Stores and theaters are randomly selected and their inside temperatures are measured in degrees Celsius. At the 5% level of significance, test the claim that theaters are warmer than stores. Use the
sample data that follow.

Stores	Theaters
$n = 40$	$n = 32$
$\bar{x} = 18.3$	$\bar{x} = 22.2$
$s = 0.8$	$s = 0.9$

8-47. Ten randomly selected volunteers test a new diet with the following results. At the 5% level of significance, test the claim that the diet is effective. All of the weights are given in kilograms.

Subject	A	B	C	D	E	F	G	H	I	J
Weight before diet	68	54	59	60	57	62	62	65	88	76
Weight after diet	65	52	52	60	58	59	60	63	78	75

EXERCISES B: TESTS COMPARING TWO MEANS

8-48. When testing hypotheses about two population means, under what conditions do we require the populations to have normal distributions, and when can we ignore that requirement? (See Figure 8-6.)

8-49. Use the x and y data of Table 8-2 to find the variance for the values of $x + y$. How does that variance compare to s_x^2 and s_y^2?

8-50. If

$$s_1^2 = \frac{\Sigma(x_1 - \bar{x}_1)^2}{n_1 - 1}$$

and

$$s_2^2 = \frac{\Sigma(x_2 - \bar{x}_2)^2}{n_2 - 1}$$

show that

$$s^2 = \frac{\Sigma(x_1 - \bar{x}_1)^2 + \Sigma(x_2 - \bar{x}_2)^2}{n_1 + n_2 - 2}$$

is equivalent to

$$s = \sqrt{\frac{(n_1 - 1)s_1^2 + (n_2 - 1)s_2^2}{n_1 + n_2 - 2}}$$

This result is used in developing the pooled estimate of σ for the test statistic found at the bottom of Figure 8-6.

8-4   TESTS COMPARING TWO PROPORTIONS

Here we consider tests of hypotheses made about two population proportions. The concepts and procedures we develop can be used to answer questions such as:

Is there a difference between the proportion of Democrats favoring capital punishment and the proportion of Republicans favoring capital punishment?

Is there a difference between the proportions of men and women in management positions?

Is there a difference between the percentage of students who passed mathematics courses and the percentage of students who passed history courses?

Throughout this section we assume that our sample data come from two independent populations and that both samples are large.

NOTATION

For Population 1 we let:

p_1 denote the population proportion

n_1 denote the size of the sample

x_1 denote the number of successes.

The corresponding meanings are attached to p_2, n_2, and x_2, which come from Population 2.

x_1/n_1 represents the sample estimate of p_1, and x_2/n_2 similarly represents the sample estimate of p_2. We know from Chapter 5 that such proportions have a distribution that is approximately normal, and the differences

$$\frac{x_1}{n_1} - \frac{x_2}{n_2}$$

also have a distribution that is approximately normal. Since the means of x_1/n_1 and x_2/n_2 are p_1 and p_2, respectively, it follows that the mean of the differences $x_1/n_1 - x_2/n_2$ will be $p_1 - p_2$. In Section 7-3, we established the fact that sample proportions x/n have a standard deviation of $\sqrt{pq/n}$, and this implies that the variance of the sample proportion x_1/n_1 is p_1q_1/n_1. By similar reasoning, the variance of the x_2/n_2 sample values is p_2q_2/n_2. In Section 8-3, we established the fact that the variance of the differences between two random variables is the sum of their individual variances, and we use this property to get

$$\sigma^2_{p_1-p_2} = \sigma^2_{p_1} + \sigma^2_{p_2} = \frac{p_1q_1}{n_1} + \frac{p_2q_2}{n_2}$$

or

$$\sigma_{p_1-p_2} = \sqrt{\frac{p_1q_1}{n_1} + \frac{p_2q_2}{n_2}}$$

However, this expression is often useless since we usually do not know the values of p_1, p_2, q_1, and q_2. But if we assume that $p_1 = p_2$, we can estimate their common value by pooling the sample data.

POLIO EXPERIMENT

*In 1954 a vast medical experiment was conducted to test the effectiveness of the Salk vaccine as a protection against the devastating effects of polio. Previously developed polio vaccines had been used, but it was discovered that some of those earlier vaccinations actually **caused** paralytic polio. Researchers justifiably developed a cautious and conservative approach to approving new vaccines for general use, and they decided to conduct a large-scale experiment of the Salk vaccine with volunteers.*

*Because of a variety of physiological factors, the vaccine could not be 100% effective, so its effectiveness had to be proved by a lowered incidence of polio among innoculated children. It was hoped that this experiment would result in a **significantly** lower incidence of polio among vaccinated children, and that the decrease would be so significant as to be overwhelmingly convincing.*

The number of children involved in this experiment was necessarily large because a small sample would not provide the conclusive evidence required by these cautious researchers. Approximately 200,000 children were injected with an ineffective salt solution, while 200,000 other children were injected with the Salk vaccine. Assignments of the real vaccine and the useless salt solution were made on a random basis. The children being injected did not know whether they were given the real vaccine or the salt solution. Even the doctors giving the injections and evaluating subsequent results did not know which injections contained the real Salk vaccine. Only 33 of the 200,000 vaccinated children later developed paralytic polio, while 115 of the 200,000 children injected with the salt solution later developed paralytic polio. Statistical analysis of these and other results led to the conclusion that the Salk vaccine was indeed effective against paralytic polio.

DEFINITION

> *The **pooled estimate of p_1 and p_2** is denoted by p and is given by*
>
> $$\bar{p} = \frac{x_1 + x_2}{n_1 + n_2}$$

DEFINITION

> $\bar{q} = 1 - \bar{p}.$

Using this pooled estimate in place of p_1 and p_2, we find that the standard deviation of the differences between the sample proportions becomes

$$\sigma_{p_1 - p_2} = \sqrt{\frac{\bar{p}\bar{q}}{n_1} + \frac{\bar{p}\bar{q}}{n_2}} = \sqrt{\bar{p}\bar{q}\left(\frac{1}{n_1} + \frac{1}{n_2}\right)}$$

We now know that the sample proportion differences $x_1/n_1 - x_2/n_2$ have a distribution that is approximately normal with mean $p_1 - p_2$ and standard deviation $\sigma_{p_1-p_2}$ as estimated. The test statistic appropriate for a test of a hypothesis made about the difference between two population proportions is therefore

$$z = \frac{\left(\dfrac{x_1}{n_1} - \dfrac{x_2}{n_2}\right) - (p_1 - p_2)}{\sqrt{\bar{p}\bar{q}\left(\dfrac{1}{n_1} + \dfrac{1}{n_2}\right)}}$$

DRUG HELPFUL IN CURING CANCER

A drug called fluorouracil has been used in conjunction with surgery in treating cancer patients. Doctors have claimed that the use of the drug has led to a significantly improved survival rate for patients afflicted with cancer of the colon and rectum. In a sample of 213 patients treated by surgery and fluorouracil, 58% were cured for at least 5 years. For patients treated by surgery alone, the usual 5 year cure rate is 24%.

We consider only hypotheses that lead to a null hypothesis of $p_1 = p_2$, or $p_1 \geq p_2$, or $p_1 \leq p_2$, so our test will be conducted under the assumption that $p_1 = p_2$ or $p_1 - p_2 = 0$. The preceding test statistic therefore simplifies to

$$z = \frac{\dfrac{x_1}{n_1} - \dfrac{x_2}{n_2}}{\sqrt{\bar{p}\bar{q}\left(\dfrac{1}{n_1} + \dfrac{1}{n_2}\right)}}$$

Example   A sample of 200 New York State voters included 88 Republicans, while a sample of 300 California voters produced 143 Republicans. At the $\alpha = 0.05$ significance level, test the claim that there is no difference between the proportions of New York Republicans and California Republicans.

The given data can be summarized as follows:

New York	California
$n_1 = 200$	$n_2 = 300$
$x_1 = 88$	$x_2 = 143$

Solution   From these data we compute

$$\frac{x_1}{n_1} = \frac{88}{200} = 0.440$$

$$\frac{x_2}{n_2} = \frac{143}{300} \overset{\circ}{=} 0.477$$

$$\bar{p} = \frac{x_1 + x_2}{n_1 + n_2} = \frac{88 + 143}{200 + 300} = \frac{231}{500} = 0.462$$

$$\bar{q} = 1 - \bar{p} = 1 - 0.462 = 0.538$$

With H_0: $p_1 = p_2$ and H_1: $p_1 \neq p_2$ and the significance level set at 0.05, we compute the test statistic based on the sample data.

$$z = \frac{\dfrac{x_1}{n_1} - \dfrac{x_2}{n_2}}{\sqrt{\bar{p}\bar{q}\left(\dfrac{1}{n_1} + \dfrac{1}{n_2}\right)}} = \frac{0.440 - 0.477}{\sqrt{(0.462)(0.538)\left(\dfrac{1}{200} + \dfrac{1}{300}\right)}} \overset{\circ}{=} -0.813$$

But with $\alpha = 0.05$ in this two-tailed test, the critical z values of 1.96 and -1.96 indicate that the test statistic is not in the critical region. As a result, we fail to reject the null hypothesis of equal proportions. The proportions of Republicans in both states appear to be equal.

The symbols x_1, x_2, n_1, n_2, $\bar{p}$, and $\bar{q}$ should have become more meaningful through this example. In particular, you should recognize that, under the assumption of equal proportions, the best estimate of the common proportion is obtained by pooling both samples into one larger sample. Then

$$\bar{p} = \frac{x_1 + x_2}{n_1 + n_2}$$

becomes a more obvious estimate of the common population proportion.

So far we have discussed only proportions in this section, but probabilities are already in decimal or fractional form, so they can directly replace proportions in the preceding discussion. Percentages can also be accommodated by using the corresponding decimal equivalents as illustrated in the following example.

Example

In a given country, 44% of the rural and suburban residents smoke, while 52% of the urbanites smoke. If that conclusion were made solely by projection of samples consisting of 150 rural and suburban residents and 400 city dwellers, test the claim that the urbanites comprise a larger proportion of smokers. Use a significance level of 0.01.

Solution

From the given data we extract the following.

Rural-Suburban	Urban
$n_1 = 150$	$n_2 = 400$
$x_1 = $ 44% of 150 = 66	$x_2 = $ 52% of 400 = 208

With $x_1/n_1 = 0.44$ and $x_2/n_2 = 0.52$, we compute $\bar{p}$ and $\bar{q}$.

$$\bar{p} = \frac{x_1 + x_2}{n_1 + n_2} = \frac{66 + 208}{150 + 400} = \frac{274}{550} = 0.498$$

and

$$\bar{q} = 1 - \bar{p} = 0.502$$

The claim of a greater proportion of smokers in the city suggests that H_1 be $p_1 < p_2$ so that H_0 becomes $p_1 \geq p_2$. With α already set at 0.01, we continue by computing the test statistic based on the sample data.

$$z = \frac{\dfrac{x_1}{n_1} - \dfrac{x_2}{n_2}}{\sqrt{\bar{p}\bar{q}\left(\dfrac{1}{n_1} + \dfrac{1}{n_2}\right)}} = \frac{0.44 - 0.52}{\sqrt{(0.498)(0.502)\left(\dfrac{1}{150} + \dfrac{1}{400}\right)}} \stackrel{\circ}{=} -1.67$$

With $\alpha = 0.01$ in this left-tailed test, we use Table A-3 to obtain the critical z value of -2.33. Since the test statistic does not fall in the critical region, we fail to reject the null hypothesis and conclude that there is insufficient evidence to support the claim that the percentage of urban smokers exceeds the percentage of rural and suburban smokers.

EXERCISES A: TESTS COMPARING TWO PROPORTIONS

8-51. In each of the following, use the given sample data to determine the values of n_1, n_2, x_1, x_2, $\bar{p}$, and $\bar{q}$.
  (a) Sample A: Of 200 voters polled, 67 are Democrats.
       Sample B: Of 400 voters polled, 148 are Democrats.
  (b) Sample A: In a sample of 250 adults, 38% smoked during the last week.
       Sample B: In a sample of 300 adults, 46% smoked during the last week.
  (c) Five hundred randomly selected subjects are divided into two equal groups. Twenty percent of the first group passes a physical fitness test, while 24% of the second group passes that same test.
  (d) In a random sample of 300 men and 400 women, exactly 53% of each group believe in life on other planets.
  (e) Forty-two percent of all United States governors favor the passage of a certain bill, while 57% of all United States senators favor passage of the same bill.

8-52. For each part of Exercise 8-51, compute the z test statistic.

8-53. Let samples from two populations be such that $x_1 = 45$, $n_1 = 100$, $x_2 = 115$, and $n_2 = 200$.
  (a) Compute the z test statistic based on the given data.
  (b) If the significance level is 0.05 and the test is two-tailed, find the critical z values.
  (c) Test the claim that the two populations have equal proportions using the significance level of $\alpha = 0.05$.

8-54. Samples taken from two populations yield the data $x_1 = 30$, $n_1 = 250$, $x_2 = 44$, $n_2 = 800$. The hypothesis of equal proportions is to be tested at the $\alpha = 0.02$ significance level.
  (a) Compute the z test statistic based on the data.
  (b) Find the critical z values.
  (c) What conclusion do you reach?

8-55. A manufacturer experiments with two production methods. The first method produces 18 defects of 275 sample items, while the second method produces 27 defects of 320 samples. At the $\alpha = 0.05$ significance level, test the claim that there is no difference between the two proportions of defects.

8-56. At a university, 40% of the 95 faculty members surveyed feel that the library facilities are adequate, while 60% of the 210 students rate the library facilities as adequate. At the $\alpha = 0.01$ significance level, test the claim that the percentage of students who feel that the library facilities are adequate exceeds the percentage for faculty members.

8-57. In a random sample of 9584 licensed drivers between the ages of 18 and 24, there were 10 fatalities due to car accidents in 1 year. In another random sample of 11,316 licensed drivers in the 25-to-34 age group, there were 8 auto fatalities. At the $\alpha = 0.01$ significance level, test the claim that drivers in the 18-to-24 age bracket have a greater proportion of fatal accidents.

8-58. An accreditation committee compares the dropout rates of two similar colleges. Of 250 of the entering freshmen in the first college, the dropout rate is found to be 22%. A random sample of 275 freshmen beginning studies at the second college produces a dropout rate of 18%. At the $\alpha = 0.05$ significance level, test the claim that both colleges have the same dropout rate.

8-59. A study in genetics involves two sample groups. Of 80 people in one group, 23 have blue eyes. Of the 120 people in the second group, 24 have blue eyes. At the $\alpha = 0.05$ significance level, test the claim that the proportion of blue-eyed people is the same in each group.

8-60. A manufacturer produces a new and inexpensive fire detection device. Advertising claims include the assertion that the percentage of defective units is not higher than the corresponding percentage for models made by competitors. Five defective units are found in a sample of 150 randomly selected new units, while the competition produces a 2% defective rate in 400 randomly selected units. At the $\alpha = 0.05$ significance level, test the claim made by the manufacturer of the new units.

8-61. As the election approaches, a candidate for a state office recognizes that she has time to campaign effectively in only one of two important counties. A poll of 600 voters in the first county reveals that 57% favor our candidate, while a poll of 800 voters in the second county shows that 49% favor her. At the $\alpha = 0.05$ significance level, test the claim that the first county has a higher percentage of voters who favor this candidate.

8-62. In the past year, two professors each taught four sections of a statistics course with 25 students in each section. If 12 students of the first professor failed and 18 students of the second professor failed, test the claim that the rate of failure is the same for the two professors. Assume a significance level of $\alpha = 0.05$.

8-63. A drug is tested on 150 randomly selected volunteers. Among the 80 men receiving this drug, there are 12 noticeable reactions. Among the 70 women who take the drug, there are 20 noticeable reactions. At the 0.05 significance level, test the claim that the proportion of noticeable reactions among women is equal to the proportion of noticeable reactions among men.

8-64. Of 5000 income tax returns reviewed from low-income families, 52 were selected for audit. Of 1000 returns reviewed from high-income families, 72 were selected for audit. At the 0.01 level of significance, test the claim that the proportions of audits from the two groups are equal.

8-65. A survey was conducted of randomly selected television viewers to determine the composition of the audience for a certain show. Of 200 people with college educations, 30 indicated that they watched the show. Of 600 people without college educations, 210 indicated that they watched the show. At the 0.01 significance level, test the claim that a greater proportion of viewers did not have a college education.

8-66. Test the claim that the sample proportion of 0.26 does not differ significantly from the sample proportion of 0.20 if both sample sizes are 50. Use a 5% level of significance.

8-67. Test the claim that the sample proportion 0.25 is significantly greater than the sample proportion of 0.20 if both sample sizes are 100. Use a 5% level of significance.

8-68. In initial tests of the Salk vaccine, 33 of 200,000 vaccinated children later developed polio. Of 200,000 children vaccinated with a placebo, 115 later developed polio. At the 1% level of significance, test the claim that the Salk vaccine is effective.

8-69. A test question is answered correctly by 80% of the 200 male respondents and 72% of the 300 female respondents. At the 5% level of significance, test the claim that the proportions of correct responses are equal.

EXERCISES B: TESTS COMPARING TWO PROPORTIONS

Sample data are randomly drawn from three independent populations. The sample sizes and the numbers of successes follow.

8-70.

Population 1	Population 2	Population 3
$n = 100$	$n = 100$	$n = 100$
$x = 40$	$x = 30$	$x = 20$

(a) At the 0.05 significance level, test the claim that $p_1 = p_2$.
(b) At the 0.05 significance level, test the claim that $p_2 = p_3$.
(c) At the 0.05 significance level, test the claim that $p_1 = p_3$.
(d) In general, if hypothesis tests lead to the decisions that $p_1 = p_2$ and $p_2 = p_3$, does it follow that the decision $p_1 = p_3$ will be reached under the same conditions?

REVIEW

In this chapter we extended to two populations the method of testing hypotheses introduced in Chapter 6. We began by developing a test for comparing two population variances (or standard deviations) that come from two independent populations having normal distributions. We began with a test for comparing two variances or standard deviations since such a test is sometimes used as part of an overall test for comparing two population means. In Section 8-2 we saw that the sampling distribution of $F = s_1^2/s_2^2$ is the F distribution for which Table A-6 was computed.

In Section 8-3 we considered various situations that can occur when we want to use a hypothesis test for comparing two means. We should begin such a test by determining whether or not the two populations are **dependent** in the sense that they are related in some way. When comparing population means that come from two dependent and normal populations, we compute the differences between corresponding pairs of values. Those differences have a mean and standard deviation denoted by $\bar{d}$ and s_d, respectively. In repeated random samplings, the values of $\bar{d}$ possess a student t distribution with mean μ_d and standard deviation $s_d/\sqrt{n}$, so our test statistic is obtained from

$$t = \frac{\bar{d} - \mu_d}{s_d/\sqrt{n}}$$

μ_d denotes the difference between the two population means.

When using hypothesis tests to compare two population means from independent populations, we encounter four situations that can be summarized best by Figure 8-6. These cases incorporate standard deviations which reflect the property that, if one random variable x has variance σ_x^2 and another random variable y has variance σ_y^2, the random variable $x - y$ will have variance $\sigma_x^2 + \sigma_y^2$.

In Section 8-4 we considered hypothesis tests that can be used to compare proportions, probabilities, or percentages that come from two independent populations. We limited our discussion to cases involving large samples only. We saw that the sample proportions have differences

$$\frac{x_1}{n_1} - \frac{x_2}{n_2}$$

which tend to have a distribution that is approximately normal with mean $p_1 - p_2$ and a standard deviation estimated by

$$\sqrt{\bar{p}\bar{q}\left(\frac{1}{n_1} + \frac{1}{n_2}\right)}$$

where $\bar{q} = 1 - \bar{p}$ and $\bar{p}$ is the pooled proportion $(x_1 + x_2)/(n_1 + n_2)$.

Figure 8-7 provides a reference chart for locating the appropriate test. One of the most difficult aspects of hypothesis testing involves the identification of the most appropriate distribution and the selection of the proper test statistic.

FIGURE 8-7

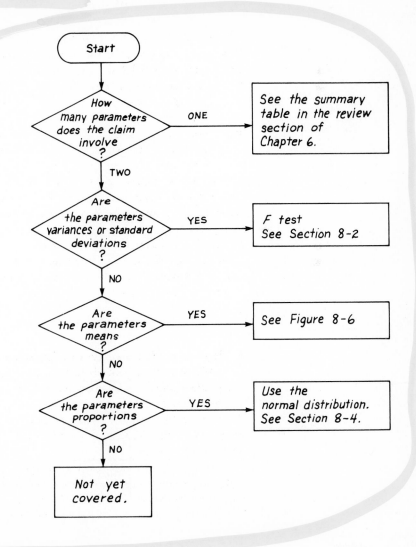

REVIEW EXERCISES

8-71. Samples of similar wires are randomly selected from two different manufacturers and tested for their breaking strengths. The results are as follows:

	Company A	Company B
n	$n = 10$	$n = 12$
$\bar{x}$	$\bar{x} = 82.0$ kilograms	$\bar{x} = 77.6$ kilograms
s	$s = 6.0$ kilograms	$s = 8.1$ kilograms

(a) At the 0.05 significance level, test the claim that the two companies produce wire having the same breaking strength.
(b) At the 0.05 significance level, test the claim that the standard deviations of the breaking strengths of both brands are equal.

8-72. The same aptitude test is given to students randomly selected from schools in two different states and the results are as follows:

	New York	Alabama
n	$n = 208$	$n = 54$
$\bar{x}$	$\bar{x} = 121.0$	$\bar{x} = 103.2$
s	$s = 9.1$	$s = 11.9$

At the 0.05 significance level, test the claim that there is no difference between the mean scores of students from both states.

8-73. A test question is considered good if it discriminates between good and poor students. The first question on a test is answered correctly by 62 of 80 good students while 23 of 50 poor students give correct answers. At the 5% level of significance, test the claim that this question is answered correctly by a greater proportion of good students.

 8-74. Automobiles are selected at random and tested for fuel economy with each of two different carburetors. The following results show the distance traveled on 1 gallon of gas. At the 5% level of significance, test the claim that both carburetors produce the same mean mileage.

Car	1	2	3	4	5	6	7	8	9
Carburetor A	16.1	21.3	19.2	14.8	29.3	20.2	18.6	19.7	16.4
Carburetor B	18.2	23.4	19.7	14.7	28.7	23.4	19.0	21.2	18.2

8-75. To test the effectiveness of a physical training program, researchers asked randomly selected participants to run as far as possible in 5 minutes. This test was conducted before and after the training program and the results follow. The numbers represent distances in meters. At the 5% level of significance, test the claim that the training program was effective.

Before course	510	620	705	590	800	1450	790	830	1220	680
After course	1130	680	810	780	1275	1410	970	1050	1380	1050

8-76. Two different production methods are used to make batteries for hearing aids. Batteries produced by both methods are randomly selected and tested for longevity. Eighteen batteries produced by the first method have a standard deviation of 42 hours, while 12 batteries produced by the second method have a standard deviation of 78 hours. At the 5% level of significance, test the claim that the two production methods yield batteries whose lives have equal standard deviations.

8-77. Two different firms manufacture garage door springs that are designed to produce a tension of 68 kilograms. Random samples are selected from each of these two suppliers and tension test results are as follows:

Firm A	Firm B
$n = 20$	$n = 32$
$\bar{x} = 66.0$ kilograms	$\bar{x} = 68.3$ kilograms
$s = 2.1$ kilograms	$s = 0.4$ kilograms

At the 5% level of significance, test the claim that the mean tension is the same.

8-78. Do Exercise 8-77 after changing the sample size for Firm A to $n = 40$.

8-79. At the 10% level of significance, test the claim that the sample proportion of 0.4 differs significantly from the sample proportion of 0.6. Assume that both sample sizes are 75.

8-80. A poll reveals that 47.0% of 1500 randomly selected voters in Ohio favor a certain candidate, while 48.0% of 500 randomly selected voters in Maryland favor that same candidate. At the 1% level of significance, test the claim that the candidate is favored by the same percentage in both states.

CORRELATION AND REGRESSION

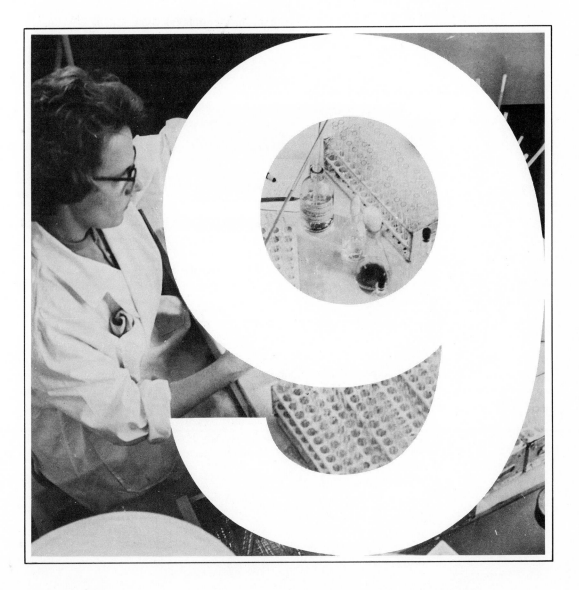

9-1 OVERVIEW

Among the topics we considered in Chapter 8 was a method for testing hypotheses about differences between the means of two populations. Here we investigate ways of analyzing the relationship between two variables. Experience has shown that such investigations are helpful in formulating answers to questions such as the following:

Is there a relationship between a college student's I.Q. and his grade point average?

Is there a relationship between the number of hours you study for a test and the subsequent test grade?

Is there a relationship between a person's age and his or her blood pressure?

What is the relationship between age and blood pressure?

We begin Section 9-2 by describing the **scatter diagram**, which serves as a graph of the sample data. We then investigate the concept of **correlation** which is used to decide whether there is a statistically significant relationship between two variables. Section 9-3 investigates **regression analysis** as we attempt to identify the exact nature of the relationship between two variables. Specifically, we show how to determine an equation that relates the two variables.

Throughout this chapter we deal only with **linear** (straight line) relationships between two variables. (Advanced texts consider more variables and nonlinear relationships.)

As a very simple example, consider the following table.

x	1	2	4	5	7	8	10
y	3	5	9	11	15	17	21

Note that each y value is one more than twice the corresponding x value. This strongly suggests that there exists a very definite relationship between the two variables. In deciding that there is a relationship, we are dealing with the concept of correlation. We can identify exactly what the relationship is. The relationship between x and y can be described by the equation $y = 1 + 2x$. We deal with the concept of regression when we determine what the relationship is.

9-2 CORRELATION

As we mentioned in the overview, this section deals with the concept of correlation and scatter diagrams as tools that help us decide whether a linear relationship exists between two variables. Since these tools are designed to analyze relationships between two variables, the sample data must be collected as paired data. We start with an example.

Is there a linear relationship between homework grades and test grades? We may suspect that those who do well on homework assignments will also do well on tests, while poor homework grades would tend to accompany poor test grades. A case study involving specific results begins with paired data representing homework and test grades for different individuals. Table 9-1 lists these scores for 10 randomly selected students. The first student (M.G.) received homework and test grades of 75 and 70, respectively. These scores of 75 and 70 represent one pair of data. The table lists 10 separate pairs of data corresponding to the 10 different students.

TABLE 9-1

Student	M.G	T.R	A.K.	D.U.	G.B.	G.J.	M.S.	J.J.	K.T.	C.D.
x (homework)	75	88	90	60	50	77	79	72	67	85
y (test)	70	80	78	65	65	92	67	67	94	94

We can often form intuitive and qualitative conclusions about paired data by constructing a **scatter diagram** similar to the one in Figure 9-1, which represents the information in Table 9-1. The points in the figure don't seem to follow any recognizable pattern, so we conclude that there probably isn't any significant linear relationship between the two variables of homework scores and test scores.

FIGURE 9-1

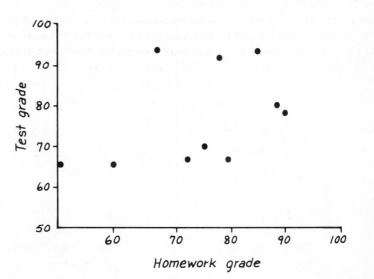

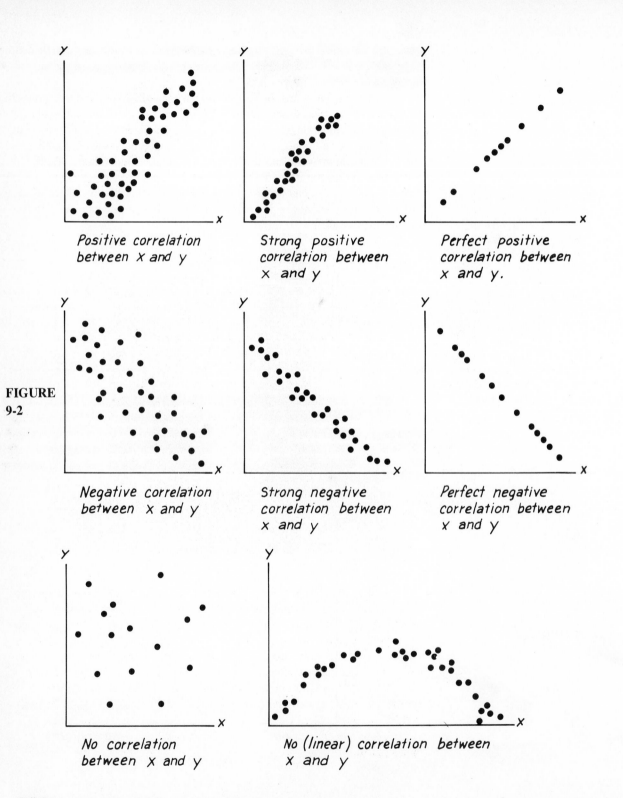

Positive correlation between x and y

Strong positive correlation between x and y

Perfect positive correlation between x and y.

Negative correlation between x and y

Strong negative correlation between x and y

Perfect negative correlation between x and y

No correlation between x and y

No (linear) correlation between x and y

FIGURE 9-2

We can use the concepts of correlation to establish that there is a relationship between the age of a child and the height of a child. Children obviously grow taller as they age, but that relationship is not perfect. The taller of the two brothers shown is one year younger than his shorter brother. The concepts of correlation accommodate exceptions of this type while enabling us to identify the dominating trend.

Using other collections of paired data we may get scatter diagrams similar to the examples illustrated in Figure 9-2. The scatter diagram can be easily plotted and doesn't require complex computations. In addition, a large collection of paired data may exhibit a pattern and become meaningful when displayed in this form. However, these advantages are often offset by the subjective and qualitative nature of the conclusions which may be drawn.

More precise and objective criteria accompany the computation of the **linear correlation coefficient**, which is denoted by *r* and is given in Formula 9-1.

Formula 9-1 $$r = \frac{n\Sigma xy - (\Sigma x)(\Sigma y)}{\sqrt{n(\Sigma x^2) - (\Sigma x)^2} \sqrt{n(\Sigma y^2) - (\Sigma y)^2}}$$

We now describe the way to compute and interpret the linear correlation coefficient r given a list of paired data. Later in this section we present the underlying theory that led to the development of this formula. Before computing the correlation coefficient r for the data of Table 9-1, we make the following notes that are relevant to the components of Formula 9-1.

n denotes the **number of pairs** of data present. In Table 9-1, for example, $n = 10$.

Σ denotes the addition of the items indicated.

Σx denotes the sum of all x scores.

Σx^2 indicates that each x score should be squared, and then those squares are added.

$(\Sigma x)^2$ indicates that the x scores should be added and then the total should be squared. It is extremely important to avoid confusion between Σx^2 and $(\Sigma x)^2$.

Σxy indicates that each x score should be multiplied by its corresponding y score. After obtaining all such products, find their sum.

For the data of Table 9-1, we compute the individual components and then use the results to determine the value of r.

Example   For the sample paired data of Table 9-1 we get;

$n = 10$ since there are 10 pairs of data.

$\Sigma x = 75 + 88 + 90 + 60 + 50 + 77 + 79 + 72 + 67 + 85 = 743$

$\Sigma y = 70 + 80 + 78 + 65 + 65 + 92 + 67 + 67 + 94 + 94 = 772$

$$\begin{aligned}
\Sigma x^2 &= 75^2 + 88^2 + 90^2 + 60^2 + 50^2 + 77^2 + 79^2 + 72^2 + 67^2 + 85^2 \\
&= 5625 + 7744 + 8100 + 3600 + 2500 + 5929 + 6241 + 5184 \\
&\quad + 4489 + 7225 \\
&= 56637
\end{aligned}$$

$$\begin{aligned}
\Sigma y^2 &= 70^2 + 80^2 + 78^2 + 65^2 + 65^2 + 92^2 + 67^2 + 67^2 + 94^2 + 94^2 \\
&= 4900 + 6400 + 6084 + 4225 + 4225 + 8464 + 4489 + 4489 \\
&\quad + 8836 + 8836 \\
&= 60948
\end{aligned}$$

$(\Sigma x)^2 = (743)^2 = 552049$

$(\Sigma y)^2 = (772)^2 = 595984$

$$\begin{aligned}
\Sigma xy &= (75 \times 70) + (88 \times 80) + (90 \times 78) + (60 \times 65) + (50 \times 65) \\
&\quad + (77 \times 92) + (79 \times 67) + (72 \times 67) + (67 \times 94) + (85 \times 94) \\
&= 5250 + 7040 + 7020 + 3900 + 3250 + 7084 + 5293 + 4824 \\
&\quad + 6298 + 7990 \\
&= 57949
\end{aligned}$$

FRANCIS GALTON WAS A BRIGHT CHILD

Sir Francis Galton developed the statistical concept of correlation. On the day before his fifth birthday, he wrote the following letter to his sister:

My dear Adele,

I am four years old and I can read any English book. I can say all the Latin substantives and adjectives and active verbs besides 52 lines of Latin poetry. I can cast up any sum in addition and can multiply by 2, 3, 4, 5, 6, 7, 8, 10. I can also say the pence table. I read French a little and know the clock.

February 15, 1827   *Francis Galton*

Galton went on to write over 200 papers and 15 books, including **Gregariousness in Cattle and Men, Statistical Enquiries into the Efficacy of Prayer,** *and* **Intelligible Signals Between Neighbouring Stars.**

Solution   Using these values, we compute r as follows:

$$r = \frac{n(\Sigma xy) - (\Sigma x)(\Sigma y)}{\sqrt{n(\Sigma x^2) - (\Sigma x)^2}\sqrt{n(\Sigma y^2) - (\Sigma y)^2}}$$

$$= \frac{10(57949) - (743)(772)}{\sqrt{10(56637) - 552049}\sqrt{10(60948) - 595984}}$$

$$= \frac{579490 - 573596}{\sqrt{566370 - 552049}\sqrt{609480 - 595984}}$$

$$= \frac{5894}{\sqrt{14321}\sqrt{13496}}$$

$$\overset{\circ}{=} 0.424$$

Terrific! Now that we have computed the value of the linear correlation coefficient r, what does it mean? The way Formula 9-1 was derived, the computed value of r must always be between -1 and $+1$, inclusive. If a computed linear correlation coefficient has a value greater than $+1$ or less than -1, an error must be present in the computations. A positive linear correlation between x and y is indicated by a value of r near $+1$. A negative linear correlation is indicated by a value of r near -1. If the linear correlation coefficient r is close to 0, we conclude that there is no significant linear correlation between x and y. These vague descriptions of close to 0 and near $+1$ or -1 are made precise by Table A-7, which lists critical values of r for various sample sizes. If the magnitude of the computed linear correlation coefficient r exceeds the critical r value found in Table A-7, we can conclude that there is a statistically significant linear relationship between the variables x and y. If the relationship is found to be significant and the computed r is positive, we say that there is a positive linear correlation. If the relationship is found to be significant and the computed r is negative, we say that there is a negative linear correlation. Table A-7 accommodates the two significance levels of $\alpha = 0.05$ and $\alpha = 0.01$.

Example

Find the critical value of the linear correlation coefficient if we have 10 pairs of data (as in Table 9-1) and the significance level is $\alpha = 0.05$.

Solution

Refer to Table A-7 and locate the critical r value of 0.632 corresponding to $n = 10$ and $\alpha = 0.05$.

Example

From 10 pairs of data, r is computed to be 0.424. What can you conclude at the significance level of $\alpha = 0.05$?

Solution

The magnitude of the computed linear correlation coefficient (0.424) does not exceed the critical r value of 0.632, so we conclude that there is no significant linear relationship between the two variables x and y.

The last example utilizes the data extracted from Table 9-1, and the results indicate that there is no significant linear relationship between homework grades and test grades. This objective and quantitative analysis supports the intuitive conclusion obtained by examining the scatter diagram of Figure 9-1. (A teacher finding such a lack of correlation between homework and test grades may benefit by reconsidering the method of selecting and correcting assignments, along with the construction of tests.)

In the next example we obtain the mean salaries for teachers in elementary and secondary schools, along with the annual per capita beer consumption (in gallons) for Americans. The results are listed in Table 9-2.

TABLE 9-2

Year	1960	1965	1970	1972	1973
Mean teacher salary	5,000	6,200	8,600	9,700	10,200
Per capita beer consumption	24.02	25.46	28.55	29.43	29.68

Figure 9-3 is the scatter diagram of Table 9-2. From Figure 9-3 we tend to develop the general impression that there is a definite positive linear correlation between teachers' salaries and per capita beer consumption. To test this impression, we compute r. From Table 9-2 we determine the following:

$$n = 5$$

$$\Sigma x = 39,700$$

$$\Sigma y = 137.14$$

$$\Sigma x^2 = 335,530,000$$

$$\Sigma y^2 = 3,787.3$$

$$(\Sigma x)^2 = 1,576,090,000$$

$$(\Sigma y)^2 = 18,807.38$$

$$\Sigma xy = 1,111,689$$

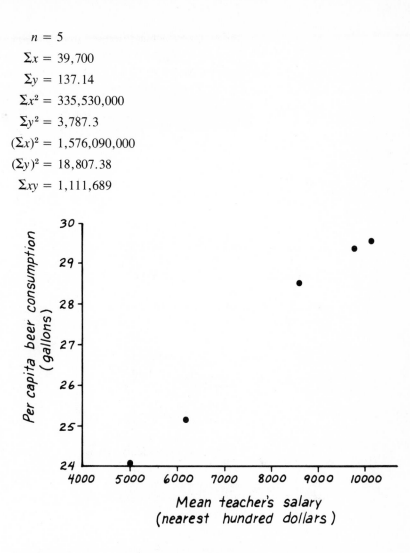

FIGURE 9-3

Now we use the components in Formula 9-1 to get:

$$r = \frac{5(1,111,689) - (39700)(137.14)}{\sqrt{5(335,530,000) - (39700)^2} \ \sqrt{5(3787.3) - (137.14)^2}} \cong 0.995$$

Even if we insist upon the strict $\alpha = 0.01$ significance level, we obtain a critical r value of 0.959 from Table A-7, so our computed r score of 0.995 does indicate a statistically significant linear correlation between teachers' salaries and per capita beer consumption.

The significance of the correlation implies that teachers are using their raises to buy more beer, right? Wrong. Perhaps increases in teachers' salaries precipitate higher taxes, which in turn cause taxpayers to drown their sorrows and forget their financial difficulties by drinking more beer. Or perhaps higher teachers' salaries and greater beer consumption are both manifestations of some other factor, such as general improvement in the standard of living. In any event, the techniques in this chapter can be used only to establish a **statistical** linear relationship. *We cannot establish the existence or absence of any inherent cause-and-effect relationship.*

A medical researcher may establish a significant correlation between the unhealthy habit of smoking and the unhealthy habit of dying. Yet such a correlation does not prove that smoking causes or hastens deaths. Perhaps people become nervous about the prospect of dying and turn to cigarettes as a way of relieving that tension. Maybe dancing puts a strain on the heart that

Medical researchers use statistics to help identify correlations between factors such as cigarette smoking and cancer. While the presence of a strong statistical correlation is not proof of a cause-effect relationship, it is an extremely valuable guide for further research and treatments.

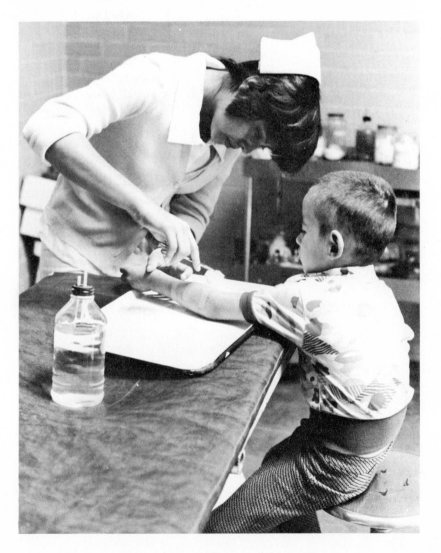

ultimately leads to death and, in the process, creates a biological urge to smoke. Statisticians cannot determine the inherent cause-and-effect relationship, but they can assist and guide the medical researcher in the analysis of the relevant physiological and biological processes.

One misuse of the correlation coefficient involves the concept of linearity. The linear correlation coefficient r, as discussed in this section, is significant only if the paired data follow a linear or straight line pattern. Consider the data of Table 9-3 along with the corresponding scatter diagram of Figure 9-4. Table 9-3 and Figure 9-4 represent nine pairs of data obtained from a physical experiment that consists of shooting an object upward and recording the height of the object at different times after its release. For example, the pair of ''2 seconds: 192 feet'' indicates that the object is at a height of 192 feet exactly 2 seconds after it is shot upward. While Figure 9-4 exhibits a clearly recognizable pattern, the relationship is not linear. The application of Formula 9-1 to the data of Table 9-3 reveals that $r = 0$, an indication that there is no **linear** relationship between the two variables. A nonlinear relationship becomes very obvious when we examine Figure 9-4. (The techniques for dealing with these nonlinear cases are beyond the scope of this text.)

TABLE 9-3

Time (seconds)	0	1	2	3	4	5	6	7	8
Distance (feet) above ground	0	112	192	240	256	240	192	112	0

FIGURE 9-4

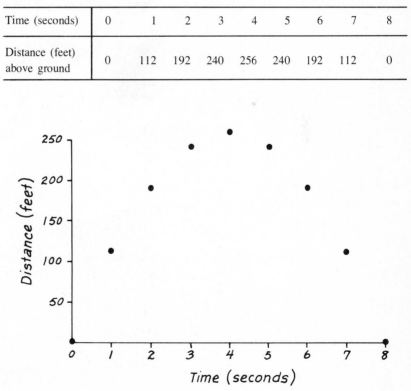

So far we have presented the formula for computing the linear correlation coefficient r, but we have given no justification for it. Formula 9-1 is a simplified form of the equivalent formula

$$r = \frac{\Sigma(x - \bar{x})(y - \bar{y})}{(n - 1)s_x s_y}$$

We could use the preceding formula to compute r, but it would then become necessary to compute $\bar{x}$, $\bar{y}$, s_x, and s_y. While this formula is equivalent to Formula 9-1, we find that Formula 9-1 is generally easier to work with, especially if we use a calculator. (There are several inexpensive calculators that are designed to compute the linear correlation coefficient r directly. The user simply enters the sample data in pairs and then presses the appropriate key to obtain r.)

Formula 9-1 is a short-cut form of the preceding formula for the linear correlation coefficient, but the following discussion refers to this formula since its form relates more directly to underlying theory. We will consider the paired data

x	1	1	2	4	7
y	4	5	8	15	23

that are depicted in the scatter diagram of Figure 9-5. Figure 9-5 includes the point $(\bar{x}, \bar{y})$, which is called the **centroid** of the sample points.

FIGURE 9-5

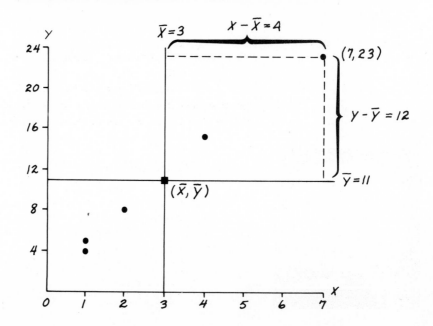

Sometimes r is called **Pearson's product moment**, and that title reflects both the fact that it was first developed by Karl Pearson and that it is based on the product of the moments $(x - \bar{x})$ and $(y - \bar{y})$. That is, Pearson based the measure of scattering on the statistic $\Sigma(x - \bar{x})(y - \bar{y})$. In any scatter diagram, vertical and horizontal lines through the centroid $(\bar{x}, \bar{y})$ divide the diagram into four quadrants (see Figure 9-5). If the points of the scatter diagram tend to approximate an uphill line (as in this figure, then individual values of $(x - \bar{x})(y - \bar{y})$ tend to be positive, since the points are predominantly found in the first and third quadrants where the products of $(x - \bar{x})$ and $(y - \bar{y})$ are positive. If the points of the scatter diagram approximate a downhill line, the points are predominantly in the second and fourth quadrants where $(x - \bar{x})$ and $(y - \bar{y})$ are opposite in sign, so $\Sigma(x - \bar{x})(y - \bar{y})$ tends to be negative. If the points follow no linear pattern, they tend to be scattered among the four quadrants, so $\Sigma(x - \bar{x})(y - \bar{y})$ tends to be close to zero.

The sum $\Sigma(x - \bar{x})(y - \bar{y})$ depends on the magnitude of the numbers used, yet r should not be affected by the particular scale used. For example, r should not change whether heights are measured in meters or centimeters. We therefore standardize the moments $(x - \bar{x})$ and $(y - \bar{y})$ by dividing them by s_x and s_y, respectively. The sum

$$\sum \frac{(x - \bar{x})(y - \bar{y})}{s_x s_y}$$

is not affected by the scales used, but it is affected by the number of pairs of data. We can average out the preceding sum by simply dividing by n, but we divide by $n - 1$ for theoretical reasons similar to those that caused us to use $n - 1$ in sample standard deviation computations. We get

$$r = \frac{\Sigma(x - \bar{x})(y - \bar{y})}{(n - 1)s_x s_y}$$

which can be algebraically manipulated into the equivalent form of Formula 9-1 (see Exercise 9-20).

We can use the linear correlation coefficient to decide whether there is a linear relationship between the two variables. When we decide that such a relationship exists, we can proceed to determine what it is.

HAVE ATOMIC TESTS CAUSED CANCER?

Conquerors was a 1954 movie made in Utah. Recently a team of investigative reporters attempted to locate members of the cast and crew. Of the 79 people they found (some living and some dead), 27 had developed cancer, including John Wayne, Susan Hayward, and Dick Powell. Two key questions arise:

1. *Is a cancer rate of 27 out of 79 significant, or could it be a coincidence?*
2. *Was the cancer caused by fallout from the nuclear tests previously conducted in the area?*

We can use statistics to answer the first question but not the second.

EXERCISES A: CORRELATION

9-1. Would you expect positive correlation, negative correlation, or no correlation for each of the following sets of paired data?
(a) The weights and the lengths of newborn babies.
(b) People's ages and blood pressures.
(c) The amounts of rainfall and vegetation growth.
(d) The weights of cars and fuel consumption rates.
(e) The hat sizes of adults and their I.Q. scores.
(f) The diameters and circumferences of circles.
(g) Hours studied preparing for tests and the resulting test scores.
(h) The numbers of absences in a course and the grades in that course.
(i) The weights and the heights of children.
(j) Annual per capita income for different nations and infant mortality rates for those nations.

In Exercises 9-2 through 9-5, use the given list of paired data to:
(a) *Construct the scatter diagram.*   (b) *Determine* n.
(c) *Find* Σx. 7   (d) *Find* Σx^2.
(e) *Find* $(\Sigma x)^2$. 49   (f) *Find* Σxy.
(g) *Find* r.

9-2. | x | 1 1 2 3 |
      |---|---------|
      | y | 1 5 4 2 |

9-3. | x | 1 2 2 3 |
      |---|---------|
      | y | 5 4 3 1 |

9-4. | x | 0 1 1 2 5 |
      |---|-----------|
      | y | 3 3 4 5 6 |

9-5. | x | 1 3 3 4 5 5 |
      |---|-------------|
      | y | 5 3 2 2 0 1 |

9-6. In each of the following, a sample of paired data produces a linear correlation coefficient r. What do you conclude in each case? Assume a significance level of $\alpha = 0.05$.
(a) $n = 20, r = 0.5$ (b) $n = 20, r = -0.5$
(c) $n = 50, r = 0.2$ (d) $n = 50, r = -0.2$
(e) $n = 37, r = 0.25$ (f) $n = 77, r = 0.35$
(g) $n = 22, r = 0.37$ (h) $n = 22, r = 0.40$
(i) $n = 22, r = -0.5$ (j) $n = 6, r = -0.8$

In Exercises 9-7 through 9-16:

(a) *Construct the scatter diagram.*
(b) *Compute the linear correlation coefficient* r.
(c) *Assume that* $\alpha = 0.05$ *and find the critical value of* r *from Table A-7.*
(d) *Based upon the results of parts (b) and (c), decide whether there is a significant positive linear correlation, a significant negative linear correlation, or no significant linear correlation. In each case, assume a significance level of* $\alpha = 0.05$.

9-7. The following table lists absences and corresponding final exam grades for 20 randomly selected students.

Absences	5	12	1	7	7	40	9	4	10	7	6	7	4	9	2	2	14	3	4	6
Final exam grades	65	60	89	72	55	0	25	60	78	65	60	88	80	85	60	68	50	60	69	84

9-8. The following table lists the grade point averages and corresponding weights of 12 randomly selected students.

Grade point averages	3.6	2.96	3.2	2.63	2.8	2.01	3.25	3.0	3.2	3.0	2.53	2.94
Weights	136	127	148	167	153	148	121	145	155	160	110	130

9-9. The following table lists per capita cigarette consumption in the United States for various years along with the percentage of the population admitted to mental institutions as psychiatric cases.

Years	1950	1955	1960	1965	1969	1970	1971	1972
Cigarette consumption	3522	3597	4171	4258	3993	3971	4042	4053
Percentage of psychiatric admissions (in percentage points)	0.20	0.22	0.23	0.29	0.31	0.33	0.33	0.32

9-10. Randomly selected subjects are given a standard I.Q. test and then tested for their receptivity to hypnosis. The results are listed in the following table.

I.Q.	103	113	119	107	78	153	114	101	103	111	105	82	110	90	92
Receptivity to hypnosis	55	55	59	64	45	72	42	63	62	46	41	49	57	52	41

9-11. A farmer notices that crickets seem to chirp faster on warm days. The accompanying table lists the number of chirps made by a cricket in 1 minute, along with the corresponding temperature (degrees Fahrenheit).

Chirps per minute	66	58	94	120	83	119	121	65	55	50
Temperature	55	53	62	70	59	69	69	55	52	51

9-12. Two different tests are designed to measure one's understanding of a certain topic. Two tests are given to 10 different subjects and the results are listed in the following table.

Test X	75	78	88	92	95	67	55	73	74	80
Test Y	81	73	85	85	89	73	66	81	81	81

9-13. A manager in a factory randomly selects 15 assembly line workers and develops scales to measure their dexterity and productivity levels. The results are listed in the following table.

Productivity	63	67	88	44	52	106	99	110	75	58	77	91	101	51	86
Dexterity	2	9	4	5	8	6	9	8	9	7	4	10	7	4	6

9-14. The values of exports and incomes on foreign investments are listed (in billions of dollars) for various years.

	1949	1955	1960	1965	1969	1970	1971
Exports	16	20	27	39	56	63	66
Incomes on foreign investments	2	3	4	7	11	11	13

9-15. The following table lists the ages and diastolic blood pressures (in millimeters of mercury) of six randomly selected subjects.

Ages	35	38	40	69	29	42
Blood pressures	89	81	100	90	59	82

9-16. In the following table, all figures are in kilograms.

	Rat	Man	Dolphin	Elephant	Chimp	Lion	Bat	Crow
Body mass	0.2	60	180	8000	60	200	0.02	0.3
Brain mass	0.0025	1.7	1.8	5	0.4	0.25	0.001	0.01

EXERCISES B: CORRELATION

9-17. Attempt to compute the linear correlation coefficient r for the data in the table and comment on the results.

x	0	3	5	5	6
y	2	2	2	2	2

9-18. Compile a list of 10 pairs of data where each pair represents the ages of married partners. Fill in the table, compute the linear correlation coefficient r, and comment on your results.

Age of husband	
Age of wife	

9-19. Compute the linear correlation coefficient for the paired data in the following table by using

$$r = \frac{\Sigma(x - \bar{x})(y - \bar{y})}{(n - 1)s_x s_y}$$

and compare your result to Exercise 9-3.

x	1	2	2	3
y	5	4	3	1

9-20. Show that

$$\frac{\Sigma(x - \bar{x})(y - \bar{y})}{(n - 1)s_x s_y} = \frac{n\Sigma xy - (\Sigma x)(\Sigma y)}{\sqrt{n(\Sigma x^2) - (\Sigma x)^2}\sqrt{n(\Sigma y^2) - (\Sigma y)^2}}$$

9-3 REGRESSION

In Section 9-2 we tested paired data for the presence or absence of the statistical relationship of a linear correlation. In this section we identify that relationship. The relationship is expressed in the form of a linear or straight line equation. Such an equation is often useful in predicting the likely value of one variable given a value of another. The straight line that concisely summarizes the relationship between the two variables is the **regression line**.

Sir Francis Galton (1822–1911), a cousin of Charles Darwin, studied the phenomenon of heredity whereby certain characteristics regress or revert to more typical values. Galton noted, for example, that children of tall parents

tend to be shorter than their parents, while short parents tend to have children taller than themselves (when fully grown, of course). These original studies of regression evolved into a fairly sophisticated branch of mathematics called regression analysis, which includes the consideration of linear and nonlinear relationships.

This section is confined to linear relationships for two basic reasons. First, the real relationship between two variables is often a linear relationship, or it can be effectively approximated by a linear relationship. Second, non-linear or curvilinear regression problems introduce complexities beyond the scope of this introductory text.

Let's begin with the paired data of Table 9-4 which lists, in billions of dollars, the values of United States exports and incomes on foreign investments for various years. The year numbers can be ignored since we are concerned only with the relationship between the exports and foreign investment incomes. Figure 9-6 is the scatter diagram that corresponds to the data of Table 9-4. The computed value of $r = 0.992$ and the scatter diagram both strongly suggest that there is a definite relationship between the two variables of exports and incomes on foreign investments. Let's denote the first variable (exports) by x and the second variable (incomes) by y. We now want the equation of the straight line that relates x and y. This equation will be of the form $y = mx + b$ where the values of m and b are computed from the paired data by using Formulas 9-2 and 9-3.

STUDENT RATINGS OF TEACHERS

A recent study compared student evaluations of teachers with the amount of material learned by the students. The researchers found a strong negative correlation between these two factors. Teachers rated highly by students seemed to induce less learning.

Many colleges equate high student ratings with good teaching, and this equation is often fostered by the fact that student evaluations are easy to administer and easy to measure.

In a related study, an audience gave a high rating to a lecturer who conveyed very little information but was interesting and entertaining.

Formula 9-2

$$m = \frac{n(\Sigma xy) - (\Sigma x)(\Sigma y)}{n(\Sigma x^2) - (\Sigma x)^2}$$

Formula 9-3

$$b = \frac{(\Sigma y)(\Sigma x^2) - (\Sigma x)(\Sigma xy)}{n(\Sigma x^2) - (\Sigma x)^2}$$

TABLE 9-4

Year	1949	1955	1960	1965	1969	1970
Exports	16	20	27	39	56	63
Incomes on foreign investments	2	3	4	7	11	11

Some inexpensive calculators accept entries of paired data and provide the m and b values directly. These formulas appear formidable, but three observations make the required computations easier. First, if the correlation coefficient r has been computed by Formula 9-1, the values of Σx, Σy, Σx^2, $(\Sigma x)^2$, and Σxy have already been computed. These values can now be used again in Formulas 9-2 and 9-3. Second, examine the denominators of the formulas for m and b and note that they are identical. This means that the computation of $n(\Sigma x^2) - (\Sigma x)^2$ need be done only once, and the resulting value can be used in both formulas. Third, the regression line always passes through the centroid $(\bar{x}, \bar{y})$, so the equation $\bar{y} = m\bar{x} + b$ must be true. This implies that $b = \bar{y} - m\bar{x}$ and it may be easier to evaluate b by computing $y - m\bar{x}$ than by using Formula 9-3.

FIGURE 9-6

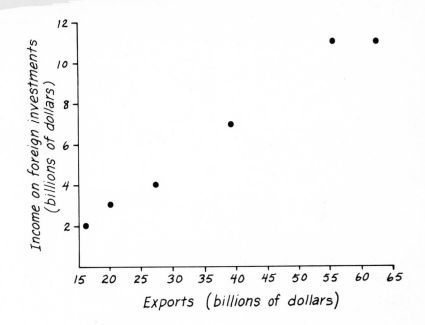

We now use the export–foreign-income data to find the equation of the regression line. From Table 9-4 we compute the following:

$$n = 6$$
$$\Sigma x = 221 \ (x \text{ represents export values})$$
$$\Sigma y = 38 \ (y \text{ represents incomes on foreign investments})$$
$$\Sigma x^2 = 10{,}011$$
$$(\Sigma x)^2 = 48{,}841$$
$$\Sigma xy = 1{,}782$$

Having determined the values of the individual components, we can now compute m and b.

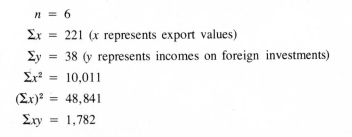

$$m = \frac{n(\Sigma xy) - (\Sigma x)(\Sigma y)}{n(\Sigma x^2) - (\Sigma x)^2} = \frac{6(1782) - (221)(38)}{6(10011) - 48841}$$

$$= \frac{10692 - 8398}{60066 - 48841} = \frac{2294}{11225} \doteq 0.20$$

$$b = \frac{(\Sigma y)(\Sigma x^2) - (\Sigma x)(\Sigma xy)}{n(\Sigma x^2) - (\Sigma x)^2} = \frac{(38)(10011) - (221)(1782)}{11225}$$

$$= \frac{380418 - 393822}{11225} = \frac{-13404}{11225} \stackrel{\circ}{=} -1.19$$

We could have found b by computing $b = \bar{y} - m\bar{x} = (38/6) - 0.20(221/6) = -1.03$. (The value of -1.19 is obtained if we use $2294/11225$ for m instead of the approximate value of 0.20.) Using these results, the general equation of $y = mx + b$ becomes $y = 0.20x - 1.19$. Those who remember the basics of graphing recognize that the straight line representing that equation is shown in Figure 9-7. Those who don't remember the basics of graphing should carefully study the following short review that pertains specifically to straight lines.

FIGURE 9-7

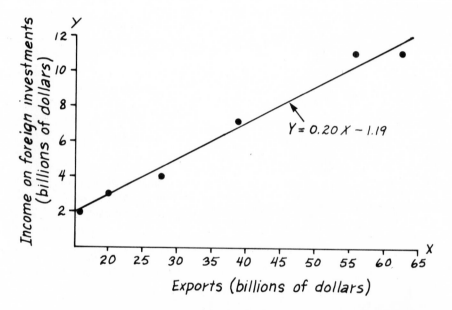

The equation $y = mx + b$ is called the **slope-intercept** form of a straight line, since the constants m and b represent slope and y-intercept, respectively. The y-intercept b is simply the value along the vertical y axis where the straight line intercepts or crosses that axis. For example, the straight line represented by the equation $y = mx + 2$ crosses the y axis at $y = 2$. The slope m represents the steepness of the straight line. Going from left to right, an uphill line has positive slope while a downhill line has a negative slope. Given any two points (x_1, y_1) and (x_2, y_2) on a straight line, the slope may be computed by $m = (y_2 - y_1)/(x_2 - x_1)$. In effect, the slope measures the changes in y values relative to the changes in x values. Here we are concerned with two basic procedures:

1. Obtain the equation of a straight line given the coordinates of different points.
2. Graph the straight line after you have determined its equation.

Formulas 9-2 and 9-3 can be used to determine the equation of the regression line, but it is often helpful (or necessary) to determine that equation by other means. One alternative that will yield approximate results begins with a rough sketch of the straight line that appears to fit the data best. Then you need to identify the x and y coordinates of two separate points on that line and compute m by evaluating $(y_2 - y_1)/(x_2 - x_1)$. After inserting the resulting value of m in the equation $y = mx + b$, continue by substituting the x and y values of either known point into the equation and solve for the only remaining unknown, b. This is all illustrated in the next example.

Example
Without using Formulas 9-2 and 9-3, estimate the equation of the straight line that best fits the paired data given in the following table.

x	1	2	3	6	8	10
y	1800	1400	1300	1000	600	500

Solution
After constructing the scatter diagram, draw the straight line that appears to best fit the given data (see Figure 9-8). The points (3, 1400) and (10, 400) appear to be on the estimated regression line, even though those pairs of data are not present in the original list given in the table. We use the coordinates of those points on the line to find the slope m.

$$m = \frac{y_2 - y_1}{x_2 - x_1} = \frac{400 - 1400}{10 - 3} = \frac{-1000}{7} = -142.9$$

The general equation $y = mx + b$ now becomes $y = -142.9x + b$. We substitute the x and y coordinates of either point in this last equation. Choosing (3, 1400), we get

$$1400 = -142.9(3) + b$$

or

$$1400 = -428.7 + b$$

or

$$1828.7 = b$$

The equation suggested by the line in Figure 9-8 is therefore $y = -142.9x + 1828.7$. For comparison, the corresponding equation obtained through application of Formulas 9-2 and 9-3 is $y = -135.9x + 1779.7$. Thus the estimated line and the resulting equation served as a fairly good approximation to the true regression line.

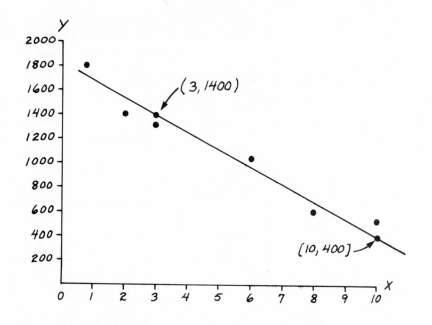

FIGURE 9-8

The second major concern of this brief review on straight lines deals with the graphing of the regression line once we have obtained its equation. Using the equation, we can substitute a value for x and then compute the corresponding y value. These values are the x and y coordinates of a single point on the line. Repeat the same process and generate a second point. Both points can be graphed and, as we all well know, two points determine a straight line. See the following example.

Example

A list of paired data result in the regression values of $m = 3$ and $b = 2$. Graph the equation $y = 3x + 2$.

Solution

Before substituting for x blindly, examine the original values of x so that you can select convenient numbers near the smallest and largest x values. Let's assume that the regression equation came from paired data in which the minimum and maximum x values were 1.8 and 6.3. Then 2 and 6 would make excellent choices for x. If $x = 2$, then $y = 3(2) + 2 = 8$. If $x = 6$, then $y = 3(6) + 2 = 20$. We have generated the pairs of coordinates (2, 8) and (6, 20) that are plotted and then connected as in Figure 9-9.

FIGURE 9-9

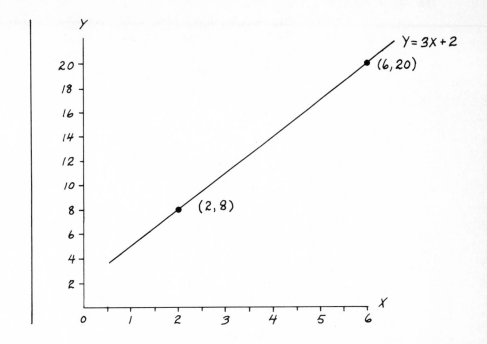

The next example incorporates concepts from this section plus the linear correlation coefficient.

Example   Two different tests are designed to measure the understanding of a certain topic and are administered to 10 subjects. The results follow. Find the linear correlation coefficient r, the equation of the regression line, and plot the scatter diagram and regression line on the same graph. Determine whether there is a significant linear correlation and then predict the likely score on Test Y for someone receiving a score of 70 on Test X.

Test X	75	78	88	92	95	67	55	73	74	80
Test Y	81	73	85	85	89	73	66	81	81	81

Solution   For the preceding experiment, we determine the following:

$$n = 10 \qquad\qquad (\Sigma x)^2 = 603,729$$
$$\Sigma x = 777 \qquad\qquad (\Sigma y)^2 = 632,025$$
$$\Sigma y = 795 \qquad\qquad \Sigma xy = 62,432$$
$$\Sigma x^2 = 61,661$$
$$\Sigma y^2 = 63,629$$

We can now compute the value of the linear correlation coefficient r.

$$r = \frac{n\Sigma xy - (\Sigma x)(\Sigma y)}{\sqrt{n(\Sigma x^2) - (\Sigma x)^2}\sqrt{n(\Sigma y^2) - (\Sigma y)^2}}$$

$$= \frac{10(62432) - (777)(795)}{\sqrt{10(61661) - 603729}\sqrt{10(63629) - 632025}} = \frac{6605}{\sqrt{12881}\sqrt{4265}} = 0.891$$

With $n = 10$, the statistic of $r = 0.891$ indicates a significant positive linear correlation since it exceeds the critical r values of 0.632 (for $\alpha = 0.05$) and 0.765 (for $\alpha = 0.01$). That is, at the 95% or 99% level of significance, we conclude that there is a significant positive linear correlation. Having determined r, we can obtain the regression values of m and b.

$$m = \frac{n\Sigma xy - (\Sigma x)(\Sigma y)}{n(\Sigma x^2) - (\Sigma x)^2} = \frac{10(62432) - (777)(795)}{10(61661) - 603729}$$

$$= \frac{6605}{12881} \stackrel{\circ}{=} 0.51$$

$$b = \bar{y} - m\bar{x} = 79.5 - 0.51(77.7) \stackrel{\circ}{=} 39.87$$

The general equation of the regression line $y = mx + b$ becomes $y = 0.51x + 39.87$. The scatter diagram appears in Figure 9-10 along with the graph of the regression line. The line $y = 0.51x + 39.87$ is plotted by generating the coordinates of two different points on that line. Specifically, when $x = 60$, then the regression equation becomes $y = 0.51(60) + 39.87 = 70.47$, so (60,

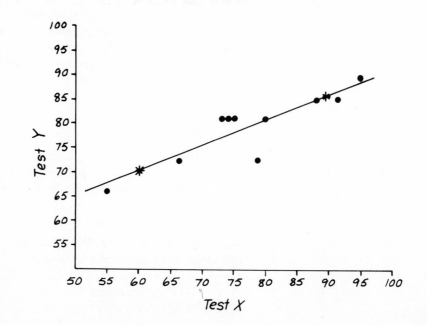

FIGURE 9-10

70.47) is on the regression line and appears as the leftmost asterisk in Figure 9-10. Letting $x = 90$ implies that $y = 0.51(90) + 39.87 = 85.77$ so that (90, 85.77) represents a second point on the regression line that corresponds to the right asterisk in the figure. Finally, we use the equation of the regression line to find the last item requested. To predict the score on Test Y given a score of 70 on Test X, we simply substitute 70 for x in the equation of the regression line. If $x = 70$, $y = 0.51x + 39.87$ becomes $y = 0.51(70) + 39.87 = 75.57$, which is our predicted score.

By examining Figure 9-10 and the computed value of the linear correlation coefficient for the given data, we conclude that the regression line fits the data reasonably well. It is therefore useful in making projections or predictions that do not go far beyond the scope of the available scores. However, if r is close to zero, even though the regression line is the best fitting line, it may not fit the data well enough. It is important to note that the value of r indicates how well the regression line actually fits the available paired data. If $r = 1$ or $r = -1$, then the regression line fits the data perfectly. If r is near $+1$ or -1, the regression line constitutes a very good approximation of the data. But if r is near zero, the regression line fits poorly. When we must estimate the value of one variable given some value of the other, **we should use the equation of the regression line only if r indicates that there is a significant linear correlation.** However, **in the absence of a significant linear correlation, we should not use the regression equation for projecting or predicting. Instead, our best estimate of the second variable is simply the sample mean of that variable**, regardless of the value assigned to the first variable. To illustrate this concept, let's suppose that we have the two samples of paired data. In both cases we want to estimate y when $x = 5$.

First collection of paired data	Second collection of paired data
$n = 100$	$n = 100$
Regression line: $y = 2x + 3$	Regression line: $y = 2x + 3$
$\bar{y} = 20$	$\bar{y} = 20$
$r = 0.95$	$r = 0.02$
To get the best estimate of y when $x = 5$, use the regression line: $y = 2(5) + 3 = 13$.	To get the best estimate of y for any value of x, simply select $\bar{y} = 20$.

In the first case, $r = 0.95$ indicates that the equation $y = 2x + 3$ will give good results, since the regression line fits the data well. Consequently, the best estimate of y for $x = 5$ can be obtained by substituting 5 for x in the equation of the regression line. The estimated y value of 13 results. However, when we consider the second collection of paired sample data, we see that the

linear correlation coefficient of 0.02 reflects a poorly fitting regression line that is useless as a predictor. In this case, the best estimate of y is 20 (the value of $\bar{y}$).

Knowing the qualitative relationship between the value of r and the goodness of fit of the regression line, we can seek some precise quantitative relationship. For paired data,

$$r = \frac{m s_x}{s_y}$$

if we adjust the computation of the sample standard deviation by replacing $n - 1$ with n. That is, let

$$s_x = \sqrt{\frac{\Sigma(x - \bar{x})^2}{n}} \quad \text{and} \quad s_y = \sqrt{\frac{\Sigma(y - \bar{y})^2}{n}}$$

when using this last expression for r (see Exercise 9-41).

Formulas 9-2 and 9-3 describe the computations necessary to obtain the regression line equation $y = mx + b$, and we now describe the criterion used to arrive at these particular formulas. To be concise, the regression line obtained through Formulas 9-2 and 9-3 is unique in that **the sum of the squares of the vertical deviations of the sample points from the regression line is the smallest sum possible.** This property is called the least-squares property and can be understood by examining Figure 9-11 and the text that follows. In Figure 9-11 we show the paired data contained in the following table.

x	1	2	4	5
y	4	24	8	32

FIGURE 9-11

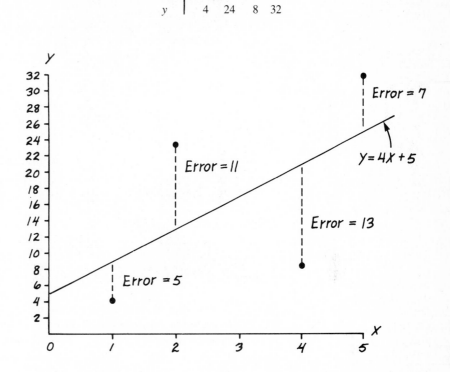

Application of the formulas for m and b results in the equation $y = 4x + 5$, which you see in Figure 9-11. The least-squares property refers to the vertical distances by which the original points miss the regression line. In this figure, these vertical deviations are shown to be 5, 11, 13, and 7 for the four pairs of data. The sum of the squares of these vertical errors is $5^2 + 11^2 + 13^2 + 7^2 = 364$, and the regression line $y = 4x + 5$ is unique in that the value of 364 is the lowest for that particular line. Any other line will yield a sum of squares that exceeds 364, and therefore produce a greater collective error. Thus, using the least-squares criterion, another line will not fit the data as well as the regression line $y = 4x + 5$.

For example, the line $y = 3x + 8$ will produce vertical errors of 7, 10, 12, and 9, so the sum of the squares of those errors is $49 + 100 + 144 + 81 = 374$. The collective error of 374 exceeds a collective error of 364. Consequently, the line $y = 4x + 5$ provides a better fit to the paired data. Fortunately, we need not deal directly with this least-squares property when we want to obtain the equation of the regression line. Calculus has been used to build the least-squares property into formulas and our calculations for m and b are a result of that property. Because Formulas 9-2 and 9-3 depend on certain methods of calculus, we have not included their development in this text.

EXERCISES A: REGRESSION

9-21. Verify that the given paired data will produce a regression line whose equation is $y = x + 2$. Plot the scatter diagram and the equation $y = x + 2$.

x	1	3	5
y	3	5	7

In Exercises 9-22 through 9-25, use the data given in the exercise named to obtain the equation of the regression line (see page 318 for the data needed in these and the following exercises).

9-22. Exercise 9-2.

9-23. Exercise 9-3.

9-24. Exercise 9-4.

9-25. Exercise 9-5.

9-26. Use the paired data in the accompanying table to find the equation of the regression line. Graph the data and the regression line on the same set of axes.

x	1	1	2	2
y	2	1	2	1

9-27. Using the data in Exercise 9-7, find the equation of the regression line.

9-28. Using the data in Exercise 9-8, find the equation of the regression line.

9-29. Using the data in Exercise 9-9, find the equation of the regression line.

9-30. Using the data in Exercise 9-10, find the equation of the regression line.

9-31. Use the data given in Exercise 9-11.
   (a) Construct the scatter diagram and sketch the line that appears to fit the data best.
   (b) Using the graph from part (a), estimate the coordinates of two different points on the estimated regression line and then use these coordinates to approximate the equation of the regression line.

(c) Determine the exact equation of the regression line by using the formulas for m and b (Formulas 9-2 and 9-3).

(d) Predict the Fahrenheit temperature when a cricket chirps 70 times in 1 minute.

9-32. Use the data in Exercise 9-12.

(a) Construct the scatter diagram and sketch the line that appears to fit the data best.

(b) Using the graph from part (a), estimate the coordinates of two different points on the estimated regression line and then use these coordinates to approximate the equation of the regression line.

(c) Determine the exact equation of the regression line by using the formulas for the regression line (Formulas 9-2 and 9-3).

(d) Predict the Test Y score of a subject who obtained a Test X score of 90.

9-33. Find the equation of the regression line for the data found in Exercise 9-13.

9-34. Using the data in Exercise 9-14, find the equation of the regression line.

9-35. Using the data in Exercise 9-15, find the equation of the regression line.

9-36. In each of the following cases, find the best estimate of y when $x = 5$. The given statistics are summarized from paired sample data.

(a) $n = 40, \bar{y} = 6, r = 0.01$, and the equation of the regression line is $y = 3x + 2$.

(b) $n = 40, \bar{y} = 6, r = 0.93$, and the equation of the regression line is $y = 3x + 2$.

(c) $n = 20, \bar{y} = 6, r = -0.654$, and the equation of the regression line is $y = -3x + 2$.

(d) $n = 20, \bar{y} = 6, r = 0.432$, and the equation of the regression line is $y = 1.2x + 3.7$.

(e) $n = 100, \bar{y} = 6, r = -0.175$, and the equation of the regression line is $y = -2.4x + 16.7$.

EXERCISES B: REGRESSION

9-37. Prove that the point $(\bar{x}, \bar{y})$ will always lie on the regression line.

9-38. Using the data in Exercise 9-3, verify that $r = ms_x/s_y$ where m, s_x, and s_y are as defined in this section and r is computed by using Formula 9-1.

9-39. Use $r = ms_x/s_y$ to prove that r and m have the same sign.

9-40. What do you know about s_x and s_y if $r = 0.500$ for paired data having the regression line $y = \frac{1}{2}x + 7.3$?

9-41. Show that

$$\frac{ms_x}{s_y} = \frac{n\Sigma xy - (\Sigma x)(\Sigma y)}{\sqrt{n(\Sigma x^2) - (\Sigma x)^2}\sqrt{n(\Sigma y^2) - (\Sigma y)^2}}$$

where s_x and s_y are both modified by replacing $n - 1$ by n.

9-42. Using the data in Exercise 9-3 and the equation of the regression line, find the sum of the squares of the vertical deviations for the given points. Show that this sum is less than the corresponding sum obtained by replacing the regression line with $y = -x + 6$.

REVIEW

In this chapter we studied the concepts of linear correlation and regression so that we could analyze paired sample data. We limited our discussion to linear relationships because consideration of nonlinear relationships requires more advanced mathematics. With correlation, we attempted to decide whether there is a significant linear relationship between the two variables. With regression, we attempted to specify what that relationship is. While a scatter diagram provides a graphic display of the paired data, the linear correlation coefficient r and the equation of the regression line serve as more precise and objective tools for analysis.

Given a list of paired data, we can compute the linear correlation coefficient r by using Formula 9-1. In Table A-7 we list critical r values that we compare to the computed r values so we can decide whether there is a significant linear relationship. The presence of a significant linear correlation does not necessarily mean that there is a direct cause-and-effect relationship between the two variables.

In Section 9-3 we developed procedures for obtaining the equation of the regression line which, by the least-squares criterion, is the straight line that best fits the paired data. When there is a significant linear correlation, the regression line can be used to predict the value of one variable when given some value of the other variable. The regression line has the form $y = mx + b$ where the constants m and b can be found by using the formulas given in this section.

For convenience, we list the key formulas of this chapter.

$$r = \frac{n\Sigma xy - (\Sigma x)(\Sigma y)}{\sqrt{n(\Sigma x^2) - (\Sigma x)^2}\sqrt{n(\Sigma y^2) - (\Sigma y)^2}}$$

$$m = \frac{n\Sigma xy - (\Sigma x)(\Sigma y)}{n(\Sigma x^2) - (\Sigma x)^2}$$

$$b = \frac{(\Sigma y)(\Sigma x^2) - (\Sigma x)(\Sigma xy)}{n(\Sigma x^2) - (\Sigma x)^2}$$

or

$$b = \bar{y} - m\bar{x}$$

$$r = \frac{ms_x}{s_y}$$

where s_x and s_y are modified by replacing $n - 1$ by n in the denominator.

REVIEW EXERCISES

9-43.   In each of the following, determine whether correlation or regression analysis is more appropriate.
   (a)   Is the value of a car related to its age?
   (b)   What is the relationship between the age of a car and annual repair costs?
   (c)   How are Celsius and Fahrenheit temperatures related?
   (d)   Is the age of a car related to annual repair costs?
   (e)   Is there a relationship between cigarette smoking and lung cancer?

9-44.   (a)   What should you conclude if 40 pairs of data produce a linear correlation coefficient of $r = -0.508$?
   (b)   What should you conclude if 10 pairs of data produce a linear correlation coefficient of $r = 0.608$?
   (c)   What should you conclude if 55 pairs of data produce a linear correlation coefficient of $r = 0.250$?
   (d)   If 10 pairs of data produce a linear correlation coefficient of $r = 0.950$, why is it impossible for $y = -3x + 4$ to be the equation of the regression line?
   (e)   The regression line always passes through the point $(\bar{x}, \bar{y})$ (true or false).

9-45.   Randomly selected subjects were given two different I.Q. tests and their scores follow.

Test A	85	97	100	76	80	116	120	105
Test B	92	109	100	74	85	118	125	90

   (a)   Plot the scatter diagram.
   (b)   Find the value of the linear correlation coefficient r.
   (c)   Assuming a 95% level of significance, find the critical value of r from Table A-7.

(d) Use the results of parts (b) and (c) to decide whether there is a significant linear correlation.

(e) Find the equation of the regression line.

(f) Plot the regression line on the scatter diagram of part (a).

(g) If someone receives a score of 90 on Test A, what is the predicted Test B score?

9-46. A pill designed to lower systolic blood pressure is administered to 10 randomly selected volunteers. The results follow.

Before pill	120	136	160	98	115	110	180	190	138	128
After pill	118	122	143	105	98	98	180	175	105	112

(a) Plot the scatter diagram.

(b) Find the value of the linear correlation coefficient r.

(c) Assuming a 95% level of significance, find the critical value of r from Table A-7.

(d) Use the results of parts (b) and (c) to decide whether there is a significant linear correlation.

(e) Find the equation of the regression line.

(f) Plot the regression line on the scatter diagram of part (a).

(g) Use the equation of the regression line to verify that the line passes through $(\bar{x}, \bar{y})$.

(h) If a subject has a blood pressure of 130 before taking the pill, what is the predicted blood pressure after taking the pill?

9-47. A college dean randomly selects several seniors and obtains their current grade point averages and their high school averages. The results follow.

High school	92	80	73	93	84	77	75	78
College	3.10	2.75	2.60	2.20	3.15	2.05	2.40	2.50

(a) Plot the scatter diagram.

(b) Find the value of the linear correlation coefficient r.

(c) Assuming a 95% level of significance, find the critical value of r from Table A-7.

(d) Use the results of parts (b) and (c) to decide whether there is a significant linear correlation.

(e) Find the equation of the regression line.

(f) Plot the regression line on the scatter diagram.

(g) If a student has a high school average of 82, what is his or her best predicted grade point average in college?

COMPARISONS AMONG SEVERAL SAMPLES

10-1   OVERVIEW

Prior to Chapter 8 we considered statistical analyses involving only one sample. In Chapter 8 we introduced methods for comparing the statistics of two different samples, and in Chapter 9 we introduced ways of analyzing paired data. In this chapter we develop methods for comparing more than two samples.

We begin by considering a method for testing a hypothesis made about several population proportions. Instead of working with the sample proportions, we deal directly with the frequencies with which the events occur. Our objective is to test for the significance of the differences between observed frequencies and the frequencies we would expect in a theoretically ideal experiment. We introduce the test statistic that measures the differences between observed frequencies and expected frequencies. In repeated large samplings, the test statistic can be approximated by the chi-square distribution discussed in Section 6-5.

We analyze tables of frequencies called **contingency tables**. In these tables the rows represent categories of one variable while the columns represent categories of another variable. We test the hypothesis that the two classification variables are independent. We again use the chi-square distribution to determine whether there is a significant difference between the observed sample frequencies and the frequencies we would expect in a theoretically ideal experiment involving independent variables. We present an example in which the variables of sex and employability are tested for independence.

Unlike the hypothesis tests discussed in previous chapters, the methods of Sections 10-2 and 10-3 do not require specific population distributions, and they do not depend on population parameters such as μ or σ. Tests that lack these requirements are called **nonparametric tests**. Section 10-4 introduces some of the basic ideas of another parametric test that depends on population parameters and requires normal distributions. We utilize a technique called **analysis of variance** to test the claim that several population means are equal. In Section 8-3 we tested the hypothesis $\mu_1 = \mu_2$, but in Section 10-4 we test claims such as $\mu_1 = \mu_2 = \mu_3$. When dealing with more than two samples, we use a fundamentally different approach that incorporates the F distribution.

10-2   MULTINOMIAL EXPERIMENTS

In Chapter 4 we introduced the binomial probability distribution and indicated that each trial must have all outcomes classified into exactly one of two categories. This feature of two categories is reflected in the prefix *bi* which begins the term binomial. In this section we consider **multinomial experiments** that require each trial to yield outcomes belonging to one of several categories. Except for this difference, binomial and multinomial experiments are essentially the same.

DEFINITION

> *A **multinomial experiment** is one in which:*
>
> 1. *There is a fixed number of trials.*
> 2. *The trials are independent.*
> 3. *Each trial must have all outcomes classified into exactly one of several categories.*
> 4. *The probabilities remain constant for each trial.*

HUGHES

Woodrow Wilson

THE POWER OF YOUR VOTE

In the system of the electoral college, the power of a voter in a large state exceeds that of a voter in a small state. When voting power is measured as the ability to affect the outcome of an election, we see that a New Yorker has 3.312 times the voting power of a resident of the District of Columbia. This result is included in an article by John Banzhof entitled "One Man, 3.312 Votes."

As an example, the outcome of the 1916 Presidential election could have been changed by shifting only 1983 votes in California. If the same number of votes were changed in a much smaller state, the resulting change in electoral votes would not have been sufficient to alter the outcome.

We have already discussed methods for testing hypotheses made about one population proportion (Section 6-4) and two population proportions (Section 8-4). However, there is often a real need to deal with more than two proportions. Suppose, for example, that an employer randomly selects 100 absence reports and obtains the sample data in Table 10-1.

TABLE 10-1

Day of week	Mon	Tues	Wed	Thurs	Fri
Number of absences	31	10	6	24	29

If absences occur with perfectly equal frequencies on different days, Table 10-1 would look like Table 10-2.

TABLE 10-2

Day of week	Mon	Tues	Wed	Thurs	Fri
Number of absences	20	20	20	20	20

We know that samples deviate naturally from ideal expectations because of chance fluctuations, so we now pose the key question: Are the differences between the real data of Table 10-1 and the theoretically ideal data of Table 10-2 attributable to chance, or are the differences significant? To answer this question we need some way of measuring the significance of the differences between the observed values and the theoretical values. **The expected frequency of an outcome is the product of the probability of that outcome and the total number of trials**. If there are five possible outcomes that are supposed to be equally likely, then the probability of each outcome is 1/5 or 0.2. For 100 trials and five equally likely outcomes, the expected frequency of each outcome is $0.2 \times 100 = 20$.

In testing for the differences among the five sample proportions (0.31, 0.10, 0.06, 0.24, 0.29), one approach might be to compare them two at a time by using the methods of Section 8-4. That approach would, however, be very inefficient and lead to severe problems related to the level of significance. There are 10 different combinations of two samples, and if each individual hypothesis test has a 0.95 probability of not leading to a type I error (rejecting a true null hypothesis), then the probability of no type I errors among the 10 tests is only 0.95^{10} or about 0.599. Instead of pairing off samples and conducting 10 separate tests, we develop one comprehensive test that is based on a statistic which measures the differences between observed values and the corresponding values that we expect in an ideal case. This statistic will incorporate the observed and expected frequencies instead of the proportions. We introduce the following notation:

NOTATION

O represents the **observed frequency** of an outcome.

E represents the theoretical or **expected frequency** of an outcome.

From Table 10-1 we see that the O values are 31, 10, 6, 24, and 29. From Table 10-2 we see that the corresponding values of E are 20, 20, 20, 20, and 20. For the given sample data we have $\Sigma O = \Sigma E = 100$ and, in general, $\Sigma O = \Sigma E = n$. The method generally used in testing for agreement between O and E values is based on the test statistic

$$\chi^2 = \Sigma \frac{(O - E)^2}{E}$$

Simply summing the differences between observed and expected frequencies would not lead to a good measure, since that sum is always zero.

$$\Sigma(O - E) = \Sigma O - \Sigma E = n - n = 0$$

Squaring the $O - E$ values provides a better statistic which does reflect the differences between observed and expected frequencies, but $\Sigma(O - E)^2$ grows larger as the sample size increases. We average out that sum of squares through division by the expected frequencies. For the data of our absentee analysis, the Monday values of Table 10-1 and Table 10-2 indicate that 31 is the observed frequency, while 20 is the expected frequency,

$$\frac{(O - E)^2}{E} = \frac{(31 - 20)^2}{20} = \frac{11^2}{20} = \frac{121}{20}$$

Proceeding in a similar manner with the remaining data we get

$$\chi^2 = \sum \frac{(O - E)^2}{E} = \frac{(31 - 20)^2}{20} + \frac{(10 - 20)^2}{20} + \frac{(6 - 20)^2}{20} + \frac{(24 - 20)^2}{20} + \frac{(29 - 20)^2}{20}$$

$$= \frac{121}{20} + \frac{100}{20} + \frac{196}{20} + \frac{16}{20} + \frac{81}{20}$$

$$= \frac{514}{20} = 25.7$$

The theoretical distribution of χ^2 is a discrete distribution, since there are a limited number of possible χ^2 values. However, extensive studies have established that, in repeated large samplings, the distribution of χ^2 can be approximated by a chi-square distribution. This approximation is generally considered to be acceptable provided that all values of E are at least 5. In Section 5-5 we saw that the continuous normal probability distribution can reasonably approximate the discrete binomial probability distribution provided that np and nq are both at least 5. We now see that the continuous chi-square distribution can reasonably approximate the discrete distribution of χ^2 provided that all values of E are at least 5. There are ways of circumventing the problem of an expected frequency that is less than 5, and one procedure requires us to combine categories so that all expected frequencies are at least 5.

When we use the chi-square distribution as an approximation to the distribution of χ^2 values, we obtain the critical test value from Table A-5 after determining the level of significance α and the number of degrees of freedom. In a multinomial experiment with k possible outcomes, the number of degrees of freedom is $k - 1$. This reflects the fact that, for n trials, the frequencies of $k - 1$ outcomes can be freely varied, but the frequency of the last outcome is determined. Our 100 absences are distributed among five categories or cells, but we can freely vary the frequencies of only four cells, since the last cell would be 100 minus the total of the first four cell frequencies. In this case, we say that the number of degrees of freedom is 4.

Note that close agreement between observed and expected values will lead to a small value of χ^2. A large value of χ^2 will indicate strong disagreement between observed and expected values. A significantly large value of χ^2 will therefore cause rejection of the null hypothesis of no difference between observed and expected frequencies. Our test is therefore right-tailed since the critical value and critical region are located at the extreme right of the distribution.

For the sample absentee data, we have determined the value of the test statistic ($\chi^2 = 25.7$) and the number of degrees of freedom (4). Let's assume a 5% level of significance so that $\alpha = 0.05$. With 4 degrees of freedom and $\alpha = 0.05$, Table A-5 indicates a critical value of 9.488 (see Figure 10-1).

Figure 10-1 indicates that our test statistic of 25.7 falls within the critical region, so the absences described in Table 10-1 are not likely to come from a population of absences that occur on the different weekdays with equal frequencies.

FIGURE 10-1

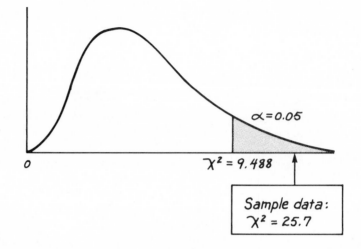

Unlike the hypothesis tests discussed in previous chapters, this method does not depend on population parameters such as μ or σ, nor does it require a particular population distribution. Consequently it is called a **nonparametric test** since it does not depend on population parameters or distributions. In Section 10-3 and in Chapter 11 we consider other nonparametric tests of hypotheses.

The first example in this section dealt with the implied null hypothesis that the frequencies of absences on the five working days were all equal. However, the theory and methods we present here can also be used in cases where the claimed frequencies are different. The next example illustrates this.

Example A car manufacturer claims that the best selling model has the following color preference rates: 30% of the buyers prefer red, 10% prefer white, 15% prefer green, 25% prefer blue, 5% prefer brown, and 15% prefer yellow. A random survey of 200 buyers produces the following results:

Color	Red	White	Green	Blue	Brown	Yellow
Number prefer	64	14	38	49	6	29
	60	20	30	50	10	30

At the $\alpha = 0.01$ significance level, test the claim that the percentages given by the manufacturer are correct.

Solution

The null hypothesis H_0 is the claim that the percentages given are correct. If the manufacturer's claim is exactly correct, then the 200 preferences would have occurred with these frequencies:

Color	Red	White	Green	Blue	Brown	Yellow
Number prefer	60	20	30	50	10	30

The first table lists observed values, while this one lists expected values. We now compute χ^2 as a measure of the disagreement between observed and expected values.

$$\chi^2 = \sum \frac{(O - E)^2}{E}$$

$$= \frac{(64 - 60)^2}{60} + \frac{(14 - 20)^2}{20} + \frac{(38 - 30)^2}{30}$$

$$+ \frac{(49 - 50)^2}{50} + \frac{(6 - 10)^2}{10} + \frac{(29 - 30)^2}{30}$$

$$\stackrel{\circ}{=} 5.853$$

This is a right-tailed test with $\alpha = 0.01$ and $6 - 1$ or 5 degrees of freedom, so the critical value from Table A-5 is 15.086. Since the computed χ^2 of 5.853 does not fall within the critical region bounded by 15.086, we fail to reject the null hypothesis and we attribute the observed deviations from the expected values to chance fluctuations. The available sample data do not warrant rejection of the manufacturer's claim.

EXERCISES A: MULTINOMIAL EXPERIMENTS

10-1. The following table is obtained from a random sample of 100 absences. At the $\alpha = 0.01$ significance level, test the claim that absences occur on the five days with equal frequency.

Day	Mon	Tues	Wed	Thurs	Fri
Number absent	27	19	22	20	12

(a) Find the χ^2 value based on the sample data.
(b) Find the critical value of χ^2.
(c) What conclusion can you draw?

10-2.   A die is rolled 60 times and the outcomes are listed in the following table. At the $\alpha = 0.05$ significance level, test the claim that the die is fair.

Outcome	1	2	3	4	5	6
Frequency	16	13	8	9	6	8

(a)   Determine the expected frequency for each outcome.
(b)   Find the χ^2 value based on the sample data.
(c)   Find the critical value of χ^2.
(d)   What conclusion can you draw?

10-3.   A die is loaded and the manufacturer claims that the outcome of 2 occurs in 30% of the rolls, while the other outcomes occur with the same frequency of 14%. Test the claim at the $\alpha = 0.05$ significance level if 600 sample rolls produce the following results.

Outcome	1	2	3	4	5	6
Frequency	102	161	100	79	65	93

10-4.   A psychology course has 10 sections with enrollments listed in the following table. At the $\alpha = 0.05$ significance level, test the claim that students enroll in the various sections with equal frequencies.

Section	1	2	3	4	5	6	7	8	9	10
Enrollment	22	14	12	25	25	20	17	15	21	19

10-5.   A biochemist conducts an experiment designed to reveal the most effective aspirin. The eight best selling brands of aspirin are tested on 120 random subjects and, for each subject, a best aspirin is selected according to certain technical criteria. The results are as follows. At the $\alpha = 0.05$ significance level, test the claim that the eight brands of aspirin are equally effective.

Brand	A	B	C	D	E	F	G	H
Number of times selected as best	10	14	16	21	17	14	15	13

10-6.   A genetics experiment involves 320 mice and is designed to determine whether Mendelian principles hold for a certain list of characteristics. The following table summarizes the actual experimental results and the expected Mendelian results for the five characteristics being considered. At the $\alpha = 0.01$ significance level, test the claim that Mendelian principles hold.

Characteristic	A	B	C	D	E
Observed frequency	30	15	58	83	134
Expected (Mendelian) frequency	20	20	40	120	120

10-7. In an experiment on perception, 50 subjects are asked to select the most pleasant of five different photographs. The results are as follows. At the $\alpha = 0.05$ significance level, test the hypothesis that the photographs are equally pleasant.

Photograph	A	B	C	D	E
Number selected	0	9	19	10	12

10-8. In an experiment on extrasensory perception, subjects were asked to correctly identify the month showing on a calendar in the next room. If the results are as shown, test the claim that months were selected with equal frequencies. Assume a significance level of 0.05.

Month	Jan	Feb	Mar	Apr	May	Jun	Jul	Aug	Sept	Oct	Nov	Dec
Number selected	8	12	9	15	6	12	4	7	11	11	5	20

10-9. The manager of an ice cream shop claims that 30% of his customers prefer vanilla, 25% prefer chocolate, 10% prefer strawberry, 20% prefer butter pecan, and 15% prefer maple walnut. A random sample of 100 customers produces the following results. At the 0.05 significance level, test the claim that the manager's percentages are correct.

Flavor	vanilla	chocolate	strawberry	butter pecan	maple walnut
Number selected	24	24	8	21	23

10-10. A politician claims that the Republican, Democratic, and Independent mayoral candidates are favored by voters at the rates of 35%, 40%, and 25%, respectively. Test the claim at the 0.05 significance level if a random survey of 30 voters produces the following results.

Candidate	Republican	Democrat	Independent
Number selected	13	11	6

10-11. A television company is told by a consulting firm that its eight leading shows are favored according to the percentages given in the following table. A separate and independent sample is obtained by another consulting firm. Do the figures agree? Assume a significance level of 0.05.

Show	A	B	C	D	E	F	G	H
First consultant	22%	18%	12%	12%	10%	9%	9%	8%
Second consultant (number of respondents favoring show)	29	30	20	16	9	17	10	19

10-12. A roulette wheel has 38 possible outcomes that are supposed to be equally likely. If we plan to test a roulette wheel for fairness using the chi-square distribution, what is the minimum number of trials we must make if we are to satisfy the requirement that each expected frequency must be at least 5? Suppose we conduct the minimum number of trials and compute χ^2 to be 49.6. What do we conclude at the 0.01 significance level?

10-13. A pair of dice has 36 possible outcomes that are supposed to be equally likely. If we plan to test a certain pair of dice for fairness using the chi-square distribution, what is the minimum number of rolls necessary if each outcome is to have an expected frequency of at least 5? Suppose we conduct the minimum number of rolls and obtain a χ^2 value of 67.2. What do we conclude at the 0.01 significance level?

10-14. A college dean expects 40% of the students to register on Monday, 45% on Tuesday, and 15% on Wednesday. Of 1850 students, 1073 register on Monday, 555 register on Tuesday, and 222 register on Wednesday. At the 1% level of significance, test the claim that the observed results are compatible with the dean's expectation.

10-15. In a certain county, 80% of the drivers have no accidents in a given year, 16% have one accident, and 4% have more than one accident. A survey of 200 randomly selected teachers from the county produced 172 with no accidents, 23 with one accident, and 5 with more than one accident. At the 5% level of significance, test the claim that the teachers exhibit the same accident rate as the countywide population.

10-16. For the past several years, the percentages of A's, B's, C's, D's, and F's for a certain statistics course have been 8%, 17%, 45%, 22%, and 8%, respectively. A new testing method was used in the last semester and there were 10 A's, 20 B's, 10 C's, 5 D's, and 5 F's. At the 5% level of significance, test the claim that the grades obtained through the new testing method are in the same proportion as before.

10-17. A teacher divides a course into four units, expecting to spend an equal amount of time on each unit. Units I, II, III, and IV required 13 classes, 15 classes, 8 classes, and 9 classes, respectively. At the 5% level of significance, test the claim that the actual number of classes agreed with the teacher's expectation.

EXERCISES B:
MULTINOMIAL EXPERIMENTS

10-18. Obtain a pair of dice and test for fairness at the 0.05 significance level by performing the minimum number of rolls necessary (see Exercise 10-13).

10-19. In testing for agreement between observed and expected frequencies in the following table, we cannot use the chi-square distribution since all expected values are not at least 5. However, we can combine some columns so that all expected values do equal or exceed 5. Use this suggestion to test the claim that the observed and expected frequencies are compatible.

Characteristic	A	B	C	D	E	F	G	H	I	J
Observed frequency	2	8	8	9	3	5	3	0	12	3
Expected frequency	4	5	8	7	4	6	5	2	9	3

10-20. Why can't we solve Exercise 8-53 using the techniques of this section?

10-3 CONTINGENCY TABLES

Suppose we collect the sample data relating to the employability of disabled adults and summarize them in Table 10-3.

TABLE 10-3 Effect of a disability on employment.

	Unemployable	Limited employment	Employable with only minor effects
Males	8	11	16
Females	18	10	12

This table indicates that 75 disabled subjects were classified according to their ability to work. Tables similar to this one are generally called **contingency tables** or **two-way tables**. In this context, the word contingency refers to dependence, and the contingency table serves as a useful medium for analyzing the dependence of one variable on another. This is only a statistical dependence that cannot be used to establish an inherent cause-and-effect relationship.

We need to test the null hypothesis that the two variables in question are independent. That is, we will test the claim that, among disabled adults, sex and employability are independent. Let's select a significance level of $\alpha = 0.05$. We can now compute a test statistic based on the data and then compare that test statistic to the appropriate critical test value. As in the previous section, we utilize the chi-square distribution where the test statistic is given by

$$\chi^2 = \sum \frac{(O - E)^2}{E}$$

This test statistic allows us to measure the degree of disagreement between the frequencies actually observed and those that we would theoretically expect when the two variables are independent. The reasons underlying the development of the χ^2 statistic in the previous section also apply here. In repeated large samplings, the distribution of the test statistic χ^2 can be approximated by the chi-square distribution provided that all expected frequencies are at least 5.

In the preceding section we knew the corresponding probabilities and could easily determine the expected values, but the typical contingency table does not come with the relevant probabilities. Consequently, we need to devise a method for obtaining the corresponding expected values. To develop

such a method, let's pretend that we know only the row and column totals and that we must fill in the cell frequencies by assuming that there is no relationship between the two variables involved (see Table 10-4).

TABLE 10-4

	Unemployable	Limited employment	Employment with only minor effects	
Males				35
Females				40
	26	21	28	

DID MENDEL FUDGE HIS DATA?

R. A. Fisher statistically analyzed the claimed results of Mendel's experiments in hybridization. Fisher noted that the data were unusually close to theoretically expected outcomes. He says that "the data have evidently been sophisticated systematically, and after examining various possibilities, I have no doubt that Mendel was deceived by a gardening assistant, who knew only too well what his principal expected from each trial made." Fisher used chi-square tests and concluded that there is only about a 0.00004 probability of such close agreement between expected observations and reported observations.

We begin with the empty cell in the upper left-hand corner that corresponds to unemployable males. Since 35 of the 75 subjects are males, if one of the subjects is randomly selected, $P(\text{male}) = 35/75$. Similarly, 26 of the 75 subjects are unemployable, so if one subject is randomly selected, $P(\text{unemployable}) = 26/75$. Since we assume that sex and employability are independent, we conclude that $P(\text{male and unemployable}) = P(\text{male}) \times P(\text{unemployable}) = 35/75 \times 26/75$. This follows from a basic principle of probability whereby $P(A \text{ and } B) = P(A) \times P(B)$ if A and B are independent events. To obtain the expected value of the upper left cell, we simply multiply the probability for that cell by the total number of subjects available to get

$$\frac{35}{75} \times \frac{26}{75} \times 75 = 12.13$$

The form of this product suggests a general way to obtain the expected frequency of a cell:

$$\text{expected frequency } E = \frac{(\text{row total})}{(\text{grand total})} \cdot \frac{(\text{column total})}{(\text{grand total})} \cdot (\text{grand total})$$

This expression can be simplified to the following form:

$$\boxed{\text{expected frequency } E = \frac{(\text{row total}) \cdot (\text{column total})}{(\text{grand total})}}$$

where grand total refers to the number of trials n. Applying this last expression to the lower right cell of Table 10-4 we get an expected frequency of

$$E = \frac{40 \cdot 28}{75} = 14.93$$

Table 10-5 reproduces Table 10-3 with the expected frequencies inserted in parentheses. As in Section 10-2, we require that all expected frequencies be at least 5 before we can conclude that the chi-square distribution serves as a suitable approximation to the distribution of χ^2 values.

TABLE 10-5

	Unemployable	Limited employment	Employable with only minor effects
Males	8 (12.13)	11 (9.8)	16 (13.07)
Females	18 (13.87)	10 (11.2)	12 (14.93)

Using the observed and expected frequencies shown in Table 10-5, we can now compute the χ^2 test statistic based on the sample data.

$$\chi^2 = \sum \frac{(O - E)^2}{E}$$

$$= \frac{(8 - 12.13)^2}{12.13} + \frac{(11 - 9.8)^2}{9.8} + \frac{(16 - 13.07)^2}{13.07}$$

$$+ \frac{(18 - 13.87)^2}{13.87} + \frac{(10 - 11.2)^2}{11.2} + \frac{(12 - 14.93)^2}{14.93}$$

$$= 1.406 + 0.147 + 0.657 + 1.230 + 0.129 + 0.575$$

$$\stackrel{\circ}{=} 4.144$$

With $\alpha = 0.05$ we proceed to Table A-5 to obtain the critical value of χ^2, but we must first know where the critical region lies and the number of degrees of freedom:

1. Tests of independence with contingency tables involve only right-tailed critical regions since small values of χ^2 support the claimed independence of the two variables. That is, χ^2 is small if observed and expected frequencies are close. Large values of χ^2 are to the right of the chi-square distribution, and they reflect significant differences between observed and expected frequencies.

2. In a contingency table with r rows and c columns, the number of degrees of freedom is $(r - 1) \cdot (c - 1)$.

Thus, Table 10-3 has $(2 - 1) \cdot (3 - 1)$ or 2 degrees of freedom. In a right-tailed test with $\alpha = 0.05$ and with 2 degrees of freedom we refer to Table A-5 to get a critical χ^2 value of 5.991. Since the calculated χ^2 value of 4.144 does not fall in the critical region bounded by $\chi^2 = 5.991$, we fail to reject the null hypothesis of independence between the two variables. Sex seems to have no effect on the impact of a disability on employment (see Figure 10-2).

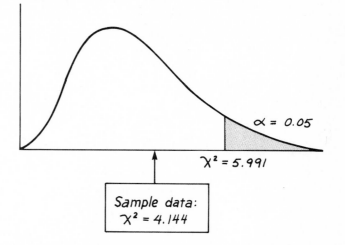

FIGURE 10-2

In the next example we test for independence between a high school student's program and the family's social class.

Example

Given the data in Table 10-6, test the claim that a high school student's program is independent of his family's social class. The table is based on a random sample of 300 high school students. Assume a significance level of $\alpha = 0.05$.

TABLE 10-6

High school program	Social class of family			
	Lower	Lower middle	Upper middle	Upper
Academic	5	10	40	30
General	20	40	50	15
Vocational	15	50	20	5

Solution

The null hypothesis is the claim that a student's high school program is independent of social class. That hypothesis will be rejected only if the sample data exhibit a significant difference between the observed frequencies (Table 10-6) and the frequencies we would expect with two independent variables. In order to evaluate $\Sigma(O - E)^2/E$, we must first determine the expected frequency E for each of the nine cells. The row and column totals are shown in Table 10-7.

TABLE 10-7

	Lower	Lower middle	Upper middle	Upper	
Academic					85
General					125
Vocational					90
	40	100	110	50	(Grand total is 300)

For each of the nine cells, we compute the expected frequency

$$E = \frac{\text{(row total)(column total)}}{\text{(grand total)}}$$

The expected frequency E for the upper left cell is computed as an illustration and it, along with the other expected frequencies, is inserted in parentheses in Table 10-8.

$$E = \frac{85 \cdot 40}{300} = 11.33$$

TABLE 10-8

	Lower	Lower middle	Upper middle	Upper
Academic	5 (11.33)	10 (28.33)	40 (31.17)	30 (14.17)
General	20 (16.67)	40 (41.67)	50 (45.83)	15 (20.83)
Vocational	15 (12)	50 (30)	20 (33)	5 (15)

Having determined the values of E, we can now compute the test statistic.

$$\chi^2 = \sum \frac{(O - E)^2}{E}$$

$$= \frac{(5 - 11.33)^2}{11.33} + \frac{(10 - 28.33)^2}{28.33} + \cdots + \frac{(5 - 15)^2}{15}$$

$$\stackrel{\circ}{=} 64.197$$

We obtain the critical value of χ^2 by noting that $\alpha = 0.05$, that the test is right-tailed, and that the number of degrees of freedom is $(r - 1)(c - 1) = (3 - 1)(4 - 1) = 6$. Since all expected frequencies are at least 5, we can approximate the distribution of χ^2 by the chi-square distribution. Referring to Table A-5, we get a critical value of 12.592. The test statistic of $\chi^2 = 64.197$ is in the critical region bounded by the critical value of 12.592, so we reject the null hypothesis and conclude that high school program and social class are *not* independent.

EXERCISES A: CONTINGENCY TABLES

10-21.   A survey on a local gun control proposal produced the data summarized in the following table. At the $\alpha = 0.05$ significance level, test the claim that voting on the bill is independent of the sex of the voter.

(a)   Assume independence of sex and voter opinion and determine the expected frequency for each of the four cells.

(b)   Compute the χ^2 test statistic based on the sample data.

(c)   Determine the critical value of χ^2.

(d)   What conclusion can you draw?

	In favor	Opposed
Males	20	30
Females	30	10

10-22.   A presidential candidate employs a polling organization to find out whether his popularity differs between men and women. The results follow. At the $\alpha = 0.05$ significance level, test the claim of no significant difference between the responses of men and women.

	In favor	Opposed
Males	5	45
Females	15	55

10-23.   A particular job in an assembly process involves considerable stress, and a study is done to determine whether there is any difference between the adjustment of men and women to the job. The results of a random survey follow. At the $\alpha = 0.01$ significance level, test the claim of no difference.

OPPOSITION TO ABORTION IS NOT A CATHOLIC MOVEMENT

Analysis of statistical data often helps dispel popular myths, such as the assumption that opposition to abortion is largely a Catholic movement. A poll by Sindinger and Company showed that 60% of all Americans oppose abortions, while 80% of the Catholics in this country oppose them. With about 220,000,000 Americans and about 40,000,000 Catholics, this means that only about 25% of those opposed to abortions are Catholics. It is therefore unfair to dismiss the antiabortion movement as a denominational lobby dominated by Catholics.

	Well adjusted	Not well adjusted
Males	5	20
Females	30	10

10-24. Two teaching methods are used to train bank employees. We want to determine whether there is any relationship between the method used and the quality of learning. The following table summarizes the sample data. At the $\alpha = 0.01$ level of significance, test the claim that the quality of learning and the teaching method are independent.

	Average	Superior
Method A	12	18
Method B	20	24

10-25. The claim has been made that males do better in mathematics courses than females. At the $\alpha = 0.01$ level of significance, use the sample data in the following table to test that claim.

Math grades

	Low	Average	High
Males	15	60	25
Females	10	35	5

10-26. Researchers studied 50 jurors to find the effect of sex on punitive attitudes. They administered psychological tests and classified each subject as having liberal, fair, or vindictive tendencies. The results follow. At the $\alpha = 0.05$ level of significance, test the claim that the presence of these tendencies is independent of the sex of the juror.

	Liberal	Fair	Vindictive	
Males	5 7.2	18 15.6	7 7.2	30
Females	7 4.8	8 10.4	5 4.8	20
	12	26	12	50

10-27. The manager of an assembly operation wants to determine whether the number of defective parts is dependent on the day of the week. She develops the following sample data. At the $\alpha = 0.05$ significance level, test the claim of the union representative that the day of the week is independent of the number of defects.

	Mon	Tues	Wed	Thurs	Fri
Acceptable products	80	100	95	93	82
Defective products	15	5	5	7	12

10-28. A candidate for national office wants to know whether there are regional differences in his popularity and conducts a survey to find out. At the $\alpha = 0.05$ level of significance, test the claim that the candidate's popularity is independent of geographic region. The sample results are given in the following table.

	In favor	Against	No opinion
Northeast	10	20	5
South	12	36	8
West	21	33	9
North Central	45	5	7

10-29. The following table summarizes sample grades for four subjects. At the $\alpha = 0.05$ level of significance, test the claim that grade distribution is independent of the subject.

	Grade				
	A	B	C	D	F
Math	13	16	13	8	9
English	6	15	15	5	6
Science	5	8	8	5	9
Psychology	9	18	18	5	6

10-30. Use the data in the following table to test the claim that the number of cigarettes smoked per day is independent of the age of the smoker. Assume a significance level of $\alpha = 0.05$.

Number of cigarettes smoked per day

Age	Under 5	6–14	15–24	25–34	35–44
17–24	7	30	42	9	5
25–34	6	18	47	12	11
35–44	6	15	42	15	15
45–54	6	15	42	13	16

EXERCISES B: CONTINGENCY TABLES

10-31. The only sample data available on SALT II is summarized as follows. We wish to test the independence of geographic region and opinion, but each cell does not have an expected value of at least 5. Combine rows in a reasonable way so that the expected value of each cell is at least 5, and then complete the test using a significance level of $\alpha = 0.05$.

	In favor	Opposed	No opinion
Northeast	4	8	3
Southeast	7	8	6
North Central	2	4	4
South Central	9	3	5
Northwest	12	20	4
Southwest	2	14	3

10-4 ANALYSIS OF VARIANCE

In Section 10-2 we developed a procedure for testing the hypothesis that the differences among several sample proportions are due to chance. In this section we develop a procedure for testing the hypothesis that differences among several sample **means** are due to chance. Unlike the previous sections of this chapter, we consider a **parametric** test which requires the populations under consideration to have a distribution that is essentially normal. The variances of the populations must be approximately equal, and the samples must be independent. The requirements of normality and equal variances are somewhat loose since the methods in this section work reasonably well unless there is a very nonnormal distribution or unless the population variances differ by very large amounts. The method we will describe is called **analysis of variance** because it is based on a comparison of two different estimates of the variance that is common to the different populations.

Let's assume that we have k different populations and that a sample of size n is drawn from each population. From the sample data we compute the sample means $\bar{x}_1, \bar{x}_2, \ldots, \bar{x}_k$ and the sample standard deviations $s_1, s_2, \ldots, s_k$. Our method of testing the null hypothesis $\mu_1 = \mu_2 = \cdots = \mu_k$ requires that we obtain two different estimates of the common population variance σ^2 and then compare those estimates by using the F distribution. Let's consider a specific example.

Example A pilot does extensive bad weather flying and decides to buy a battery-powered radio as an independent backup for his regular radios which depend on his airplane's electrical system. He has a choice of three brands of rechargeable batteries that vary in cost. He obtains the sample data in the following table. He randomly selects five batteries for each brand, and tests them for the operating time (in hours) before recharging is necessary. Do the three brands have the same mean usable time before recharging is required?

Brand X	Brand Y	Brand Z
26.0	29.0	30.0
28.5	28.8	26.3
27.3	27.6	29.2
25.9	28.1	27.1
28.2	27.0	29.8
$\bar{x}_1 = 27.18$	$\bar{x}_2 = 28.10$	$\bar{x}_3 = 28.48$
$s_1^2 = 1.46$	$s_2^2 = 0.69$	$s_3^2 = 2.81$

Solution Using these sample results we develop two separate estimates of σ^2 where σ^2 is assumed to be the population variance common to all three brands. We employ the F distribution in the comparison of the two separate estimates. In Section 8.2, we discussed the use of the F distribution in testing for equality of two variances. We saw that when two independent samples were drawn from populations that were approximately normally distributed, the sampling distribution of $F = s_1^2/s_2^2$ was the F distribution. Here we again use the F distribution, but we replace the two sample variances by two different estimates of the population variance common to all of the populations under consideration. Specifically, we use

$$F = \frac{\text{variance between samples}}{\text{variance within samples}} = \frac{ns_{\bar{x}}^2}{\mu_{s_i^2}}$$

The variance *between* samples is obtained by finding the product of n and $s_{\bar{x}}^2$; n is the number of scores in each sample and $s_{\bar{x}}^2$ is the variance of the sample means. For the data in the table, $n = 5$ since each sample consists of 5 scores, and $s_{\bar{x}}^2 = 0.45$ is found by computing the variance of the sample means (27.18, 28.10, 28.48). Thus $ns_{\bar{x}}^2 = (5)(0.45) = 2.25$. The expression $ns_{\bar{x}}^2$ is justified as an estimate of σ^2 since, by the Central Limit Theorem $\sigma_{\bar{x}} = \sigma/\sqrt{n}$. Squaring both sides of this last expression and solving for σ^2, we get $\sigma^2 = n\sigma_{\bar{x}}^2$ which indicates that σ^2 can be estimated by $ns_{\bar{x}}^2$.

The variance *within* samples is obtained by computing $\mu_{s_i^2}$, which denotes the mean of the sample variances. Since we have three sets of samples, $\mu_{s_i^2}$ becomes

$$\frac{s_1^2 + s_2^2 + s_3^2}{3} = \frac{1.46 + 0.69 + 2.81}{3} = 1.65$$

This approach seems reasonable since we assume that equal population variances imply that representative samples yield sample variances which, when pooled, provide a good estimate of the value of the common population variance.

We can now proceed to include both estimates of σ^2 in the determination of the value of the F statistic based on the data.

$$F = \frac{\text{variance between samples}}{\text{variance within samples}} = \frac{ns_{\bar{x}}^2}{\mu_{s_i^2}} = \frac{2.25}{1.65} \stackrel{\circ}{=} 1.36$$

If the two estimates of variance are close, the calculated value of F will be close to 1 and we would conclude that there are no significant differences among the sample means. But if the value of F is excessively *large*, then we would reject the claim of equal means. The estimate of variance in the denominator ($\mu_{s_i^2}$) depends only on the sample variances and is not affected by differences among the sample means. However, if there are extreme differences among the sample means, the numerator will be larger so the value of F will be larger. In general, as the sample means move farther apart, the value of $ns_{\bar{x}}^2$ grows larger and the value of F itself grows larger.

The critical value of F which separates excessive values from acceptable values is found in Table A-6 where α is again the level of significance and the numbers of degrees of freedom are as follows. Assuming that there are k sets of separate samples with n scores in each set,

> numerator degrees of freedom $= k - 1$
>
> denominator degrees of freedom $= k(n - 1)$

Our battery example involves $k = 3$ separate sets of samples, and there are $n = 5$ scores in each set, so

$$\text{numerator degrees of freedom } = k - 1 = 3 - 1 = 2$$

$$\text{denominator degrees of freedom } = k(n - 1) = 3(5 - 1) = 12$$

With $\alpha = 0.05$, the critical F value corresponding to these degrees of freedom is 3.8853. The computed test statistic of $F = 1.36$ does not fall within the right-tailed critical region bounded by $F = 3.8853$, so we fail to reject the null hypothesis of equal population means. That is, there is no significant difference among the means of the three given sets of samples. The pilot should purchase the least expensive battery.

MORE POLICE, FEWER CRIMES?

Does an increase in the number of police officers result in lower crime rates? That question was studied in a New York City experiment that involved a 40% increase in police officers in one precinct while adjacent precincts maintained a constant level of officers. Statistical analysis of the crime records showed that crimes visible from the street (such as auto thefts) did decrease, but crimes not visible from the street (such as burglaries) were not significantly affected.

The preceding analysis led to a right-tailed critical region, and every other similar situation will also involve a right-tailed critical region. F is always positive since the individual components of n, $s_{\bar{x}}^2$, and $\mu_{s_i}^2$ are all positive, and we know that values of F near 1 correspond to relatively close sample means and close sample variances. The only values of the F test statistic that are indicative of significant differences among the sample means are those F values that exceed 1 beyond the critical value obtained from Table A-6.

It may seem strange that we are testing equality of several means by analyzing only variances, but the sample means directly affect the value of the test statistic F. To illustrate this important point, we add 10 to each score listed under Brand X, but we leave the Brand Y and Brand Z values unchanged. The revised sample statistics follow.

Brand X	Brand Y	Brand Z
$\bar{x}_1 = 37.18$	$\bar{x}_2 = 28.10$	$\bar{x}_3 = 28.48$
$s_1^2 = 1.46$	$s_2^2 = 0.69$	$s_3^2 = 2.81$

We again assume a significance level of $\alpha = 0.05$ and test the claim that the three brands have the same population mean. In this way, we can see the effect of increasing the Brand X scores by 10. The F statistic based on the revised sample data follows.

$$F = \frac{\text{variance between samples}}{\text{variance within samples}} = \frac{ns_{\bar{x}}^2}{\mu_{s_i^2}} = \frac{(5)(26.38)}{(1.46 + 0.69 + 2.81)/3}$$

$$= \frac{131.90}{1.65} = 79.94$$

We get $n = 5$ since each brand yields 5 sample values. The value of $s_{\bar{x}}^2 = 26.38$ is obtained by calculating the variance of the sample means 37.18, 28.10, and 28.48. $\mu_{s_i}^2$ is found to be 1.65 by computing the mean of the three sample variances. With three samples of five scores each, we have $3 - 1$ or 2 degrees of freedom for the numerator and $3(5 - 1)$ or 12 degrees of freedom for the denominator. With $\alpha = 0.05$ we therefore obtain a critical F value of 3.8853. Since the computed F test statistic of 79.94 far exceeds the critical F value of 3.8853, we reject the null hypothesis of equal population means.

Before adding 10 to each score for Brand X, we obtained a test statistic of $F = 1.36$, but the increased Brand X values cause F to become 79.94. Note that, in both cases, the Brand X variance is 1.46, so the difference in the F test statistic is attributable only to the change in $\bar{x}_1$. By changing $\bar{x}_1$ from 27.18 to 37.18 and retaining the same values of $\bar{x}_2$, $\bar{x}_3$, s_1^2, s_2^2, and s_3^2, we find that the

F test statistic changes from 1.36 to 79.94. This illustrates that the F statistic is very sensitive to sample means even though it is obtained through two different estimates of the common population variance.

The method described in this section is part of a much larger branch of statistics called **analysis of variance**, well named since we do analyze the variance to determine whether the sample data vary because of chance or because they came from populations with different characteristics. The method we consider in this section has two basic limitations:

1. All samples must consist of the exact same number of values n.

2. Only one variable may be used as the basis for classifying the individual values into their respective samples.

Further study of analysis of variance techniques would reveal that both of these limitations can be circumvented. Test statistics accommodating samples of unequal sizes have been developed, and methods for incorporating any number of classification variables have been developed. The collection of all of these concepts provides a powerful tool that can be used in the analysis of complicated data that involve several variables.

EXERCISES A: ANALYSIS OF VARIANCE

10-32. Three production methods are used in making fuses, and five randomly selected fuses are selected from each process and tested for their breaking points. We want to test for equality of the three population means. What can you conclude in each case if the computed F statistic and significance level α are as follows?

(a) $F = 4.0$, $\alpha = 0.05$
(b) $F = 7.1$, $\alpha = 0.01$
(c) $F = 3.2$, $\alpha = 0.05$
(d) $F = 6.7$, $\alpha = 0.01$
(e) $F = 5.2$, $\alpha = 0.05$
(f) $F = 2.2$, $\alpha = 0.05$
(g) $F = 3.3$, $\alpha = 0.01$

10-33. Five car models are studied in a test that involves four of each model. For each of the four cars in each of the five samples, exactly 1 gallon of gas is placed in the tank and the car is driven until the gas is used up. The results follow. At the $\alpha = 0.05$ significance level, test the claim that the five population means are all equal.

Distance Traveled in Miles

A	B	C	D	E
16	18	18	19	15
22	23	18	21	16
17	15	20	22	20
17	20	20	22	17

10-34. A sociologist randomly selects subjects from three types of family structure and obtains the I.Q. scores that follow. At the $\alpha = 0.05$ level of significance, test the claim that the three population means are equal.

A	B	C
110	115	90
105	105	120
100	110	125
95	130	100
120	105	105

10-35. Do Exercise 10-34 after adding 30 to each score in group A.

10-36. A unit on elementary algebra is taught to five different classes of randomly selected students with the same academic backgrounds. A different method of teaching is used in each class, and the final averages of the 20 students in each class are compiled. The results yield the following data. At the $\alpha = 0.05$ level of significance, test the claim that the five population means are equal.

Traditional	Programmed	Audio	Audio-Visual	Visual
$n = 20$	$n = 20$	$n = 20$	$n = 20$	$n = 20$
$\bar{x} = 76$	$\bar{x} = 74$	$\bar{x} = 70$	$\bar{x} = 75$	$\bar{x} = 74$
$s^2 = 60$	$s^2 = 50$	$s^2 = 100$	$s^2 = 36$	$s^2 = 40$

10-37. The dean of a college wants to compare grade point averages of resident, commuting, and part-time students. A random sample of each group is selected and the results are as follows. At the $\alpha = 0.05$ level of significance, test the claim that the three populations have equal means.

Residents	Commuters	Part-Time
$n = 15$	$n = 15$	$n = 15$
$\bar{x} = 2.60$	$\bar{x} = 2.55$	$\bar{x} = 2.30$
$s^2 = 0.30$	$s^2 = 0.25$	$s^2 = 0.16$

10-38. Repeat Exercise 10-37 after changing the sample mean grade point average for resident students from 2.60 to 3.00.

10-39. Readability studies are conducted to determine the clarity of four different texts, and the sample scores follow. At the $\alpha = 0.05$ level of significance, test the claim that the four texts produce the same mean readability score.

Text A	Text B	Text C	Text D
50	59	48	60
51	60	51	65
53	58	47	62
58	57	49	68
53	61	50	70

10-40. An introductory calculus course is taken by students with varying high school records. The sample results of six students from each of three groups follow. The values given are the final numerical averages in the calculus course. At the $\alpha = 0.05$ level of significance, test the claim that the mean scores are equal in the three groups.

Good high school record	Fair high school record	Poor high school record
90	80	60
86	70	60
88	61	55
93	52	62
80	73	50
96	65	70

10-41. A preliminary study is conducted to determine whether there is any relationship between education and income. The sample results are as follows. The figures represent, in thousands of dollars, the life-time incomes of five randomly selected workers from each category.

At the $\alpha = 0.05$ level of significance, test the claim that the samples come from populations with equal means.

		Years of Education		
8 years or less	9–11 years	12 years	13–15 years	16 or more years
300	270	400	420	570
210	330	430	480	640
260	380	370	510	590
330	310	390	390	700
290	340	420	470	620

10-42. Three car models are studied in tests that involve several cars of each model. In each case, the car is run on exactly 1 gallon of gas until the fuel supply is exhausted, and the distances traveled are as follows. At the $\alpha = 0.05$ significance level, test the claim that the three population means are equal.

A	B	C
16	14	20
20	16	21
18	16	19
18	17	22
15	15	18
17	18	24
17	13	18
19	16	20

10-43. Five socioeconomic classes are being studied by a sociologist, and sample I.Q. scores are obtained from each group and summarized as follows. At the $\alpha = 0.05$ level of significance, test the claim that the five populations have equal means.

A	B	C	D	E
$n_1 = 10$	$n_2 = 10$	$n_3 = 10$	$n_4 = 10$	$n_5 = 10$
$\bar{x}_1 = 103$	$\bar{x}_2 = 97$	$\bar{x}_3 = 102$	$\bar{x}_4 = 100$	$\bar{x}_5 = 110$
$s_1^2 = 230$	$s_2^2 = 75$	$s_3^2 = 200$	$s_4^2 = 150$	$s_5^2 = 100$

EXERCISES B:
ANALYSIS OF VARIANCE

10-44. A study is made of three police precincts to determine the time required for a police car to be dispatched after a crime is reported. Sample results are as follows.

Precinct 1	Precinct 2	Precinct 3
$n_1 = 50$	$n_2 = 50$	$n_3 = 50$
$\bar{x}_1 = 170$ seconds	$\bar{x}_2 = 202$ seconds	$\bar{x}_3 = 165$ seconds
$s_1 = 18$ seconds	$s_2 = 20$ seconds	$s_3 = 23$ seconds

(a) At the 5% level of significance, test the claim that $\mu_1 = \mu_2$. Use the methods discussed in Chapter 8.

(b) At the 5% level of significance, test the claim that $\mu_2 = \mu_3$. Use the methods discussed in Chapter 8.

(c) At the 5% level of significance, test the claim that $\mu_1 = \mu_3$. Use the methods discussed in Chapter 8.

(d) At the 5% level of significance, test the claim that $\mu_1 = \mu_2 = \mu_3$. Use analysis of variance.

(e) Compare the methods and results of parts (a), (b), and (c) to part (d).

10-45. Five independent samples of 50 scores are randomly drawn from populations that are normally distributed with equal variances. We wish to test the claim that $\mu_1 = \mu_2 = \mu_3 = \mu_4 = \mu_5$.

(a) If we use only the methods of Chapter 8, we would test the individual claims $\mu_1 = \mu_2, \mu_1 = \mu_3, \ldots, \mu_4 = \mu_5$. What is the number of such claims? That is, how many ways can we pair off five means?

(b) Assume that for each test of equality between two means, there is a 0.95 probability of not making a type I error. If all possible pairs of means are tested for equality, what is the probability of making no type I errors?

(c) If we use analysis of variance to test the claim $\mu_1 = \mu_2 = \mu_3 = \mu_4 = \mu_5$ at the 5% level of significance, what is the probability of a type I error?

(d) Compare the results of parts (b) and (c).

10-46. Five independent samples of 50 scores are randomly drawn from populations that are normally distributed with equal variances, and the values of n, $\bar{x}$, and s are obtained in each case. Analysis of variance is then used to test the claim that $\mu_1 = \mu_2 = \mu_3 = \mu_4 = \mu_5$.

(a) If a constant is added to each of the five sample means, how is the value of the test statistic affected?

(b) If each of the five means is multiplied by a constant, how is the value of the test statistic affected?

REVIEW

We began this chapter by developing methods for testing hypotheses made about more than two population proportions. For **multinomial experiments** we tested for agreement between observed and expected frequencies by using the test statistic

$$\chi^2 = \sum \frac{(O - E)^2}{E}$$

In repeated large samplings, the distribution of the preceding χ^2 statistic can be approximated by the chi-square distribution. This approximation is generally considered acceptable as long as all expected frequencies are at least 5. In a multinomial experiment with k cells or categories, the number of degrees of freedom is $k - 1$.

In Section 10-3 we used the preceding sample χ^2 statistic to measure disagreement between observed and expected frequencies in *contingency tables*. A contingency table contains frequencies; the rows correspond to categories of one variable while the columns correspond to categories of another variable. With contingency tables, we test the hypothesis that the two variables of classification are independent. The test statistic is obtained by use of this expression for χ^2, and we can approximate the sampling distribution of that statistic by the chi-square distribution as long as all expected frequencies are at least 5. In a contingency table with r rows and c columns, the number of degrees of freedom is $(r - 1)(c - 1)$. The value of the expected frequency E for any particular cell is given by

$$E = \frac{(\text{row total})(\text{column total})}{(\text{grand total})}$$

With these additional applications of the chi-square distribution, we developed **nonparametric** tests that do not require specific population distributions and do not depend upon population parameters.

In Section 10-4 we used **analysis of variance** to determine whether differences among three or more sample means are due to chance fluctuations or whether the differences are significant. This method is **parametric** since it requires normally distributed populations with equal variances and it is based on estimated values of population parameters. Our comparison of sample means is based on two different estimates of the common population variance. The test statistic is

$$F = \frac{\text{variance between samples}}{\text{variance within samples}} = \frac{ns_{\bar{x}}^2}{\mu_{s_1^2}}$$

In repeated samplings, the distribution of this F statistic can be approximated by the F distribution which has critical values given in Table A-6. With k sets of different samples and n values in each sample, we have $k - 1$ degrees of freedom for the numerator and $k(n - 1)$ degrees of freedom for the denominator.

REVIEW EXERCISES

10-47. The owner of a new grocery store records the number of customers arriving on the different days for one week. The results are as follows. At the $\alpha = 0.05$ significance level, test the claim that customers arrive on the different days with equal frequencies.

Sun	Mon	Tues	Wed	Thurs	Fri	Sat
97	72	55	68	70	88	110

10-48. A lawyer is studying punishments for a certain crime and wants to compare the sentences imposed by three different judges. Randomly selected results follow. At the $\alpha = 0.05$ level of significance, test the claim that the three judges impose sentences that have the same mean.

Judge A	Judge B	Judge C
$n = 36$	$n = 36$	$n = 36$
$\bar{x} = 5.2$ years	$\bar{x} = 4.1$ years	$\bar{x} = 5.5$ years
$s = 1.4$ years	$s = 1.1$ years	$s = 1.5$ years

10-49. Five years ago, a survey was made of a group of students. The same survey was made of a group of students this year. The first question dealt with sexual permissiveness, and there were five possible responses. At the $\alpha = 0.01$ level of significance, test the claim that the responses are independent of the student group.

	A	B	C	D	E
Student group five years ago	25	32	18	10	5
Student group today	40	73	44	28	20

10-50. In a certain region, a survey is made of companies that officially declared bankruptcy during the past year. Seventy-two of the 120 small bankrupt businesses advertised in weekly newspapers. Twenty-five of the 65 medium-sized bankrupt businesses advertised in weekly newspapers, while 8 of the 15 large bankrupt businesses did so. At the 5% level of significance, test the claim that the three proportions of bankrupt businesses which used weekly newspaper ads are equal.

10-51. A psychologist conducted studies on the relationship between forgetfulness and I.Q. scores, and the results are as follows. At the 0.05 level of significance, test the claim that I.Q. scores and levels of forgetfulness are independent.

	Forgets infrequently	Forgets occasionally	Forgets often
Low I.Q.	15	10	5
Medium I.Q.	20	30	10
High I.Q.	15	25	15

10-52. Three teaching methods are used with three groups of randomly selected students and the results follow. At the 0.05 level of significance, test the claim that the samples came from populations with equal means.

Method A	Method B	Method C
$n = 20$	$n = 20$	$n = 20$
$\bar{x} = 72.0$	$\bar{x} = 76.0$	$\bar{x} = 71.0$
$s = 9.0$	$s = 10.0$	$s = 12.0$

10-53. (a) What is the difference between parametric and nonparametric tests of hypotheses?
(b) What is a multinomial experiment?
(c) If a contingency table has five rows and six columns, what is the number of degrees of freedom?
(d) What assumptions are necessary for the use of analysis of variance as discussed in this chapter?
(e) In testing the claim $\mu_1 = \mu_2 = \mu_3$, why would we use analysis of variance instead of using the normal distribution to test $\mu_1 = \mu_2$, $\mu_2 = \mu_3$, and $\mu_1 = \mu_3$?

NONPARAMETRIC STATISTICS

11-1 OVERVIEW

Prior to Chapter 10, most of the methods of inferential statistics could be classified as **parametric** methods because they involved parameters such as means, standard deviations, or variances. In addition, those methods usually required some strict prerequisites, such as the requirement that the sample data be extracted from a population having a normal distribution. Nonparametric techniques are advantageous because they transcend these limitations by circumventing dependence on population parameters like μ and σ, and by removing the strict requirements about the population distributions. Another advantage of nonparametric methods is that they can accommodate qualitative or attribute data that lack numerical meaning. If all of these advantages could be accrued without any significant disadvantages, we could ignore parametric methods and enjoy much less complicated procedures. However, the balance of nature seems to work against this as the advantages are offset by some major disadvantages. In general, nonparametric methods tend to be wasteful of information, since specific numerical data are often reduced to a qualitative form. For example, one nonparametric test requires that a weight loss of 75 pounds be recorded simply as a negative sign. A weight loss of only 1 pound would receive the same representation. (This is almost an insult to those on diets.) Nonparametric tests usually lack the sensitivity of parametric tests, with the result that the null hypothesis H_0 is not rejected as often as it is in similar parametric tests. The likelihood of type II errors is therefore higher.

In Chapter 10 we introduced nonparametric tests which used the chi-square distribution in analyses of multinomial experiments and contingency tables. In this chapter we introduce the **sign test** and the **rank-sum** test as additional examples of nonparametric methods. Through these tests the advantages and disadvantages of nonparametric methods become more apparent.

11-2 SIGN TEST

One example from Section 8-3 involves an educational service that offered a course designed to improve scores on I.Q. tests. The sample results for ten randomly selected subjects appear in Table 11-1.

In Section 8-3 we proceeded with the parametric student t test, but in this section we apply the nonparametric sign test which can be used to test for equality between two medians. The key concept underlying the sign test is this: **if there is no difference between the two sets of data, the number of plus signs should be approximately equal to the number of minus signs.** For the data in Table 11-1, we can conclude that the course is effective if there are significantly more plus signs than minus signs.

TABLE 11-1

Subject	I.Q. before course	I.Q. after course	Sign of change from before to after
A	96	99	+
B	110	112	+
C	98	107	+
D	113	110	−
E	88	88	0
F	92	101	+
G	106	107	+
H	119	123	+
I	100	91	−
J	97	99	+

The sign test requires that we exclude ties (represented by zeros), so we are left with this specific question: Do the seven plus signs in Table 11-1 significantly outnumber the two minus signs? The answer to this question depends on the level of significance we require, so let's use $\alpha = 0.05$ as we did in Section 8-3. When we assume the null hypothesis of no increase in I.Q. scores, we assume that plus signs and minus signs occur with equal frequency, so $P(\text{plus sign}) = P(\text{minus sign}) = 0.5$. (The null hypothesis of no increase also includes the possibility of a decrease, but we continue to assume that plus signs and minus signs are equally likely.)

Since our results fall into two categories (positive, negative) and we have a fixed number of independent subjects, we can use the binomial probability distribution to determine the likelihood of getting seven or more plus signs among the nine subjects. From Table A-2 with $n = 9$ and $p = 0.5$, we see that the probability of seven or more plus signs is $0.070 + 0.018 + 0.002 = 0.090$. But with $\alpha = 0.05$, we fail to reject the null hypothesis of no increase in I.Q. scores since the probability of seven or more plus signs is 0.090. Since the sample results are associated with a probability greater than 0.05, we conclude that they can easily occur by chance. We consider as significant only those cases that have a probability less than 0.05.

This sign test involves simple computations but leads to the same conclusion we obtained in Section 8-3. However, this sign test is wasteful of information in that it ignores the magnitudes of the changes.

To appreciate the consequences resulting from the wasted quantitative data, suppose that the sample data correspond to Table 11-2. The sign test again involves seven plus signs and two minus signs, so once again we fail to reject the null hypothesis of no increase in I.Q. scores. But if we employ the student t test as in Section 8-3, we get

$$t = \frac{\overline{d}}{s_d/\sqrt{n}} = \frac{22.8}{18.8/\sqrt{10}} \stackrel{\circ}{=} 3.835$$

which causes *rejection* of the null hypothesis because the critical t value is found to be 1.833. Using the data of Table 11-2, we fail to reject H_0 if we use the sign test, but we reject H_0 if we use the student t test.

TABLE 11-2

Subject	I.Q. before course	I.Q. after course	Difference	Sign of change
A	96	129	+33	+
B	110	142	+32	+
C	98	137	+39	+
D	113	110	−3	−
E	88	88	0	0
F	92	131	+39	+
G	106	137	+31	+
H	119	153	+34	+
I	100	91	−9	−
J	97	129	+32	+

An intuitive analysis of Table 11-2 suggests that the course seems to be effective, but the sign test is blind to the magnitude of the changes and therefore fails to reject the null hypothesis which is probably false. This illustrates the previous assertion that nonparametric tests lack the sensitivity of parametric tests with the resulting tendency that null hypotheses are not rejected so often as they are in parametric tests.

The next example illustrates the fact that nonparametric methods can accommodate attribute or qualitative data.

Example

Use the World Series victories from 1903 to 1978 to test the claim that the American and National Leagues are equal in their ability to win. Use a significance level of $\alpha = 0.05$.

1903 AL	1914 NL	1925 NL	1936 AL	1947 AL	1958 AL	1969 NL
1904 No series	1915 AL	1926 NL	1937 AL	1948 AL	1959 NL	1970 AL
1905 NL	1916 AL	1927 AL	1938 AL	1949 AL	1960 NL	1971 NL
1906 AL	1917 AL	1928 AL	1939 AL	1950 AL	1961 AL	1972 AL
1907 NL	1918 AL	1929 AL	1940 NL	1951 AL	1962 AL	1973 AL
1908 NL	1919 NL	1930 AL	1941 AL	1952 AL	1963 NL	1974 AL
1909 NL	1920 AL	1931 NL	1942 NL	1953 AL	1964 NL	1975 NL
1910 AL	1921 NL	1932 AL	1943 AL	1954 NL	1965 NL	1976 NL
1911 AL	1922 NL	1933 NL	1944 NL	1955 NL	1966 AL	1977 AL
1912 AL	1923 AL	1934 NL	1945 AL	1956 AL	1967 NL	1978 AL
1913 AL	1924 AL	1935 AL	1946 NL	1957 NL	1968 AL	

Solution

If we denote the American League (AL) victories by + and the National League (NL) victories by −, we have 45 plus signs and 30 minus signs. With

$p = 0.5$ and $n = 75$, we use the normal distribution as an approximation to the binomial distribution to determine that the probability of 45 or more plus signs is 0.0526. With $\alpha = 0.05$, we reject the null hypothesis of equal abilities only if the plus signs occur so frequently (or infrequently) that the probability is computed to be less than 0.05 in one-tailed tests or less than 0.025 in two-tailed tests. The test under discussion involves two tails since a significantly high or low number of plus signs would cause us to reject the claim of equal winning abilities. But the computed probability of 0.0526 indicates that the 45 plus signs among the 75 contests could easily occur by chance. We fail to reject the null hypothesis of equal winning abilities.

In determining the probability of getting x or more plus signs among n signs, we can use Table A-2 provided $n \leq 25$. When $n > 25$, we use the normal distribution as an approximation to the binomial distribution with

$$\mu = n \cdot p = n \cdot 0.5$$

$$\sigma = \sqrt{n \cdot p \cdot q} = \sqrt{n \cdot 0.5 \cdot 0.5} = \frac{\sqrt{n}}{2}$$

$$z = \frac{x - \mu}{\sigma} = \frac{x - (n \cdot 0.5)}{\sqrt{n}/2}$$

where x is adjusted for the continuity correction factor as discussed in Section 5-5. For the preceding example we had $n = 75$ and $x = 45$ and we sought the probability of getting at least 45 successes in 75 trials. We determine that

$$z = \frac{x - (n \cdot 0.5)}{\sqrt{n}/2} = \frac{44.5 - 37.5}{\sqrt{75}/2} = 1.62$$

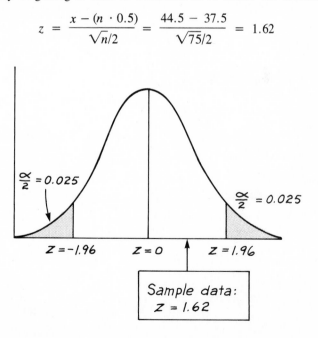

FIGURE 11-1

$\frac{\alpha}{2} = 0.025$

$\frac{\alpha}{2} = 0.025$

$Z = -1.96$ $Z = 0$ $Z = 1.96$

Sample data:
$Z = 1.62$

Monkeys have become extremely important as experimental subjects because there is often a strong correlation between their behavior and human behavior. Monkeys have been used successfully in studies of insomnia, fatigue, ulcers, sexual motivation, and effects of various drugs. Statistics is used to analyze the significance of the experimental results.

Note that use of the continuity correction factor causes us to represent the discrete value of 45 by the interval from 44.5 to 45.5. We obtain the critical z values of 1.96 and -1.96 from Table A-3. Since the test statistic of $z = 1.62$ does not fall in the critical region, we fail to reject the null hypothesis of equal winning abilities (see Figure 11-1).

The previous examples involved application of the sign test to a comparison of *two* sets of data, but we can sometimes use the sign test to investigate a claim made about one set of data. The next example illustrates this.

Example

Use the sign test to test the claim that the median I.Q. of pilots is at least 100 if a sample of 50 pilots contained exactly 22 members with I.Q.'s of 100 or higher.

Solution

The null hypothesis is the claim that the median is equal to or greater than 100; the alternate hypothesis is the claim that the median is less than 100. We select $\alpha = 0.05$ and we denote by $+$ each I.Q. score that is at least 100, so we have 22 plus signs and 28 minus signs. We can now determine the significance of getting 22 plus signs out of a possible 50. Since we have a fixed number of independent trials (50) with constant probabilities ($p = 0.5$) and only two outcomes (plus, minus), we use the binomial probability distribution. With $n = 50$ we are not able to use Table A-2 so we use the normal distribution as an approximation to the binomial distribution and compute the z test statistic as follows.

$$z = \frac{x - (n \cdot 0.5)}{\sqrt{n}/2} = \frac{22.5 - (50 \cdot 0.5)}{\sqrt{50}/2} = -0.71$$

From Figure 11-2 we can see that the computed z value of -0.71 does not fall within the critical region. Therefore we fail to reject the null hypothesis. A corresponding parametric test may or may not lead to the same conclusion, depending on the specific values of the 50 sample scores.

FIGURE 11-2

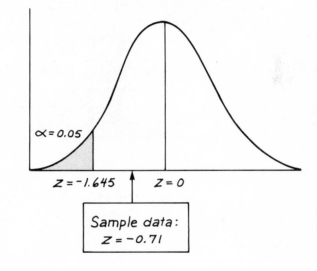

EXERCISES A: SIGN TEST

In Exercises 11-1 through 11-6, use the sign test for the exercise named (see pages (285-291). Note that you will be testing for equality between two medians instead of two means.

11-1. Exercise 8-21.   11-2. Exercise 8-22.

11-3. Exercise 8-31.   11-4. Exercise 8-36.

11-5. Exercise 8-38.   11-6. Exercise 8-47.

11-7. Thirteen department heads of a college are asked if they understand the purpose of the Department Heads Council. Nine respond affirmatively and 4 respond negatively. At the 0.05 level of significance, use the sign test to test the claim that most (more than half) department heads feel that they understand the purpose of their council. Assume that the 13 department heads comprise a sample drawn from a larger population.

11-8. A standardized aptitude test yields a mathematics score M and a verbal score V for each person. Among 15 male subjects, $M - V$ is positive in 12 cases, negative in two cases, and zero in one case. At the 0.05 level of significance, use the sign test to test the claim that males do better on the mathematics portion of the test.

11-9. Referring to the aptitude test of Exercise 11-8, of 15 girls taking the test there are 10 girls for whom $M - V$ is negative and 5 girls for whom $M - V$ is positive. At the 0.05 level of significance, use the sign test to test the claim that girls do better on the verbal portion of the test.

11-10. A political party preference poll is taken among 20 randomly selected voters. If seven prefer the Republican Party while 13 prefer the Democratic Party, apply the sign test to test the claim that both parties are preferred equally. Use a 0.05 level of significance.

11-11. A television commercial advertises that "7 out of 10 dentists surveyed prefer Covariant toothpaste over the leading competitor." Assume that 10 dentists are surveyed and 7 do prefer Covariant while 3 favor the other brand. Is this a reasonable basis for making the claim that most (more than half) dentists favor Covariant toothpaste? Use the sign test with a significance level of 0.05.

11-12. One-hundred randomly selected high school seniors are given a college aptitude test. After a period of intensive training, another similar test is given to the same students, and 59 receive higher grades, 36 receive lower grades, and 5 students receive the same grades. At the 0.05 level of significance, use the sign test to test the claim that the training is effective.

11-13. A company is experimenting with a new fertilizer at 50 different locations. In 32 of the locations there is an increase in production, while in 18 locations there is a decrease. At the 0.05 level of significance, use the sign test to test the claim that production is increased by the new fertilizer.

11-14. A new diet is designed to lower cholesterol levels. Sixty subjects go on this diet. In 6 months, 36 subjects have lower cholesterol levels while 22 have slightly higher levels. (Two subjects registered no change.) At the 0.05 level of significance, use the sign test to test the claim that the diet produces no change in cholesterol levels.

11-15. Thirty drivers are tested for reaction times. They are then given two drinks and tested again, with the result that 22 have slower reaction times while 6 have faster reaction times. (Two drivers received the same scores before and after the drinks.) At the 0.01 significance level, use the sign test to test the claim that the drinks had no effect on the reaction times.

11-16. Forty police academy students use two different pistols in target practice. Analysis of the scores shows that 24 students get higher scores with the more expensive pistol, while 16 students get better scores with the less expensive pistol. At the 0.05 level of significance, use the sign test to test the claim that both pistols are equally effective.

11-17. Of 50 voters surveyed, exactly 28 favor a tax revision bill before Congress, while all the others are opposed. At the 0.10 level of significance, use the sign test to test the claim that the majority (more than half) of voters favor the bill.

11-18. Use the sign test to test the claim that the median I.Q. score of Philadelphians is at least 100 if a sample of 200 Philadelphians contains exactly 86 with I.Q.'s of 100 or more. Assume a significance level of 0.01.

11-19. Use the sign test to test the claim that the median life of a battery is at least 40 hours if a random sample of 75 includes exactly 32 that last 40 hours or more. Assume a significance level of 0.05.

EXERCISES B: SIGN TEST

11-20. Fifty-seven of the voters surveyed favor passage of a certain bill and they constitute a majority. At the 0.05 significance level, we apply the sign test and reject the claim that voters are equally split on the bill. Given the preceding information, what is the largest sample size possible?

11-21. Of n subjects tested for high blood pressure, a majority of exactly 50 provided negative results. (That is, their blood pressure is not high.) This is sufficient for us to apply the sign test and reject (at the 0.01 level of significance) the claim that the median blood pressure level is high. Find the smallest value n can assume.

11-3 RANK-SUM TEST

As we have just said, the sign test is a very simple way of testing for equality of two population medians when we have samples of matched pairs, but it has the one very serious disadvantage of totally disregarding the magnitudes of the differences between the paired observations. This disadvantage is largely overcome by the **rank-sum test*** for dealing with samples

---

*This test is also called the Wilcoxon rank-sum test and is equivalent to the Mann-Whitney test.

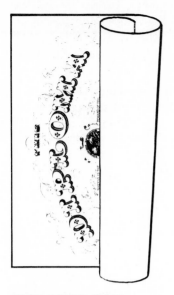

NEW YORK CITY SCHOOL DROPOUT RATE RISES

The New York State Board of Regents recently noted that "in New York City, less than 50% of the public high school students graduate." The actual dropout rate could be either 53.1% or 13.2%, depending upon which definition of dropout is used. In any case, that rate appears to be increasing. An analysis of the relevant statistics is intended to be a first step in identifying and correcting the problem of allowing students to leave schools without employable skills.

of matched pairs. Here we consider a rank-sum test that can be used when the two samples are not paired.

Example

Random samples of teachers' salaries from New York State and Florida are as follows. At the $\alpha = 0.05$ significance level, test the claim that the salaries of teachers are the same in both states.

New York	Florida
$23,300 (26)	$13,500 (8)
14,500 (12)	14,200 (10)
17,700 (18)	12,800 (5.5)
19,600 (24)	14,000 (9)
18,100 (21)	15,300 (13)
18,000 (20)	11,800 (1)
18,200 (22)	12,300 (4)
20,800 (25)	15,400 (14)
15,500 (15)	14,400 (11)
16,200 (16)	12,800 (5.5)
17,900 (19)	13,300 (7)
	17,000 (17)
	19,300 (23)
	12,200 (3)
	12,100 (2)

Solution

We may be tempted to use the student t test to compare the means of two independent samples (as in Section 8-3), but we cannot meet the prerequisite of having normally distributed populations since salaries are not normally distributed. We therefore require a nonparametric method and the rank-sum test is appropriate here.

We begin by ranking the 26 salaries beginning with a rank of 1 (assigned to the lowest salary of $11,800), a rank of 2 (assigned to the second lowest salary of $12,100), and so forth. The ranks corresponding to the various salaries are shown in parentheses in the preceding table. Note that there is a tie between the fifth and sixth scores. When ties occur, we compute the mean of the ranks involved in the tie and assign that mean rank to each of the tying values. In the case of our tie between the fifth and sixth scores, we assign the rank of 5.5 to each of those two salaries. We denote by R the sum of the ranks for one of the two samples. If we choose the New York salaries we get

$$R = 26 + 12 + 18 + 24 + 21 + 20 + 22 + 25 + 15 + 16 + 19$$

$$= 218$$

With a null hypothesis of equal means and with both sample sizes at least 10, the sampling distribution of R is approximately normal. Denoting the mean of the sample R values by μ_R and the standard deviation of the sample R values by σ_R, we are able to use the test statistic

$$z = \frac{R - \mu_R}{\sigma_R}$$

where

$$\mu_R = \frac{n_1(n_1 + n_2 + 1)}{2}$$

$$\sigma_R = \sqrt{\frac{n_1 n_2(n_1 + n_2 + 1)}{12}}$$

n_1 = size of the sample from which the rank sum R is found

n_2 = size of the other sample

The expression for μ_R is a variation of a result of mathematical induction which states that the sum of the first n positive integers is given by $1 + 2 + 3 + \cdots + n = n(n + 1)/2$ and the expression for σ_R is a variation of a result which states that the integers $1, 2, 3, \ldots, n$ have standard deviation $\sqrt{n(n + 1)/12}$.

For the salary data given in the table we have already found that the rank sum R for the New York salaries is 218. Since there are 11 New York salaries present, we have $n_1 = 11$. Also, $n_2 = 15$ since there are 15 Florida salaries. We can now determine the values of μ_R, σ_R, and z.

$$\mu_R = \frac{n_1(n_1 + n_2 + 1)}{2} = \frac{11(11 + 15 + 1)}{2} = 148.50$$

$$\sigma_R = \sqrt{\frac{n_1 n_2(n_1 + n_2 + 1)}{12}} = \sqrt{\frac{(11)(15)(11 + 15 + 1)}{12}} = 19.27$$

$$z = \frac{R - \mu_R}{\sigma_R} = \frac{218 - 148.50}{19.27} = 3.61$$

The significance of the test statistic z can now be treated in the same manner as in previous chapters. We are now testing (with $\alpha = 0.05$) the hypothesis that the two population means are equal, so we have a two-tailed test with critical z values of 1.96 and -1.96. The test statistic of $z = 3.61$ falls within the critical region and we therefore reject the null hypothesis that the salaries are the same in both states.

In general, relatively large values of R reflect the situation where the ranks of the first sample are larger than those of the second sample. These relatively large R values lead to positive z test statistics. Conversely, relatively small R values reflect ranks of the first sample which are lower than the ranks of the other sample, and this condition leads to a negative z test statistic. A value of R near μ_R indicates that the sample data closely follow the pattern we would expect for similar populations.

You can verify that, if we interchange the two sets of salaries, we will find that $R = 133$, $\mu_R = 202.5$, $\sigma_R = 19.27$, and $z = -3.61$, so the same conclusion will be reached.

We have stated that this rank-sum test considers the relative magnitudes of the sample data whereas the sign test does not. In the sign test, a weight loss of 1 pound or 50 pounds receives the same sign so the actual magnitude of the loss is ignored. While rank-sum tests do not directly involve quantitative differences between data from two samples, changes in magnitude do cause changes in rank, and these in turn affect the value of the test statistic.

For example, if we change the Florida salary of $13,500 to $20,000, then the value of the rank-sum R will change and the value of the z test statistic will also change.

There are ways of dealing with a rank-sum test when either sample contains less than 10 scores, and there are ways of applying a rank-sum test to two samples of matched pairs of data.

We have attempted to convey the general spirit of the nonparametric approach through illustrations involving the sign test and rank-sum test. For more complete and detailed treatments of these and other nonparametric methods, you can consult texts such as *Elements of Nonparametric Statistics* by G. E. Noether, *Practical Nonparametric Statistics* by W.J. Conover, and *Nonparametric Statistics* by Sydney Siegel.

EXERCISES A: RANK-SUM TEST

11-22. The following scores were randomly selected from last year's college entrance examination scores. Use the rank-sum approach to test the claim that the performance of New Yorkers equals that of Californians. Assume a significance level of 0.05.

New York	520 490 571 398 602 475 557 621 737 403 511 598
California	508 563 385 617 704 401 409 527 393 478 521 536

11-23. The following scores represent the reaction times (in seconds) of randomly selected subjects from two age groups. Use the rank-sum approach at the 0.05 level of significance to test the claim that both groups have the same reaction times.

| 18 years of age: | 1.96 | 0.94 | 0.96 | 1.51 | 1.36 | 1.41 | 1.03 | 1.12 |
| | 2.12 | 0.86 | 0.79 | 1.17 | 1.13 | 1.00 | 1.01 | |

| 50 years of age: | 1.03 | 1.42 | 1.75 | 2.01 | 0.93 | 1.92 | 2.00 | 1.87 |
| | 2.09 | 1.73 | 1.49 | 1.82 | | | | |

11-24. A class of statistics students rated the president's performance using several different criteria and the composite scores follow. Use the rank-sum test at the 0.05 level of significance to test the claim that the ratings of both sexes are the same.

Males	27	36	42	57	88	92	60	60	43	29	76	79
Females	21	43	38	40	40	60	87	72	73	10		

11-25. Job applicants from two cultural backgrounds are tested to determine the number of trials they need to learn a certain task. The results are as follows. Use the rank-sum test at the 0.05 level of significance to test the claim that the number of trials is the same for both groups.

Group A	7	12	18	15	13	14	22	9	11	10	10	10
Group B	21	19	17	8	16	16	20	24	6	19		

11-26. An auto parts supplier must send many shipments from the central warehouse to the city in which the assembly takes place, and she wants to determine the faster of two railroad routes. A search of past records provides the following data. (The shipment times are in hours.) At the 0.05 level of significance, use the rank-sum test to determine whether or not there is a significant difference between the routes.

Route A	98	102	83	117	128	92	112	108	108	100	93	72	95	91
Route B	96	132	121	87	106	102	116	95	99	76	97	104	115	114

11-27. A large city police department offers a refresher course on arrest procedures. The effectiveness of this course is examined by testing 15 randomly selected officers who have completed the course. The same test is given to 15 randomly selected officers who have not had the refresher course. The results are as follows. At the 0.05 level of significance, test the claim that the course has no effect on the test grades by using the rank-sum approach.

Group completing the course	173 141 219 157 163 165 178 200
	154 189 192 201 157 168 181
Group without the course	159 124 170 148 135 133 137 189 181 111
	144 127 138 151 162

11-28. A study is conducted to determine whether a drug affects eye movements. A standardized scale is developed and the drug is administered to one group, while a control group is given a placebo that produces no effects. The eye movement ratings of the subjects are as follows. At the 0.01 level of significance, test the claim that the drug has no effect on eye movements. Use the rank-sum approach.

| Drugged group | 652 512 711 621 508 603 787 747 516 624 627 777 729 |
| Control group | 674 676 821 830 565 821 837 652 549 668 772 563 703 789 800 711 598 |

11-29. Two vending machines that dispense coffee are studied to determine whether they distribute the same amounts. Samples are obtained and the contents (in liters) are as follows. At the 0.05 level of significance, use the rank-sum approach to test the claim that the machines distribute the same amount.

| Machine A | 0.210 0.213 0.206 0.195 0.180 0.250 0.212 0.217 0.213 0.222 0.201 0.205 0.209 |
| Machine B | 0.229 0.224 0.221 0.247 0.270 0.233 0.237 0.235 0.238 0.200 0.198 0.216 0.241 0.273 0.205 |

11-30. Groups of randomly selected men and women are given questionnaires designed to measure their attitudes towards capital punishment, and the results are as follows. At the 0.05 level of significance, test the claim that there is no difference between the attitudes of men and women concerning capital punishment. Use the rank-sum approach.

| Men | 67 72 48 30 92 15 5 87 91 54 66 72 98 97 75 74 |
| Women | 20 40 37 42 51 15 68 35 12 31 85 |

EXERCISES B: RANK-SUM TEST

11-31.   (a)   The *ranks* for Group A are 1, 2, . . . , 10 and the *ranks* for Group B are 11, 12, . . . , 20. At the 0.05 level of significance, use the rank-sum test to test the claim that both groups come from the same population.

      (b)   The *ranks* for Group A are 1, 3, 5, 7, . . . , 19 and the *ranks* for Group B are 2, 4, 6, . . . , 20. At the 0.05 level of significance, use the rank-sum test to test the claim that both groups come from the same population.

      (c)   Compare parts (a) and (b).

      (d)   What changes occur when the rankings of the two groups in part (a) are interchanged?

      (e)   Use the two groups in part (a) and interchange the ranks of 1 and 20 and then note the changes that occur.

REVIEW

In this chapter we examined two tests of statistics that are categorized as nonparametric. Besides excluding involvement with the usual population parameters such as μ and σ, nonparametric methods are not encumbered by many of the restrictions placed on parametric methods. But nonparametric methods generally lack the sensitivity of their parametric counterparts, and this results in a greater chance of the mistake of failing to reject a false null hypothesis.

The sign test is used to compare the medians of two dependent samples or to test the median value of a single sample. The test statistic is obtained through the simple process of counting plus or minus signs. The null hypothesis of equal medians leads to the assumption that both signs are equally likely and have a probability of 0.5. A significantly large number of either sign is evidence against the null hypothesis. Specific cutoff significance levels are obtained by using Table A-2 or the normal approximation to the binomial distribution with

$$\mu = n \cdot 0.5$$

$$\sigma = \sqrt{n}/2$$

$$z = \frac{x - \mu}{\sigma}$$

Another nonparametric method is the rank-sum test, which can be used when the two samples are not paired. If the two samples come from the same population, the ranks of their scores should be distributed evenly between the two samples. But if the higher ranks are concentrated in one sample, this is evidence that the samples come from different populations. Letting R represent the sum of the ranks for either sample we compute

$$z = \frac{R - \mu_R}{\sigma_R}$$

where

$$\mu_R = \frac{n_1(n_1 + n_2 + 1)}{2}$$

$$\sigma_R = \sqrt{\frac{n_1 n_2(n_1 + n_2 + 1)}{12}}$$

and n_1 is the size of the sample from which R is found, while n_2 is the size of the other sample. After obtaining the test statistic z, we proceed with the hypothesis test as we did in previous chapters.

REVIEW EXERCISES

11-32. Randomly selected cars are tested for fuel consumption and then re-tested after a tune-up. The measures of fuel consumption follow. At the 0.05 significance level, use the sign test to evaluate the claim that the tune-up has no effect on fuel consumption.

Before tune-up	16	23	12	13	7	31	27	18	19	19	19	11	9	15
After tune-up	18	23	16	17	8	29	31	21	19	20	24	13	14	18

11-33. A police academy gives an entrance exam. Sample results for applicants from two different counties follow. Use the rank-sum approach to test the claim that there is no difference between the scores from the two counties. Assume a significance level of 0.05.

Kings County	76	52	39	27	88	73	75	92	99	83	79		
Queens County	67	48	52	40	53	91	30	20	18	23	61	63	62

11-34. Use the sign test to test the claim that the median age of a teacher in Montana is 38 years. The ages of a sample group of teachers from Montana are 35, 31, 27, 42, 38, 39, 56, 64, 61, 33, 35, 24, 25, 28, 37, 36, 40, 43, 54, and 50. Use a 0.05 significance level.

11-35. An annual art award was won by a woman in 30 out of 40 presentations. At the 0.05 significance level, use the sign test to test the claim that men and women are equal in their abilities to win this award.

DESIGN, SAMPLING, AND REPORT WRITING

This chapter summarizes the procedure for setting up a statistical experiment.

12-1 IDENTIFYING OBJECTIVES

We must begin by determining exactly what question we want answered. Beginning researchers are often overcome by enthusiasm as they set out to collect facts without considering *precisely* what they are investigating and which facts are truly relevant. The original statement of the problem is often too vague or too broad. When this happens, too many different directions are sometimes pursued. For example, a social scientist may want to know how attitudes toward racial integration have changed in the past 30 years. This broad problem must be condensed to some specific areas that can be objectively measured and analyzed. One method is to repeat earlier surveys that were used to determine the percentages of people with prointegration responses to questions such as:

Do you think white students and black students should attend separate schools?

Do you think marriages between blacks and whites should be prohibited by law?

It is important at this stage of the experiment to identify the population that will be considered. If we are measuring changing attitudes toward racial integration, we must identify exactly the group whose attitudes we want to know. The population may be adult Americans over the age of 18, Southern whites, middle-class adult white males, etc.

12-2 DESIGNING THE EXPERIMENT

In order to obtain meaningful data in an efficient way, we must develop a complete plan for collecting data *before* the collection is actually begun. Researchers are often frustrated and discouraged if they learn that the method of collection or the data themselves cannot be used to answer their questions. Will the experiment be conducted on the entire population (in which case we will use descriptive statistics) or will a sample be drawn from the population (in which case we will use inferential statistics)? The population size usually requires us to make inferences on the basis of sample data. We need to determine the size of the sample and the method of sampling in the very beginning.

Some aspects of sample size determination were discussed in Chapter 7, and we discuss some types of sampling later in this chapter. In addition to

M vs *Q*
L vs *S*

THE CASE OF COKE VERSUS PEPSI

In an advertising war between Coca-Cola and Pepsi-Cola, a television commercial showed a taste test in which the majority of regular Coke drinkers preferred Pepsi (labeled M*) over coke (labeled* Q*). The Coke camp responded with research results indicating that people prefer the letter* M *over the letter* Q*. One Coke commercial has an announcer saying "Here's a fascinating report. Two glasses, one marked* M *and the other marked* Q*. Both glasses contain the same thing. Coca-Cola. We asked people to pick the one that tasted better. Most of them picked* M *even though the drinks were the same. You know what that proves? It proves that people will pick* M *more often than* Q*, So* M *has an advantage." After Pepsi switched letters to* L *(for Pepsi) and* S *(for Coke), another Coke commercial had the taster saying that "when it comes to letters, I will pick* L *every time over* S*. L stands for liberty and lunch . . . all the things that really made our country great. Could we do numbers now? My favorite number is six."*

determining sample size and sampling plan, we need to obtain a list that will serve as the source from which our samples are drawn. This source may be a telephone directory, voter registration files, class roster, car registration list, etc.

Finally, we should describe how we intend to analyze the data after sampling is completed. We should note which statistics are to be computed and which formulas will be used in making estimates and conducting tests of hypotheses.

12-3 SAMPLING AND COLLECTING DATA

Sampling and data collection usually require the most time, effort, and money. Careful planning will help to minimize the expenditure of those precious resources. Take care to ensure that the sampling is done according to plan and that the data are recorded in a complete and accurate manner. Some different methods of sampling follow.

Random Sampling

In random sampling, each member of the population has an equal chance of being selected. Random sampling is also called representative or proportionate sampling since all groups of the population should be proportionately represented in the sample.

Random sampling is not the same as haphazard or unsystematic sampling. Much effort and planning must be invested in order to ensure that any bias will be avoided. For example, if a list of all elements from the population is available, then the names of those elements can be placed in capsules and put in a bowl; those capsules can then be mixed and samples selected.

Another approach involves numbering the list and using a table of random numbers to determine which specific elements are to be selected.

A major problem with these approaches is the difficulty of finding a complete list of *all* elements in the population.

Stratified Sampling

After classifying the population into at least two different strata (or classes) that share the same characteristics, we draw a sample from each stratum. In surveying views on an equal rights amendment to the Constitution, we might use sex as a basis for creating two strata. After obtaining a list of men and a list of women, we use some suitable method (such as random sampling) to select a certain number of people from each list. If it should happen that some strata are not represented in the proper proportion, then the results can be adjusted or weighted accordingly. Stratified sampling is often the most efficient of the various types.

Systematic Sampling

In systematic sampling we select some starting point and then select every nth element. For example, suppose we use a telephone directory of 10,000 names as our population, and we must choose 200 of those names. We can randomly select one of the first 50 names and then choose every 50th name after that. This method is simple and used frequently.

Cluster Sampling

In cluster sampling we first divide the population area into sections and then randomly select a few of those sections. We can then choose all of the members of those chosen sections or we can randomly select members from those sections. For example, in conducting a pre-election poll, we could randomly select 30 election precincts and then survey 50 people from each of those chosen precincts. This would be much more efficient than selecting one person from each of 1500 precincts. We can again adjust or weight the results to correct for any disproportionate representations of groups. Cluster sampling is used extensively by the government and by private research organizations.

Importance of Sampling

Even experienced and reputable research organizations sometimes get erroneous results due to biased sampling or poor methodology. One classic example is the 1936 telephone survey which indicated that Landon would defeat Roosevelt in the presidential election. The samples were drawn from a population of telephone owners and, in 1936, that population consisted of a disproportionately large number of people with high incomes. It was not representative of the population of all voters.

In 1948 the Gallup poll was wrong when it foresaw Truman losing to Dewey. That mistake led to a revision of Gallup's polling methods. A quota system had been used to obtain the opinions of a proportionate number of men, women, rich, poor, Catholics, Protestants, Jews, etc. After the 1948 fiasco, Gallup abandoned the quota system and instituted random sampling based on clusters of interviews in several hundred areas throughout the nation.

From these examples you should be able to see the importance of the sample.

12-4   ANALYZING DATA AND DRAWING CONCLUSIONS

The data must be analyzed and the conclusions drawn according to the methods specified when the experiment was designed. After the data have been analyzed and the conclusions have been formed, it is important to note the level of confidence used in any hypothesis test or estimation.

12-5   WRITING THE REPORT

In writing the final report, the author should consider the people who will read the report. Statisticians will expect specific and detailed results that are accompanied by fairly complete descriptions of methodology. However, a lay audience would not benefit from this type of report so a different approach is necessary. A suggested outline for the written report follows.

1. Front matter
    (a) Title page
    (b) Table of contents
    (c) List of tables and illustrations
    (d) Preface
    (e) Summary of results and conclusions

2. Body of report
    (a) Statement of objectives
    (b) Description of procedure and methods
    (c) Analysis of data
    (d) Conclusions drawn from data

3. Supplementary material
    (a) Appendices
    (b) Bibliography
    (c) Glossary of terms requiring definition
    (d) Index

12-6   AN ACTUAL SURVEY

The Gallup organization conducted a nationwide survey for the Zenith Radio Corporation to determine the extent of brand loyalty among color television owners. We can gain some insight into experimental design by examining some aspects of the approach used by this reputable company. The survey involved a nationwide sample of 9532 telephone interviews conducted with male or female heads of households which had color television sets.

Sample Design

The survey was designed to be representative of U.S. households owning a color television set. The sample employed the Gallup national probability sample of interviewing areas or localities. These localities were drawn in the following manner.

The United States was divided into four size-of-community strata: (a) cities of population 1,000,000 or over, (b) 250,000 to 999,999, (c) 50,000 to 249,999, (d) less than 50,000.

Within each of these strata, the population was further stratified by seven regions: New England, Middle Atlantic, East Central, West Central, South, Mountain, and Pacific.

Within each size of community-regional stratum, the population was arrayed in geographic order by states. In each of the first three strata, the cities were listed in alphabetical order within states. In the fourth stratum, counties were arrayed in geographic order within states.

Population in the fourth stratum was arrayed by counties, with each county's population classified into these four substrata: urban fringe around cities in the first three strata, cities of 2,500 to 49,999, rural towns and villages, and rural open-country areas.

From this array a national sample of locations was drawn, with the probability of a location's selection proportional to its size in the 1970 census. This sample of locations also represents a sample of all U.S. counties.

In the first three city-size strata, cities over 50,000 population, a working bank of telephone numbers in each city in the sample was located. In the fourth stratum, a working bank of numbers in each county in the sample was located. The working banks were obtained from national personal interviewing surveys conducted by The Gallup Organization in May of 1973. As part of these interviews, respondents were asked for their telephone numbers. In each sample location about eight telephone starts were used, drawn from these personal interview surveys. A telephone start consisted of the telephone prefix plus the next two digits. From that start, a set of numbers to be used in the survey was developed. For example, a start obtained from the survey might be 921-4720. All numbers 921-4720 to 921-4799 might then be taken excluding the original number of 921-4720.

In each location, interviewers were instructed to call about 225 telephone numbers. At each number the interviewer asked questions to "screen out" numbers which were:

1. Nonresidential telephone numbers.

2. Telephone numbers of households outside of a city or county, depending upon the stratum of the location.

3. Households that did not own a color television.

Interviewers were instructed to make up to four calls (an original call plus three call backs) to complete an interview or determine that the household was not qualified.

At qualifying households, the interviewer asked to speak to the male or female household head. A quota of about equal male and female assignments was made. A total of 9532 interviews were completed with color television owners.

After the completion of interviewing, survey results were examined by known population and market characteristics, and no statistical adjustments were deemed necessary on that basis. However, the data were adjusted statistically to correct for the fact that the interviewing localities were drawn on the basis of population size and not telephone population, and to adjust for the fact that in some interviewing localities more numbers were contacted than in others.

Composition of the Sample

	Percent
All Color Television Owners	100.0
Sex	
Men	47.5
Women	52.3
Undesignated	0.2
	100.0
Age	
18–29 years	19.8
30–39 years	20.5
40–49 years	19.4
50 and over	34.4
Undesignated	5.9
	100.0
Annual Family Income	
Under $10,000	30.1
$10,000–$15,000	32.0
Over $15,000	27.0
Undesignated	10.9
	100.0

Education
College	35.6
High school	52.1
Grade school	9.6
Undesignated	2.7
	100.0

Region of Country

East:	Maine, New Hampshire, Rhode Island, Connecticut, Vermont, Massachusetts, New York, New Jersey, Pennsylvania, West Virginia, Delaware, Maryland, District of Columbia	27.5
Midwest:	Ohio, Indiana, Illinois, Michigan, Minnesota, Wisconsin, Iowa, North Dakota, South Dakota, Kansas, Nebraska, Missouri	29.4
South:	Kentucky, Tennessee, Virginia, North Carolina, South Carolina, Georgia, Florida, Alabama, Mississippi, Texas, Arkansas, Oklahoma, Louisiana	26.7
West:	Arizona, New Mexico, Colorado, Nevada, Montana, Idaho, Wyoming, Utah, California, Washington, Oregon, Alaska, Hawaii	16.4
		100.0

Percent of owners of major brands* who would purchase brand again.

	Would buy again**	Would buy other brand	Not sure what brand would buy	Total	Number of interviews
	%	%	%	%	
All Color Television Owners	61	22	17	100	(9532)
Brand of Color Television Most Recently Purchased:					
Zenith	82	8	10	100	(2118)
RCA	70	16	14	100	(2140)
Sony	69	17	14	100	(200)

	Would buy again**	Would buy other brand	Not sure what brand would buy	Total	Number of interviews
	%	%	%	%	
Motorola	66	19	15	100	(645)
Magnavox	63	23	14	100	(721)
Sylvania	56	24	20	100	(396)
Admiral	51	28	21	100	(396)
Sears	48	34	18	100	(639)
Philco	48	32	20	100	(330)
Panasonic	47	33	20	100	(161)
General Electric	45	36	19	100	(528)
All other brands	40	38	22	100	(973)

*Brand most recently purchased, if more than one brand owned.

**Table is read: 82% of those who own Zenith named it as the brand that would be their first choice for purchase today.

EXERCISES A: DESIGN, SAMPLING, AND REPORT WRITING

12-1.   What is random sampling?

12-2.   What is stratified sampling?

12-3.   What is systematic sampling?

12-4.   What is cluster sampling?

12-5.   Identify the type of sampling used in each of the following:
  (a)   A teacher selects every fifth student in the class for a test.
  (b)   A teacher writes the name of each student on a card, shuffles the cards, and then draws five names.
  (c)   A teacher selects five students from each of twelve classes.
  (d)   A teacher selects five men and five women from each of four classes.

12-6.   You plan to conduct a survey to determine the percentage of students who favor a pub on campus. Describe the population and a method of sampling that should give reasonable results.

12-7.   You plan to estimate the mean weight of all passenger cars used in the United States. Is there universal agreement as to what a "passenger car" is? Are there any factors which might lead to regional differences among the weights of passenger cars? How can you obtain a sample?

12-8.   Describe in detail a statistical experiment that can be used to test the effectiveness of a drug for birth control.

EXERCISES B: DESIGN, SAMPLING, AND REPORT WRITING

12-9.   After developing some hypothesis, test that hypothesis by conducting a complete statistical experiment. Develop a written report that follows the outline in this chapter. Be sure to identify the sample size and type of sampling. Consult your instructor for possible suggestions.

12-10. Find an article that deals with some hypothesis test in a professional journal. Use the given information to write a report that follows, as closely as possible, the outline for a written report. Include the name of the journal, the author, and the title of the article.

APPENDIX A

TABLES

Table A-1. Factorials of Numbers 1 to 20.

n	n!
0	1
1	1
2	2
3	6
4	24
5	120
6	720
7	5040
8	40320
9	362880
10	3628800
11	39916800
12	479001600
13	6227020800
14	87178291200
15	1307674368000
16	20922789888000
17	355687428096000
18	6402373705728000
19	121645100408832000
20	2432902008176640000

Table A-2. Binomial Probabilities.

n	x	.01	.05	.10	.20	.30	.40	.50	.60	.70	.80	.90	.95	.99	x
2	0	980	902	810	640	490	360	250	160	090	040	010	002	0+	0
	1	020	095	180	320	420	480	500	480	420	320	180	095	020	1
	2	0+	002	010	040	090	160	250	360	490	640	810	902	980	2
3	0	970	857	729	512	343	216	125	064	027	008	001	0+	0+	0
	1	029	135	243	384	441	432	375	288	189	096	027	007	0+	1
	2	0+	007	027	096	189	288	375	432	441	384	243	135	029	2
	3	0+	0+	001	008	027	064	125	216	343	512	729	857	970	3
4	0	961	815	656	410	240	130	062	026	008	002	0+	0+	0+	0
	1	039	171	292	410	412	346	250	154	076	026	004	0+	0+	1
	2	001	014	049	154	265	346	375	346	265	154	049	014	001	2
	3	0+	0+	004	026	076	154	250	346	412	410	292	171	039	3
	4	0+	0+	0+	002	008	026	062	130	240	410	656	815	961	4
5	0	951	774	590	328	168	078	031	010	002	0+	0+	0+	0+	0
	1	048	204	328	410	360	259	156	077	028	006	0+	0+	0+	1
	2	001	021	073	205	309	346	312	230	132	051	008	001	0+	2
	3	0+	001	008	051	132	230	312	346	309	205	073	021	001	3
	4	0+	0+	0+	006	028	077	156	259	360	410	328	204	048	4
	5	0+	0+	0+	0+	002	010	031	078	168	328	590	774	951	5
6	0	941	735	531	262	118	047	016	004	001	0+	0+	0+	0+	0
	1	057	232	354	393	303	187	094	037	010	002	0+	0+	0+	1
	2	001	031	098	246	324	311	234	138	060	015	001	0+	0+	2
	3	0+	002	015	082	185	276	312	276	185	082	015	002	0+	3
	4	0+	0+	001	015	060	138	234	311	324	246	098	031	001	4
	5	0+	0+	0+	002	010	037	094	187	303	393	354	232	057	5
	6	0+	0+	0+	0+	001	004	016	047	118	262	531	735	941	6
7	0	932	698	478	210	082	028	008	002	0+	0+	0+	0+	0+	0
	1	066	257	372	367	247	131	055	017	004	0+	0+	0+	0+	1
	2	002	041	124	275	318	261	164	077	025	004	0+	0+	0+	2
	3	0+	004	023	115	227	290	273	194	097	029	003	0+	0+	3
	4	0+	0+	003	029	097	194	273	290	227	115	023	004	0+	4
	5	0+	0+	0+	004	025	077	164	261	318	275	124	041	002	5
	6	0+	0+	0+	0+	004	017	055	131	247	367	372	257	066	6
	7	0+	0+	0+	0+	0+	002	008	028	082	210	478	698	932	7
8	0	923	663	430	168	058	017	004	001	0+	0+	0+	0+	0+	0
	1	075	279	383	336	198	090	031	008	001	0+	0+	0+	0+	1
	2	003	051	149	294	296	209	109	041	010	001	0+	0+	0+	2
	3	0+	005	033	147	254	279	219	124	047	009	0+	0+	0+	3
	4	0+	0+	005	046	136	232	273	232	136	046	005	0+	0+	4
	5	0+	0+	0+	009	047	124	219	279	254	147	033	005	0+	5
	6	0+	0+	0+	001	010	041	109	209	296	294	149	051	003	6
	7	0+	0+	0+	0+	001	008	031	090	198	336	383	279	075	7
	8	0+	0+	0+	0+	0+	001	004	017	058	168	430	663	923	8

The column headers span under the label *p*.

Table A-2 (continued). Binomial Probabilities.

n	x	.01	.05	.10	.20	.30	.40	.50	.60	.70	.80	.90	.95	.99	x
9	0	914	630	387	134	040	010	002	0+	0+	0+	0+	0+	0+	0
	1	083	299	387	302	156	060	018	004	0+	0+	0+	0+	0+	1
	2	003	063	172	302	267	161	070	021	004	0+	0+	0+	0+	2
	3	0+	008	045	176	267	251	164	074	021	003	0+	0+	0+	3
	4	0+	001	007	066	172	251	246	167	074	017	001	0+	0+	4
	5	0+	0+	001	017	074	167	246	251	172	066	007	001	0+	5
	6	0+	0+	0+	003	021	074	164	251	267	176	045	008	0+	6
	7	0+	0+	0+	0+	004	021	070	161	267	302	172	063	003	7
	8	0+	0+	0+	0+	0+	004	018	060	156	302	387	299	083	8
	9	0+	0+	0+	0+	0+	0+	002	010	040	134	387	630	914	9
10	0	904	599	349	107	028	006	001	0+	0+	0+	0+	0+	0+	0
	1	091	315	387	268	121	040	010	002	0+	0+	0+	0+	0+	1
	2	004	075	194	302	233	121	044	011	001	0+	0+	0+	0+	2
	3	0+	010	057	201	267	215	117	042	009	001	0+	0+	0+	3
	4	0+	001	011	088	200	251	205	111	037	006	0+	0+	0+	4
	5	0+	0+	001	026	103	201	246	201	103	026	001	0+	0+	5
	6	0+	0+	0+	006	037	111	205	251	200	088	011	001	0+	6
	7	0+	0+	0+	001	009	042	117	215	267	201	057	010	0+	7
	8	0+	0+	0+	0+	001	011	044	121	233	302	194	075	004	8
	9	0+	0+	0+	0+	0+	002	010	040	121	268	387	315	091	9
	10	0+	0+	0+	0+	0+	0+	001	006	028	107	349	599	904	10
11	0	895	569	314	086	020	004	0+	0+	0+	0+	0+	0+	0+	0
	1	099	329	384	236	093	027	005	001	0+	0+	0+	0+	0+	1
	2	005	087	213	295	200	089	027	005	001	0+	0+	0+	0+	2
	3	0+	014	071	221	257	177	081	023	004	0+	0+	0+	0+	3
	4	0+	001	016	111	220	236	161	070	017	002	0+	0+	0+	4
	5	0+	0+	002	039	132	221	226	147	057	010	0+	0+	0+	5
	6	0+	0+	0+	010	057	147	226	221	132	039	002	0+	0+	6
	7	0+	0+	0+	002	017	070	161	236	220	111	016	001	0+	7
	8	0+	0+	0+	0+	004	023	081	177	257	221	071	014	0+	8
	9	0+	0+	0+	0+	001	005	027	089	200	295	213	087	005	9
	10	0+	0+	0+	0+	0+	001	005	027	093	236	384	329	099	10
	11	0+	0+	0+	0+	0+	0+	0+	004	020	086	314	569	895	11
12	0	886	540	282	069	014	002	0+	0+	0+	0+	0+	0+	0+	0
	1	107	341	377	206	071	017	003	0+	0+	0+	0+	0+	0+	1
	2	006	099	230	283	168	064	016	002	0+	0+	0+	0+	0+	2
	3	0+	017	085	236	240	142	054	012	001	0+	0+	0+	0+	3
	4	0+	002	021	133	231	213	121	042	008	001	0+	0+	0+	4
	5	0+	0+	004	053	158	227	193	101	029	003	0+	0+	0+	5
	6	0+	0+	0+	016	079	177	226	177	079	016	0+	0+	0+	6
	7	0+	0+	0+	003	029	101	193	227	158	053	004	0+	0+	7
	8	0+	0+	0+	001	008	042	121	213	231	133	021	002	0+	8
	9	0+	0+	0+	0+	001	012	054	142	240	236	085	017	0+	9

Table A-2 (continued). Binomial Probabilities.

n	x	.01	.05	.10	.20	.30	.40	p .50	.60	.70	.80	.90	.95	.99	x
12	10	0+	0+	0+	0+	0+	002	016	064	168	283	230	099	006	10
	11	0+	0+	0+	0+	0+	0+	003	017	071	206	377	341	107	11
	12	0+	0+	0+	0+	0+	0+	0+	002	014	069	282	540	886	12
13	0	878	513	254	055	010	001	0+	0+	0+	0+	0+	0+	0+	0
	1	115	351	367	179	054	011	002	0+	0+	0+	0+	0+	0+	1
	2	007	111	245	268	139	045	010	001	0+	0+	0+	0+	0+	2
	3	0+	021	100	246	218	111	035	006	001	0+	0+	0+	0+	3
	4	0+	003	028	154	234	184	087	024	003	0+	0+	0+	0+	4
	5	0+	0+	006	069	180	221	157	066	014	001	0+	0+	0+	5
	6	0+	0+	001	023	103	197	209	131	044	006	0+	0+	0+	6
	7	0+	0+	0+	006	044	131	209	197	103	023	001	0+	0+	7
	8	0+	0+	0+	001	014	066	157	221	180	069	006	0+	0+	8
	9	0+	0+	0+	0+	003	024	087	184	234	154	028	003	0+	9
	10	0+	0+	0+	0+	001	006	035	111	218	246	100	021	0+	10
	11	0+	0+	0+	0+	0+	001	010	045	139	268	245	111	007	11
	12	0+	0+	0+	0+	0+	0+	002	011	054	179	367	351	115	12
	13	0+	0+	0+	0+	0+	0+	0+	001	010	055	254	513	878	13
14	0	869	488	229	044	007	001	0+	0+	0+	0+	0+	0+	0+	0
	1	123	359	356	154	041	007	001	0+	0+	0+	0+	0+	0+	1
	2	008	123	257	250	113	032	006	001	0+	0+	0+	0+	0+	2
	3	0+	026	114	250	194	085	022	003	0+	0+	0+	0+	0+	3
	4	0+	004	035	172	229	155	061	014	001	0+	0+	0+	0+	4
	5	0+	0+	008	086	196	207	122	041	007	0+	0+	0+	0+	5
	6	0+	0+	001	032	126	207	183	092	023	002	0+	0+	0+	6
	7	0+	0+	0+	009	062	157	209	157	062	009	0+	0+	0+	7
	8	0+	0+	0+	002	023	092	183	207	126	032	001	0+	0+	8
	9	0+	0+	0+	0+	007	041	122	207	196	086	008	0+	0+	9
	10	0+	0+	0+	0+	001	014	061	155	229	172	035	004	0+	10
	11	0+	0+	0+	0+	0+	003	022	085	194	250	114	026	0+	11
	12	0+	0+	0+	0+	0+	001	006	032	113	250	257	123	008	12
	13	0+	0+	0+	0+	0+	0+	001	007	041	154	356	359	123	13
	14	0+	0+	0+	0+	0+	0+	0+	001	007	044	229	488	869	14
15	0	860	463	206	035	005	0+	0+	0+	0+	0+	0+	0+	0+	0
	1	130	366	343	132	031	005	0+	0+	0+	0+	0+	0+	0+	1
	2	009	135	267	231	092	022	003	0+	0+	0+	0+	0+	0+	2
	3	0+	031	129	250	170	063	014	002	0+	0+	0+	0+	0+	3
	4	0+	005	043	188	219	127	042	007	001	0+	0+	0+	0+	4
	5	0+	001	010	103	206	186	092	024	003	0+	0+	0+	0+	5
	6	0+	0+	002	043	147	207	153	061	012	001	0+	0+	0+	6
	7	0+	0+	0+	014	081	177	196	118	035	003	0+	0+	0+	7
	8	0+	0+	0+	003	035	118	196	177	081	014	0+	0+	0+	8
	9	0+	0+	0+	001	012	061	153	207	147	043	002	0+	0+	9

Table A-2 (continued). Binomial Probabilities.

n	x	.01	.05	.10	.20	.30	.40	p .50	.60	.70	.80	.90	.95	.99	x
15	10	0+	0+	0+	0+	003	024	092	186	206	103	010	001	0+	10
	11	0+	0+	0+	0+	001	007	042	127	219	188	043	005	0+	11
	12	0+	0+	0+	0+	0+	002	014	063	170	250	129	031	0+	12
	13	0+	0+	0+	0+	0+	0+	003	022	092	231	267	135	009	13
	14	0+	0+	0+	0+	0+	0+	0+	005	031	132	343	366	130	14
	15	0+	0+	0+	0+	0+	0+	0+	0+	005	035	206	463	860	15
16	0	851	440	185	028	003	0+	0+	0+	0+	0+	0+	0+	0+	0
	1	138	371	329	113	023	003	0+	0+	0+	0+	0+	0+	0+	1
	2	010	146	275	211	073	015	002	0+	0+	0+	0+	0+	0+	2
	3	0+	036	142	246	146	047	009	001	0+	0+	0+	0+	0+	3
	4	0+	006	051	200	204	101	028	004	0+	0+	0+	0+	0+	4
	5	0+	001	014	120	210	162	067	014	001	0+	0+	0+	0+	5
	6	0+	0+	003	055	165	198	122	039	006	0+	0+	0+	0+	6
	7	0+	0+	0+	020	101	189	175	084	019	001	0+	0+	0+	7
	8	0+	0+	0+	006	049	142	196	142	049	006	0+	0+	0+	8
	9	0+	0+	0+	001	019	084	175	189	101	020	0+	0+	0+	9
	10	0+	0+	0+	0+	006	039	122	198	165	055	003	0+	0+	10
	11	0+	0+	0+	0+	001	014	067	162	210	120	014	001	0+	11
	12	0+	0+	0+	0+	0+	004	028	101	204	200	051	006	0+	12
	13	0+	0+	0+	0+	0+	001	009	047	146	246	142	036	0+	13
	14	0+	0+	0+	0+	0+	0+	002	015	073	211	275	146	010	14
	15	0+	0+	0+	0+	0+	0+	0+	003	023	113	329	371	138	15
	16	0+	0+	0+	0+	0+	0+	0+	0+	003	028	185	440	851	16
17	0	843	418	167	023	002	0+	0+	0+	0+	0+	0+	0+	0+	0
	1	145	374	315	096	017	002	0+	0+	0+	0+	0+	0+	0+	1
	2	012	158	280	191	058	010	001	0+	0+	0+	0+	0+	0+	2
	3	001	041	156	239	125	034	005	0+	0+	0+	0+	0+	0+	3
	4	0+	008	060	209	187	080	018	002	0+	0+	0+	0+	0+	4
	5	0+	001	017	136	208	138	047	008	001	0+	0+	0+	0+	5
	6	0+	0+	004	068	178	184	094	024	003	0+	0+	0+	0+	6
	7	0+	0+	001	027	120	193	148	057	009	0+	0+	0+	0+	7
	8	0+	0+	0+	008	064	161	185	107	028	002	0+	0+	0+	8
	9	0+	0+	0+	002	028	107	185	161	064	008	0+	0+	0+	9
	10	0+	0+	0+	0+	009	057	148	193	120	027	001	0+	0+	10
	11	0+	0+	0+	0+	003	024	094	184	178	068	004	0+	0+	11
	12	0+	0+	0+	0+	001	008	047	138	208	136	017	001	0+	12
	13	0+	0+	0+	0+	0+	002	018	080	187	209	060	008	0+	13
	14	0+	0+	0+	0+	0+	0+	005	034	125	239	156	041	001	14
	15	0+	0+	0+	0+	0+	0+	001	010	058	191	280	158	012	15
	16	0+	0+	0+	0+	0+	0+	0+	002	017	096	315	374	145	16
	17	0+	0+	0+	0+	0+	0+	0+	0+	002	023	167	418	843	17

Table A-2 (continued). Binomial Probabilities.

n	x	.01	.05	.10	.20	.30	.40	p .50	.60	.70	.80	.90	.95	.99	x
18	0	835	397	150	018	002	0+	0+	0+	0+	0+	0+	0+	0+	0
	1	152	376	300	081	013	001	0+	0+	0+	0+	0+	0+	0+	1
	2	013	168	284	172	046	007	001	0+	0+	0+	0+	0+	0+	2
	3	001	047	168	230	105	025	003	0+	0+	0+	0+	0+	0+	3
	4	0+	009	070	215	168	061	012	001	0+	0+	0+	0+	0+	4
	5	0+	001	022	151	202	115	033	004	0+	0+	0+	0+	0+	5
	6	0+	0+	005	082	187	166	071	015	001	0+	0+	0+	0+	6
	7	0+	0+	001	035	138	189	121	037	005	0+	0+	0+	0+	7
	8	0+	0+	0+	012	081	173	167	077	015	001	0+	0+	0+	8
	9	0+	0+	0+	003	039	128	185	128	039	003	0+	0+	0+	9
	10	0+	0+	0+	001	015	077	167	173	081	012	0+	0+	0+	10
	11	0+	0+	0+	0+	005	037	121	189	138	035	001	0+	0+	11
	12	0+	0+	0+	0+	001	015	071	166	187	082	005	0+	0+	12
	13	0+	0+	0+	0+	0+	004	033	115	202	151	022	001	0+	13
	14	0+	0+	0+	0+	0+	001	012	061	168	215	070	009	0+	14
	15	0+	0+	0+	0+	0+	0+	003	025	105	230	168	047	001	15
	16	0+	0+	0+	0+	0+	0+	001	007	046	172	284	168	013	16
	17	0+	0+	0+	0+	0+	0+	0+	001	013	081	300	376	152	17
	18	0+	0+	0+	0+	0+	0+	0+	0+	002	018	150	397	835	18
19	0	826	377	135	014	001	0+	0+	0+	0+	0+	0+	0+	0+	0
	1	159	377	285	068	009	001	0+	0+	0+	0+	0+	0+	0+	1
	2	014	179	285	154	036	005	0+	0+	0+	0+	0+	0+	0+	2
	3	001	053	180	218	087	017	002	0+	0+	0+	0+	0+	0+	3
	4	0+	011	080	218	149	047	007	001	0+	0+	0+	0+	0+	4
	5	0+	002	027	164	192	093	022	002	0+	0+	0+	0+	0+	5
	6	0+	0+	007	095	192	145	052	008	001	0+	0+	0+	0+	6
	7	0+	0+	001	044	153	180	096	024	002	0+	0+	0+	0+	7
	8	0+	0+	0+	017	098	180	144	053	008	0+	0+	0+	0+	8
	9	0+	0+	0+	005	051	146	176	098	022	001	0+	0+	0+	9
	10	0+	0+	0+	001	022	098	176	146	051	005	0+	0+	0+	10
	11	0+	0+	0+	0+	008	053	144	180	098	017	0+	0+	0+	11
	12	0+	0+	0+	0+	002	024	096	180	153	044	001	0+	0+	12
	13	0+	0+	0+	0+	001	008	052	145	192	095	007	0+	0+	13
	14	0+	0+	0+	0+	0+	002	022	093	192	164	027	002	0+	14
	15	0+	0+	0+	0+	0+	001	007	047	149	218	080	011	0+	15
	16	0+	0+	0+	0+	0+	0+	002	017	087	218	180	053	001	16
	17	0+	0+	0+	0+	0+	0+	0+	005	036	154	285	179	014	17
	18	0+	0+	0+	0+	0+	0+	0+	001	009	068	285	377	159	18
	19	0+	0+	0+	0+	0+	0+	0+	0+	001	014	135	377	826	19
20	0	818	358	122	012	001	0+	0+	0+	0+	0+	0+	0+	0+	0
	1	165	377	270	058	007	0+	0+	0+	0+	0+	0+	0+	0+	1
	2	016	189	285	137	028	003	0+	0+	0+	0+	0+	0+	0+	2
	3	001	060	190	205	072	012	001	0+	0+	0+	0+	0+	0+	3
	4	0+	013	090	218	130	035	005	0+	0+	0+	0+	0+	0+	4

Table A-2 (continued). Binomial Probabilities.

n	x	.01	.05	.10	.20	.30	.40	p .50	.60	.70	.80	.90	.95	.99	x
20	5	0+	002	032	175	179	075	015	001	0+	0+	0+	0+	0+	5
	6	0+	0+	009	109	192	124	037	005	0+	0+	0+	0+	0+	6
	7	0+	0+	002	055	164	166	074	015	001	0+	0+	0+	0+	7
	8	0+	0+	0+	022	114	180	120	035	004	0+	0+	0+	0+	8
	9	0+	0+	0+	007	065	160	160	071	012	0+	0+	0+	0+	9
	10	0+	0+	0+	002	031	117	176	117	031	002	0+	0+	0+	10
	11	0+	0+	0+	0+	012	071	160	160	065	007	0+	0+	0+	11
	12	0+	0+	0+	0+	004	035	120	180	114	022	0+	0+	0+	12
	13	0+	0+	0+	0+	001	015	074	166	164	055	002	0+	0+	13
	14	0+	0+	0+	0+	0+	005	037	124	192	109	009	0+	0+	14
	15	0+	0+	0+	0+	0+	001	015	075	179	175	032	002	0+	15
	16	0+	0+	0+	0+	0+	0+	005	035	130	218	090	013	0+	16
	17	0+	0+	0+	0+	0+	0+	001	012	072	205	190	060	001	17
	18	0+	0+	0+	0+	0+	0+	0+	003	028	137	285	189	016	18
	19	0+	0+	0+	0+	0+	0+	0+	0+	007	058	270	377	165	19
	20	0+	0+	0+	0+	0+	0+	0+	0+	001	012	122	358	818	20
21	0	810	341	109	009	001	0+	0+	0+	0+	0+	0+	0+	0+	0
	1	172	376	255	048	005	0+	0+	0+	0+	0+	0+	0+	0+	1
	2	017	198	284	121	022	002	0+	0+	0+	0+	0+	0+	0+	2
	3	001	066	200	192	058	009	001	0+	0+	0+	0+	0+	0+	3
	4	0+	016	100	216	113	026	003	0+	0+	0+	0+	0+	0+	4
	5	0+	003	038	183	164	059	010	001	0+	0+	0+	0+	0+	5
	6	0+	0+	011	122	188	105	026	003	0+	0+	0+	0+	0+	6
	7	0+	0+	003	065	172	149	055	009	0+	0+	0+	0+	0+	7
	8	0+	0+	001	029	129	174	097	023	002	0+	0+	0+	0+	8
	9	0+	0+	0+	010	080	168	140	050	006	0+	0+	0+	0+	9
	10	0+	0+	0+	003	041	134	168	089	018	001	0+	0+	0+	10
	11	0+	0+	0+	001	018	089	168	134	041	003	0+	0+	0+	11
	12	0+	0+	0+	0+	006	050	140	168	080	010	0+	0+	0+	12
	13	0+	0+	0+	0+	002	023	097	174	129	029	001	0+	0+	13
	14	0+	0+	0+	0+	0+	009	055	149	172	065	003	0+	0+	14
	15	0+	0+	0+	0+	0+	003	026	105	188	122	011	0+	0+	15
	16	0+	0+	0+	0+	0+	001	010	059	164	183	038	003	0+	16
	17	0+	0+	0+	0+	0+	0+	003	026	113	216	100	016	0+	17
	18	0+	0+	0+	0+	0+	0+	001	009	058	192	200	066	001	18
	19	0+	0+	0+	0+	0+	0+	0+	002	022	121	284	198	017	19
	20	0+	0+	0+	0+	0+	0+	0+	0+	005	048	255	376	172	20
	21	0+	0+	0+	0+	0+	0+	0+	0+	001	009	109	341	810	21
22	0	802	324	098	007	0+	0+	0+	0+	0+	0+	0+	0+	0+	0
	1	178	375	241	041	004	0+	0+	0+	0+	0+	0+	0+	0+	1
	2	019	207	281	107	017	001	0+	0+	0+	0+	0+	0+	0+	2
	3	001	073	208	178	047	006	0+	0+	0+	0+	0+	0+	0+	3
	4	0+	018	110	211	096	019	002	0+	0+	0+	0+	0+	0+	4

Table A-2 (continued). Binomial Probabilities.

n	x	.01	.05	.10	.20	.30	.40	.50	.60	.70	.80	.90	.95	.99	x
22	5	0+	003	044	190	149	046	006	0+	0+	0+	0+	0+	0+	5
	6	0+	001	014	134	181	086	018	001	0+	0+	0+	0+	0+	6
	7	0+	0+	004	077	177	131	041	005	0+	0+	0+	0+	0+	7
	8	0+	0+	001	036	142	164	076	014	001	0+	0+	0+	0+	8
	9	0+	0+	0+	014	095	170	119	034	003	0+	0+	0+	0+	9
	10	0+	0+	0+	005	053	148	154	066	010	0+	0+	0+	0+	10
	11	0+	0+	0+	001	025	107	168	107	025	001	0+	0+	0+	11
	12	0+	0+	0+	0+	010	066	154	148	053	005	0+	0+	0+	12
	13	0+	0+	0+	0+	003	034	119	170	095	014	0+	0+	0+	13
	14	0+	0+	0+	0+	001	014	076	164	142	036	001	0+	0+	14
	15	0+	0+	0+	0+	0+	005	041	131	177	077	004	0+	0+	15
	16	0+	0+	0+	0+	0+	001	018	086	181	134	014	001	0+	16
	17	0+	0+	0+	0+	0+	0+	006	046	149	190	044	003	0+	17
	18	0+	0+	0+	0+	0+	0+	002	019	096	211	110	018	0+	18
	19	0+	0+	0+	0+	0+	0+	0+	006	047	178	208	073	001	19
	20	0+	0+	0+	0+	0+	0+	0+	001	017	107	281	207	019	20
	21	0+	0+	0+	0+	0+	0+	0+	0+	004	041	241	375	178	21
	22	0+	0+	0+	0+	0+	0+	0+	0+	0+	007	098	324	802	22
23	0	794	307	089	006	0+	0+	0+	0+	0+	0+	0+	0+	0+	0
	1	184	372	226	034	003	0+	0+	0+	0+	0+	0+	0+	0+	1
	2	020	215	277	093	013	001	0+	0+	0+	0+	0+	0+	0+	2
	3	001	079	215	163	038	004	0+	0+	0+	0+	0+	0+	0+	3
	4	0+	021	120	204	082	014	001	0+	0+	0+	0+	0+	0+	4
	5	0+	004	051	194	133	035	004	0+	0+	0+	0+	0+	0+	5
	6	0+	001	017	145	171	070	012	001	0+	0+	0+	0+	0+	6
	7	0+	0+	005	088	178	113	029	003	0+	0+	0+	0+	0+	7
	8	0+	0+	001	044	153	151	058	009	0+	0+	0+	0+	0+	8
	9	0+	0+	0+	018	109	168	097	022	002	0+	0+	0+	0+	9
	10	0+	0+	0+	006	065	157	136	046	005	0+	0+	0+	0+	10
	11	0+	0+	0+	002	033	123	161	082	014	0+	0+	0+	0+	11
	12	0+	0+	0+	0+	014	082	161	123	033	002	0+	0+	0+	12
	13	0+	0+	0+	0+	005	046	136	157	065	006	0+	0+	0+	13
	14	0+	0+	0+	0+	002	022	097	168	109	018	0+	0+	0+	14
	15	0+	0+	0+	0+	0+	009	058	151	153	044	001	0+	0+	15
	16	0+	0+	0+	0+	0+	003	029	113	178	088	005	0+	0+	16
	17	0+	0+	0+	0+	0+	001	012	070	171	145	017	001	0+	17
	18	0+	0+	0+	0+	0+	0+	004	035	133	194	051	004	0+	18
	19	0+	0+	0+	0+	0+	0+	001	014	082	204	120	021	0+	19
	20	0+	0+	0+	0+	0+	0+	0+	004	038	163	215	079	001	20
	21	0+	0+	0+	0+	0+	0+	0+	001	013	093	277	215	020	21
	22	0+	0+	0+	0+	0+	0+	0+	0+	003	034	226	372	184	22
	23	0+	0+	0+	0+	0+	0+	0+	0+	0+	006	089	307	794	23

Table A-2 (continued). Binomial Probabilities.

n	x	.01	.05	.10	.20	.30	.40	p .50	.60	.70	.80	.90	.95	.99	x
24	0	786	292	080	005	0+	0+	0+	0+	0+	0+	0+	0+	0+	0
	1	190	369	213	028	002	0+	0+	0+	0+	0+	0+	0+	0+	1
	2	022	223	272	081	010	001	0+	0+	0+	0+	0+	0+	0+	2
	3	002	086	221	149	031	003	0+	0+	0+	0+	0+	0+	0+	3
	4	0+	024	129	196	069	010	001	0+	0+	0+	0+	0+	0+	4
	5	0+	005	057	196	118	027	003	0+	0+	0+	0+	0+	0+	5
	6	0+	001	020	155	160	056	008	0+	0+	0+	0+	0+	0+	6
	7	0+	0+	006	100	176	096	021	002	0+	0+	0+	0+	0+	7
	8	0+	0+	001	053	160	136	044	005	0+	0+	0+	0+	0+	8
	9	0+	0+	0+	024	122	161	078	014	001	0+	0+	0+	0+	9
	10	0+	0+	0+	009	079	161	117	032	003	0+	0+	0+	0+	10
	11	0+	0+	0+	003	043	137	149	061	008	0+	0+	0+	0+	11
	12	0+	0+	0+	001	020	099	161	099	020	001	0+	0+	0+	12
	13	0+	0+	0+	0+	008	061	149	137	043	003	0+	0+	0+	13
	14	0+	0+	0+	0+	003	032	117	161	079	009	0+	0+	0+	14
	15	0+	0+	0+	0+	001	014	078	161	122	024	0+	0+	0+	15
	16	0+	0+	0+	0+	0+	005	044	136	160	053	001	0+	0+	16
	17	0+	0+	0+	0+	0+	002	021	096	176	100	006	0+	0+	17
	18	0+	0+	0+	0+	0+	0+	008	056	160	155	020	001	0+	18
	19	0+	0+	0+	0+	0+	0+	003	027	118	196	057	005	0+	19
	20	0+	0+	0+	0+	0+	0+	001	010	069	196	129	024	0+	20
	21	0+	0+	0+	0+	0+	0+	0+	003	031	149	221	086	002	21
	22	0+	0+	0+	0+	0+	0+	0+	001	010	081	272	223	022	22
	23	0+	0+	0+	0+	0+	0+	0+	0+	002	028	213	369	190	23
	24	0+	0+	0+	0+	0+	0+	0+	0+	0+	005	080	292	786	24
25	0	778	277	072	004	0+	0+	0+	0+	0+	0+	0+	0+	0+	0
	1	196	365	199	024	001	0+	0+	0+	0+	0+	0+	0+	0+	1
	2	024	231	266	071	007	0+	0+	0+	0+	0+	0+	0+	0+	2
	3	002	093	226	136	024	002	0+	0+	0+	0+	0+	0+	0+	3
	4	0+	027	138	187	057	007	0+	0+	0+	0+	0+	0+	0+	4
	5	0+	006	065	196	103	020	002	0+	0+	0+	0+	0+	0+	5
	6	0+	001	024	163	147	044	005	0+	0+	0+	0+	0+	0+	6
	7	0+	0+	007	111	171	080	014	001	0+	0+	0+	0+	0+	7
	8	0+	0+	002	062	165	120	032	003	0+	0+	0+	0+	0+	8
	9	0+	0+	0+	029	134	151	061	009	0+	0+	0+	0+	0+	9
	10	0+	0+	0+	012	092	161	097	021	001	0+	0+	0+	0+	10
	11	0+	0+	0+	004	054	147	133	043	004	0+	0+	0+	0+	11
	12	0+	0+	0+	001	027	114	155	076	011	0+	0+	0+	0+	12
	13	0+	0+	0+	0+	011	076	155	114	027	001	0+	0+	0+	13
	14	0+	0+	0+	0+	004	043	133	147	054	004	0+	0+	0+	14
	15	0+	0+	0+	0+	001	021	097	161	092	012	0+	0+	0+	15
	16	0+	0+	0+	0+	0+	009	061	151	134	029	0+	0+	0+	16
	17	0+	0+	0+	0+	0+	003	032	120	165	062	002	0+	0+	17
	18	0+	0+	0+	0+	0+	001	014	080	171	111	007	0+	0+	18
	19	0+	0+	0+	0+	0+	0+	005	044	147	163	024	001	0+	19
	20	0+	0+	0+	0+	0+	0+	002	020	103	196	065	006	0+	20
	21	0+	0+	0+	0+	0+	0+	0+	007	057	187	138	027	0+	21
	22	0+	0+	0+	0+	0+	0+	0+	002	024	136	226	093	002	22
	23	0+	0+	0+	0+	0+	0+	0+	0+	007	071	266	231	024	23
	24	0+	0+	0+	0+	0+	0+	0+	0+	001	024	199	365	196	24
	25	0+	0+	0+	0+	0+	0+	0+	0+	0+	004	072	277	778	25

Frederick Mosteller/Robert E. K. Rourke/George B. Thomas, Jr., *Probability with Statistical Applications*, Second Edition. © 1961 and 1970, Addison-Wesley, Reading, Massachusetts. Reprinted with permission of the publisher.

Table A-3. The Standard Normal (z) Distribution.

Z	.00	.01	.02	.03	.04	.05	.06	.07	.08	.09
0.0	.0000	.0040	.0080	.0120	.0160	.0199	.0239	.0279	.0319	.0359
0.1	.0398	.0438	.0478	.0517	.0557	.0596	.0636	.0675	.0714	.0753
0.2	.0793	.0832	.0871	.0910	.0948	.0987	.1026	.1064	.1103	.1141
0.3	.1179	.1217	.1255	.1293	.1331	.1368	.1406	.1443	.1480	.1517
0.4	.1554	.1591	.1628	.1664	.1700	.1736	.1772	.1808	.1844	.1879
0.5	.1915	.1950	.1985	.2019	.2054	.2088	.2123	.2157	.2190	.2224
0.6	.2257	.2291	.2324	.2357	.2389	.2422	.2454	.2486	.2517	.2549
0.7	.2580	.2611	.2642	.2673	.2704	.2734	.2764	.2794	.2823	.2852
0.8	.2881	.2910	.2939	.2967	.2995	.3023	.3051	.3078	.3106	.3133
0.9	.3159	.3186	.3212	.3238	.3264	.3289	.3315	.3340	.3365	.3389
1.0	.3413	.3438	.3461	.3485	.3508	.3531	.3554	.3577	.3599	.3621
1.1	.3643	.3665	.3686	.3708	.3729	.3749	.3770	.3790	.3810	.3830
1.2	.3849	.3869	.3888	.3907	.3925	.3944	.3962	.3980	.3997	.4015
1.3	.4032	.4049	.4066	.4082	.4099	.4115	.4131	.4147	.4162	.4177
1.4	.4192	.4207	.4222	.4236	.4251	.4265	.4279	.4292	.4306	.4319
1.5	.4332	.4345	.4357	.4370	.4382	.4394	.4406	.4418	.4429	.4441
1.6	.4452	.4463	.4474	.4484	.4495	.4505	.4515	.4525	.4535	.4545
1.7	.4554	.4564	.4573	.4582	.4591	.4599	.4608	.4616	.4625	.4633
1.8	.4641	.4649	.4656	.4664	.4671	.4678	.4686	.4693	.4699	.4706
1.9	.4713	.4719	.4726	.4732	.4738	.4744	.4750	.4756	.4761	.4767
2.0	.4772	.4778	.4783	.4788	.4793	.4798	.4803	.4808	.4812	.4817
2.1	.4821	.4826	.4830	.4834	.4838	.4842	.4846	.4850	.4854	.4857
2.2	.4861	.4864	.4868	.4871	.4875	.4878	.4881	.4884	.4887	.4890
2.3	.4893	.4896	.4898	.4901	.4904	.4906	.4909	.4911	.4913	.4916
2.4	.4918	.4920	.4922	.4925	.4927	.4929	.4931	.4932	.4934	.4936
2.5	.4938	.4940	.4941	.4943	.4945	.4946	.4948	.4949	.4951	.4952
2.6	.4953	.4955	.4956	.4957	.4959	.4960	.4961	.4962	.4963	.4964
2.7	.4965	.4966	.4967	.4968	.4969	.4970	.4971	.4972	.4973	.4974
2.8	.4974	.4975	.4976	.4977	.4977	.4978	.4979	.4979	.4980	.4981
2.9	.4981	.4982	.4982	.4983	.4984	.4984	.4985	.4985	.4986	.4986
3.0	.4987	.4987	.4987	.4988	.4988	.4989	.4989	.4989	.4990	.4990

Frederick Mosteller/Robert E. K. Rourke, *Sturdy Statistics*. © 1973, Addison-Wesley, Reading, Massachusetts. Table A-1. Reprinted with permission.

Table A-4. t Distribution.

α

degrees of freedom	.005 (one tail) .01 (two tails)	.01 (one tail) .02 (two tails)	.025 (one tail) .05 (two tails)	.05 (one tail) .10 (two tails)	.10 (one tail) .20 (two tails)	.25 (one tail) .50 (two tails)
1	63.657	31.821	12.706	6.314	3.078	1.000
2	9.925	6.965	4.303	2.920	1.886	0.816
3	5.841	4.541	3.182	2.353	1.638	0.765
4	4.604	3.747	2.776	2.132	1.533	0.741
5	4.032	3.365	2.571	2.015	1.476	0.727
6	3.707	3.143	2.447	1.943	1.440	0.718
7	3.500	2.998	2.365	1.895	1.415	0.711
8	3.355	2.896	2.306	1.860	1.397	0.706
9	3.250	2.821	2.262	1.833	1.383	0.703
10	3.169	2.764	2.228	1.812	1.372	0.700
11	3.106	2.718	2.201	1.796	1.363	0.697
12	3.054	2.681	2.179	1.782	1.356	0.696
13	3.012	2.650	2.160	1.771	1.350	0.694
14	2.977	2.625	2.145	1.761	1.345	0.692
15	2.947	2.602	2.132	1.753	1.341	0.691
16	2.921	2.584	2.120	1.746	1.337	0.690
17	2.898	2.567	2.110	1.740	1.333	0.689
18	2.878	2.552	2.101	1.734	1.330	0.688
19	2.861	2.540	2.093	1.729	1.328	0.688
20	2.845	2.528	2.086	1.725	1.325	0.687
21	2.831	2.518	2.080	1.721	1.323	0.686
22	2.819	2.508	2.074	1.717	1.321	0.686
23	2.807	2.500	2.069	1.714	1.320	0.685
24	2.797	2.492	2.064	1.711	1.318	0.685
25	2.787	2.485	2.060	1.708	1.316	0.684
26	2.779	2.479	2.056	1.706	1.315	0.684
27	2.771	2.473	2.052	1.703	1.314	0.684
28	2.763	2.467	2.048	1.701	1.313	0.683
29	2.756	2.462	2.045	1.699	1.311	0.683
Large	2.575	2.327	1.960	1.645	1.282	0.675

t DISTRIBUTION:

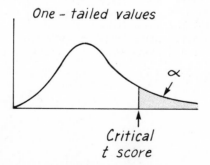

One - tailed values

Critical
t score

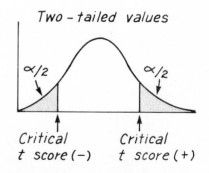

Two - tailed values

Critical
t score (−)

Critical
t score (+)

Table A-5. The Chi-Square (χ^2) Distribution

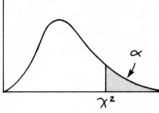

df	0.995	0.99	0.975	0.95	0.90	0.10	0.05	0.025	0.01	0.005
1	–	–	0.001	0.004	0.016	2.706	3.841	5.024	6.635	7.879
2	0.010	0.020	0.051	0.103	0.211	4.605	5.991	7.378	9.210	10.597
3	0.072	0.115	0.216	0.352	0.584	6.251	7.815	9.348	11.345	12.838
4	0.207	0.297	0.484	0.711	1.064	7.779	9.488	11.143	13.277	14.860
5	0.412	0.554	0.831	1.145	1.610	9.236	11.071	12.833	15.086	16.750
6	0.676	0.872	1.237	1.635	2.204	10.645	12.592	14.449	16.812	18.548
7	0.989	1.239	1.690	2.167	2.833	12.017	14.067	16.013	18.475	20.278
8	1.344	1.646	2.180	2.733	3.490	13.362	15.507	17.535	20.090	21.955
9	1.735	2.088	2.700	3.325	4.168	14.684	16.919	19.023	21.666	23.589
10	2.156	2.558	3.247	3.940	4.865	15.987	18.307	20.483	23.209	25.188
11	2.603	3.053	3.816	4.575	5.578	17.275	19.675	21.920	24.725	26.757
12	3.074	3.571	4.404	5.226	6.304	18.549	21.026	23.337	26.217	28.299
13	3.565	4.107	5.009	5.892	7.042	19.812	22.362	24.736	27.688	29.819
14	4.075	4.660	5.629	6.571	7.790	21.064	23.685	26.119	29.141	31.319
15	4.601	5.229	6.262	7.261	8.547	22.307	24.996	27.488	30.578	32.801
16	5.142	5.812	6.908	7.962	9.312	23.542	26.296	28.845	32.000	34.267
17	5.697	6.408	7.564	8.672	10.085	24.769	27.587	30.191	33.409	35.718
18	6.265	7.015	8.231	9.390	10.865	25.989	28.869	31.526	34.805	37.156
19	6.844	7.633	8.907	10.117	11.651	27.204	30.144	32.852	36.191	38.582
20	7.434	8.260	9.591	10.851	12.443	28.412	31.410	34.170	37.566	39.997
21	8.034	8.897	10.283	11.591	13.240	29.615	32.671	35.479	38.932	41.401
22	8.643	9.542	10.982	12.338	14.042	30.813	33.924	36.781	40.289	42.796
23	9.260	10.196	11.689	13.091	14.848	32.007	35.172	38.076	41.638	44.181
24	9.886	10.856	12.401	13.848	15.659	33.196	36.415	39.364	42.980	45.559
25	10.520	11.524	13.120	14.611	16.473	34.382	37.652	40.646	44.314	46.928
26	11.160	12.198	13.844	15.379	17.292	35.563	38.885	41.923	45.642	48.290
27	11.808	12.879	14.573	16.151	18.114	36.741	40.113	43.194	46.963	49.645
28	12.461	13.565	15.308	16.928	18.939	37.916	41.337	44.461	48.278	50.993
29	13.121	14.257	16.047	17.708	19.768	39.087	42.557	45.722	49.588	52.336
30	13.787	14.954	16.791	18.493	20.599	40.256	43.773	46.979	50.892	53.672
40	20.707	22.164	24.433	26.509	29.051	51.805	55.758	59.342	63.691	66.766
50	27.991	29.707	32.357	34.764	37.689	63.167	67.505	71.420	76.154	79.490
60	35.534	37.485	40.482	43.188	46.459	74.397	79.082	83.298	88.379	91.952
70	43.275	45.442	48.758	51.739	55.329	85.527	90.531	95.023	100.425	104.215
80	51.172	53.540	57.153	60.391	64.278	96.578	101.879	106.629	112.329	116.321
90	59.196	61.754	65.647	69.126	73.291	107.565	113.145	118.136	124.116	128.299
100	67.328	70.065	74.222	77.929	82.358	118.498	124.342	129.561	135.807	140.169

Donald B. Owen, *Handbook of Statistical Tables,* © 1962, U.S. Department of Energy. Published by Addison-Wesley, Reading, Massachusetts. Reprinted with permission of the publisher.

Table A-6. *F* Distribution.

($\alpha = 0.01$ in the right tail)

Numerator degrees of freedom

df_1 / df_2	1	2	3	4	5	6	7	8	9
1	4052.2	4999.5	5403.4	5624.6	5763.6	5859.0	5928.4	5981.1	6022.5
2	98.503	99.000	99.166	99.249	99.299	99.333	99.356	99.374	99.388
3	34.116	30.817	29.457	28.710	28.237	27.911	27.672	27.489	27.345
4	21.198	18.000	16.694	15.977	15.522	15.207	14.976	14.799	14.659
5	16.258	13.274	12.060	11.392	10.967	10.672	10.456	10.289	10.158
6	13.745	10.925	9.7795	9.1483	8.7459	8.4661	8.2600	8.1017	7.9761
7	12.246	9.5466	8.4513	7.8466	7.4604	7.1914	6.9928	6.8400	6.7188
8	11.259	8.6491	7.5910	7.0061	6.6318	6.3707	6.1776	6.0289	5.9106
9	10.561	8.0215	6.9919	6.4221	6.0569	5.8018	5.6129	5.4671	5.3511
10	10.044	7.5594	6.5523	5.9943	5.6363	5.3858	5.2001	5.0567	4.9424
11	9.6460	7.2057	6.2167	5.6683	5.3160	5.0692	4.8861	4.7445	4.6315
12	9.3302	6.9266	5.9525	5.4120	5.0643	4.8206	4.6395	4.4994	4.3875
13	9.0738	6.7010	5.7394	5.2053	4.8616	4.6204	4.4410	4.3021	4.1911
14	8.8616	6.5149	5.5639	5.0354	4.6950	4.4558	4.2779	4.1399	4.0297
15	8.6831	6.3589	5.4170	4.8932	4.5556	4.3183	4.1415	4.0045	3.8948
16	8.5310	6.2262	5.2922	4.7726	4.4374	4.2016	4.0259	3.8896	3.7804
17	8.3997	6.1121	5.1850	4.6690	4.3359	4.1015	3.9267	3.7910	3.6822
18	8.2854	6.0129	5.0919	4.5790	4.2479	4.0146	3.8406	3.7054	3.5971
19	8.1849	5.9259	5.0103	4.5003	4.1708	3.9386	3.7653	3.6305	3.5225
20	8.0960	5.8489	4.9382	4.4307	4.1027	3.8714	3.6987	3.5644	3.4567
21	8.0166	5.7804	4.8740	4.3688	4.0421	3.8117	3.6396	3.5056	3.3981
22	7.9454	5.7190	4.8166	4.3134	3.9880	3.7583	3.5867	3.4530	3.3458
23	7.8811	5.6637	4.7649	4.2636	3.9392	3.7102	3.5390	3.4057	3.2986
24	7.8229	5.6136	4.7181	4.2184	3.8951	3.6667	3.4959	3.3629	3.2560
25	7.7698	5.5680	4.6755	4.1774	3.8550	3.6272	3.4568	3.3239	3.2172
26	7.7213	5.5263	4.6366	4.1400	3.8183	3.5911	3.4210	3.2884	3.1818
27	7.6767	5.4881	4.6009	4.1056	3.7848	3.5580	3.3882	3.2558	3.1494
28	7.6356	5.4529	4.5681	4.0740	3.7539	3.5276	3.3581	3.2259	3.1195
29	7.5977	5.4204	4.5378	4.0449	3.7254	3.4995	3.3303	3.1982	3.0920
30	7.5625	5.3903	4.5097	4.0179	3.6990	3.4735	3.3045	3.1726	3.0665
40	7.3141	5.1785	4.3126	3.8283	3.5138	3.2910	3.1238	2.9930	2.8876
60	7.0771	4.9774	4.1259	3.6490	3.3389	3.1187	2.9530	2.8233	2.7185
120	6.8509	4.7865	3.9491	3.4795	3.1735	2.9559	2.7918	2.6629	2.5586
∞	6.6349	4.6052	3.7816	3.3192	3.0173	2.8020	2.6393	2.5113	2.4073

From "Tables of Percentage Points of the Inverted Beta (*F*) Distribution," *Biometrika, 33*, 1943, 80–84, Maxine Merrington and Catherine M. Thompson. Reproduced by permission of Professor E. S. Pearson.

Denominator degrees of freedom

Table A-6 (continued). F Distribution.

($\alpha = 0.01$ in the right tail)

Numerator degrees of freedom

df_2 \ df_1	10	12	15	20	24	30	40	60	120	∞
1	6055.8	6106.3	6157.3	6208.7	6234.6	6260.6	6286.8	6313.0	6339.4	6365.9
2	99.399	99.416	99.433	99.449	99.458	99.466	99.474	99.482	99.491	99.499
3	27.229	27.052	26.872	26.690	26.598	26.505	26.411	26.316	26.221	26.125
4	14.546	14.374	14.198	14.020	13.929	13.838	13.745	13.652	13.558	13.463
5	10.051	9.8883	9.7222	9.5526	9.4665	9.3793	9.2912	9.2020	9.1118	9.0204
6	7.8741	7.7183	7.5590	7.3958	7.3127	7.2285	7.1432	7.0567	6.9690	6.8800
7	6.6201	6.4691	6.3143	6.1554	6.0743	5.9920	5.9084	5.8236	5.7373	5.6495
8	5.8143	5.6667	5.5151	5.3591	5.2793	5.1981	5.1156	5.0316	4.9461	4.8588
9	5.2565	5.1114	4.9621	4.8080	4.7290	4.6486	4.5666	4.4831	4.3978	4.3105
10	4.8491	4.7059	4.5581	4.4054	4.3269	4.2469	4.1653	4.0819	3.9965	3.9090
11	4.5393	4.3974	4.2509	4.0990	4.0209	3.9411	3.8596	3.7761	3.6904	3.6024
12	4.2961	4.1553	4.0096	3.8584	3.7805	3.7008	3.6192	3.5355	3.4494	3.3608
13	4.1003	4.9603	3.8154	3.6646	3.5868	3.5070	3.4253	3.3413	3.2548	3.1654
14	3.9394	3.8001	3.6557	3.5052	3.4274	3.3476	3.2656	3.1813	3.0942	3.0040
15	3.8049	3.6662	3.5222	3.3719	3.2940	3.2141	3.1319	3.0471	2.9595	2.8684
16	3.6909	3.5527	3.4089	3.2587	3.1808	3.1007	3.0182	2.9330	2.8447	2.7528
17	3.5931	3.4552	3.3117	3.1615	3.0835	3.0032	2.9205	2.8348	2.7459	2.6530
18	3.5082	3.3706	3.2273	3.0771	2.9990	2.9185	2.8354	2.7493	2.6597	2.5660
19	3.4338	3.2965	3.1533	3.0031	2.9249	2.8442	2.7608	2.6742	2.5839	2.4893
20	3.3682	3.2311	3.0880	2.9377	2.8594	2.7785	2.6947	2.6077	2.5168	2.4212
21	3.3098	3.1730	3.0300	2.8796	2.8010	2.7200	2.6359	2.5484	2.4568	2.3603
22	3.2576	3.1209	2.9779	2.8274	2.7488	2.6675	2.5831	2.4951	2.4029	2.3055
23	3.2106	3.0740	2.9311	2.7805	2.7017	2.6202	2.5355	2.4471	2.3542	2.2558
24	3.1681	3.0316	2.8887	2.7380	2.6591	2.5773	2.4923	2.4035	2.3100	2.2107
25	3.1294	2.9931	2.8502	2.6993	2.6203	2.5383	2.4530	2.3637	2.2696	2.1694
26	3.0941	2.9578	2.8150	2.6640	2.5848	2.5026	2.4170	2.3273	2.2325	2.1315
27	3.0618	2.9256	2.7827	2.6316	2.5522	2.4699	2.3840	2.2938	2.1985	2.0965
28	3.0320	2.8959	2.7530	2.6017	2.5223	2.4397	2.3535	2.2629	2.1670	2.0642
29	3.0045	2.8685	2.7256	2.5742	2.4946	2.4118	2.3253	2.2344	2.1379	2.0342
30	2.9791	2.8431	2.7002	2.5487	2.4689	2.3860	2.2992	2.2079	2.1108	2.0062
40	2.8005	2.6648	2.5216	2.3689	2.2880	2.2034	2.1142	2.0194	1.9172	1.8047
60	2.6318	2.4961	2.3523	2.1978	2.1154	2.0285	1.9360	1.8363	1.7263	1.6006
120	2.4721	2.3363	2.1915	2.0346	1.9500	1.8600	1.7628	1.6557	1.5330	1.3805
∞	2.3209	2.1847	2.0385	1.8783	1.7908	1.6964	1.5923	1.4730	1.3246	1.0000

Denominator degrees of freedom

Table A-6 (continued). F Distribution.

($\alpha = 0.025$ in the right tail)

Numerator degrees of freedom

df_1 / df_2	1	2	3	4	5	6	7	8	9
1	647.79	799.50	864.16	899.58	921.85	937.11	948.22	956.66	963.28
2	38.506	39.000	39.165	39.248	39.298	39.331	39.335	39.373	39.387
3	17.443	16.044	15.439	15.101	14.885	14.735	14.624	14.540	14.473
4	12.218	10.649	9.9792	9.6045	9.3645	9.1973	9.0741	8.9796	8.9047
5	10.007	8.4336	7.7636	7.3879	7.1464	6.9777	6.8531	6.7572	6.6811
6	8.8131	7.2599	6.5988	6.2272	5.9876	5.8198	5.6955	5.5996	5.5234
7	8.0727	6.5415	5.8898	5.5226	5.2852	5.1186	4.9949	4.8993	4.8232
8	7.5709	6.0595	5.4160	5.0526	4.8173	4.6517	4.5286	4.4333	4.3572
9	7.2093	5.7147	5.0781	4.7181	4.4844	4.3197	4.1970	4.1020	4.0260
10	6.9367	5.4564	4.8256	4.4683	4.2361	4.0721	3.9498	3.8549	3.7790
11	6.7241	5.2559	4.6300	4.2751	4.0440	3.8807	3.7586	3.6638	3.5879
12	6.5538	5.0959	4.4742	4.1212	3.8911	3.7283	3.6065	3.5118	3.4358
13	6.4143	4.9653	4.3472	3.9959	3.7667	3.6043	3.4827	3.3880	3.3120
14	6.2979	4.8567	4.2417	3.8919	3.6634	3.5014	3.3799	3.2853	3.2093
15	6.1995	4.7650	4.1528	3.8043	3.5764	3.4147	3.2934	3.1987	3.1227
16	6.1151	4.6867	4.0768	3.7294	3.5021	3.3406	3.2194	3.1248	3.0488
17	6.0420	4.6189	4.0112	3.6648	3.4379	3.2767	3.1556	3.0610	2.9849
18	5.9781	4.5597	3.9539	3.6083	3.3820	3.2209	3.0999	3.0053	2.9291
19	5.9216	4.5075	3.9034	3.5587	3.3327	3.1718	3.0509	2.9563	2.8801
20	5.8715	4.4613	3.8587	3.5147	3.2891	3.1283	3.0074	2.9128	2.8365
21	5.8266	4.4199	3.8188	3.4754	3.2501	3.0895	2.9686	2.8740	2.7977
22	5.7863	4.3828	3.7829	3.4401	3.2151	3.0546	2.9338	2.8392	2.7628
23	5.7498	4.3492	3.7505	3.4083	3.1835	3.0232	2.9023	2.8077	2.7313
24	5.7166	4.3187	3.7211	3.3794	3.1548	2.9946	2.8738	2.7791	2.7027
25	5.6864	4.2909	3.6943	3.3530	3.1287	2.9685	2.8478	2.7531	2.6766
26	5.6586	4.2655	3.6697	3.3289	3.1048	2.9447	2.8240	2.7293	2.6528
27	5.6331	4.2421	3.6472	3.3067	3.0828	2.9228	2.8021	2.7074	2.6309
28	5.6096	4.2205	3.6264	3.2863	3.0626	2.9027	2.7820	2.6872	2.6106
29	5.5878	4.2006	3.6072	3.2674	3.0438	2.8840	2.7633	2.6686	2.5919
30	5.5675	4.1821	3.5894	3.2499	3.0265	2.8667	2.7460	2.6513	2.5746
40	5.4239	4.0510	3.4633	3.1261	2.9037	2.7444	2.6238	2.5289	2.4519
60	5.2856	3.9253	3.3425	3.0077	2.7863	2.6274	2.5068	2.4117	2.3344
120	5.1523	3.8046	3.2269	2.8943	2.6740	2.5154	2.3948	2.2994	2.2217
∞	5.0239	3.6889	3.1161	2.7858	2.5665	2.4082	2.2875	2.1918	2.1136

Table A-6 (continued). F Distribution.

($\alpha = 0.025$ in the right tail)

Numerator degrees of freedom

df_2 \ df_1	10	12	15	20	24	30	40	60	120	∞
1	968.63	976.71	984.87	993.10	997.25	1001.4	1005.6	1009.8	1014.0	1018.3
2	39.398	39.415	39.431	39.448	39.456	39.465	39.473	39.481	39.490	39.498
3	14.419	14.337	14.253	14.167	14.124	14.081	14.037	13.992	13.947	13.902
4	8.8439	8.7512	8.6565	8.5599	8.5109	8.4613	8.4111	8.3604	8.3092	8.2573
5	6.6192	6.5245	6.4277	6.3286	6.2780	6.2269	6.1750	6.1225	6.0693	6.0153
6	5.4613	5.3662	5.2687	5.1684	5.1172	5.0652	5.0125	4.9589	4.9044	4.8491
7	4.7611	4.6658	4.5678	4.4667	4.4150	4.3624	4.3089	4.2544	4.1989	4.1423
8	4.2951	4.1997	4.1012	3.9995	3.9472	3.8940	3.8398	3.7844	3.7279	3.6702
9	3.9639	3.8682	3.7694	3.6669	3.6142	3.5604	3.5055	3.4493	3.3918	3.3329
10	3.7168	3.6209	3.5217	3.4185	3.3654	3.3110	3.2554	3.1984	3.1399	3.0798
11	3.5257	3.4296	3.3299	3.2261	3.1725	3.1176	3.0613	3.0035	2.9441	2.8828
12	3.3736	3.2773	3.1772	3.0728	3.0187	2.9633	2.9063	2.8478	2.7874	2.7249
13	3.2497	3.1532	3.0527	2.9477	2.8932	2.8372	2.7797	2.7204	2.6590	2.5955
14	3.1469	3.0502	2.9493	2.8437	2.7888	2.7324	2.6742	2.6142	2.5519	2.4872
15	3.0602	2.9633	2.8621	2.7559	2.7006	2.6437	2.5850	2.5242	2.4611	2.3953
16	2.9862	2.8890	2.7875	2.6808	2.6252	2.5678	2.5085	2.4471	2.3831	2.3163
17	2.9222	2.8249	2.7230	2.6158	2.5598	2.5020	2.4422	2.3801	2.3153	2.2474
18	2.8664	2.7689	2.6667	2.5590	2.5027	2.4445	2.3842	2.3214	2.2558	2.1869
19	2.8172	2.7196	2.6171	2.5089	2.4523	2.3937	2.3329	2.2696	2.2032	2.1333
20	2.7737	2.6758	2.5731	2.4645	2.4076	2.3486	2.2873	2.2234	2.1562	2.0853
21	2.7348	2.6368	2.5338	2.4247	2.3675	2.3082	2.2465	2.1819	2.1141	2.0422
22	2.6998	2.6017	2.4984	2.3890	2.3315	2.2718	2.2097	2.1446	2.0760	2.0032
23	2.6682	2.5699	2.4665	2.3567	2.2989	2.2389	2.1763	2.1107	2.0415	1.9677
24	2.6396	2.5411	2.4374	2.3273	2.2693	2.2090	2.1460	2.0799	2.0099	1.9353
25	2.6135	2.5149	2.4110	2.3005	2.2422	2.1816	2.1183	2.0516	1.9811	1.9055
26	2.5896	2.4908	2.3867	2.2759	2.2174	2.1565	2.0928	2.0257	1.9545	1.8781
27	2.5676	2.4688	2.3644	2.2533	2.1946	2.1334	2.0693	2.0018	1.9299	1.8527
28	2.5473	2.4484	2.3438	2.2324	2.1735	2.1121	2.0477	1.9797	1.9072	1.8291
29	2.5286	2.4295	2.3248	2.2131	2.1540	2.0923	2.0276	1.9591	1.8861	1.8072
30	2.5112	2.4120	2.3072	2.1952	2.1359	2.0739	2.0089	1.9400	1.8664	1.7867
40	2.3882	2.2882	2.1819	2.0677	2.0069	1.9429	1.8752	1.8028	1.7242	1.6371
60	2.2702	2.1692	2.0613	1.9445	1.8817	1.8152	1.7440	1.6668	1.5810	1.4821
120	2.1570	2.0548	1.9450	1.8249	1.7597	1.6899	1.6141	1.5299	1.4327	1.3104
∞	2.0483	1.9447	1.8326	1.7085	1.6402	1.5660	1.4835	1.3883	1.2684	1.0000

Denominator degrees of freedom

Table A-6 (continued). F Distribution.

($\alpha = 0.05$ in the right tail)

Numerator degrees of freedom

df_1 df_2	1	2	3	4	5	6	7	8	9
1	161.45	199.50	215.71	224.58	230.16	233.99	236.77	238.88	240.54
2	18.513	19.000	19.164	19.247	19.296	19.330	19.353	19.371	19.385
3	10.128	9.5521	9.2766	9.1172	9.0135	8.9406	8.8867	8.8452	8.8123
4	7.7086	9.9443	6.5914	6.3882	6.2561	6.1631	6.0942	6.0410	6.9988
5	6.6079	5.7861	5.4095	5.1922	5.0503	4.9503	4.8759	4.8183	4.7725
6	5.9874	5.1433	4.7571	4.5337	4.3874	4.2839	4.2067	4.1468	4.0990
7	5.5914	4.7374	4.3468	4.1203	3.9715	3.8660	3.7870	3.7257	3.6767
8	5.3177	4.4590	4.0662	3.8379	3.6875	3.5806	3.5005	3.4381	3.3881
9	5.1174	4.2565	3.8625	3.6331	3.4817	3.3738	3.2927	3.2296	3.1789
10	4.9646	4.1028	3.7083	3.4780	3.3258	3.2172	3.1355	3.0717	3.0204
11	4.8443	3.9823	3.5874	3.3567	3.2039	3.0946	3.0123	2.9480	2.8962
12	4.7472	3.8853	3.4903	3.2592	3.1059	2.9961	2.9134	2.8486	2.7964
13	4.6672	3.8056	3.4105	3.1791	3.0254	2.9153	2.8321	2.7669	2.7144
14	4.6001	3.7389	3.3439	3.1122	2.9582	2.8477	2.7642	2.6987	2.6458
15	4.5431	3.6823	3.2874	3.0556	2.9013	2.7905	2.7066	2.6408	2.5876
16	4.4940	3.6337	3.2389	3.0069	2.8524	2.7413	2.6572	2.5911	2.5377
17	4.4513	3.5915	3.1968	2.9647	2.8100	2.6987	2.6143	2.5480	2.4943
18	4.4139	3.5546	3.1599	2.9277	2.7729	2.6613	2.5767	2.5102	2.4563
19	4.3807	3.5219	3.1274	2.8951	2.7401	2.6283	2.5435	2.4768	2.4227
20	4.3512	3.4928	3.0984	2.8661	2.7109	2.5990	2.5140	2.4471	2.3928
21	4.3248	3.4668	3.0725	2.8401	2.6848	2.5727	2.4876	2.4205	2.3660
22	4.3009	3.4434	3.0491	2.8167	2.6613	2.5491	2.4638	2.3965	2.3419
23	4.2793	3.4221	3.0280	2.7955	2.6400	2.5277	2.4422	2.3748	2.3201
24	4.2597	3.4028	3.0088	2.7763	2.6207	2.5082	2.4226	2.3551	2.3002
25	4.2417	3.3852	2.9912	2.7587	2.6030	2.4904	2.4047	2.3371	2.2821
26	4.2252	3.3690	2.9752	2.7426	2.5868	2.4741	2.3883	2.3205	2.2655
27	4.2100	3.3541	2.9604	2.7278	2.5719	2.4591	2.3732	2.3053	2.2501
28	4.1960	3.3404	2.9467	2.7141	2.5581	2.4453	2.3593	2.2913	2.2360
29	4.1830	3.3277	2.9340	2.7014	2.5454	2.4324	2.3463	2.2783	2.2229
30	4.1709	3.3158	2.9223	2.6896	2.5336	2.4205	2.3343	2.2662	2.2107
40	4.0847	3.2317	2.8387	2.6060	2.4495	2.3359	2.2490	2.1802	2.1240
60	4.0012	3.1504	2.7581	2.5252	2.3683	2.2541	2.1665	2.0970	2.0401
120	3.9201	3.0718	2.6802	2.4472	2.2899	2.1750	2.0868	2.0164	1.9588
∞	3.8415	2.9957	2.6049	2.3719	2.2141	2.0986	2.0096	1.9384	1.8799

Denominator degrees of freedom

Table A-6 (continued). *F* Distribution.

(α = 0.05 in the right tail)

Numerator degrees of freedom

df_1 / df_2	10	12	15	20	24	30	40	60	120	∞
1	241.88	243.91	245.95	248.01	249.05	250.10	251.14	252.20	253.25	254.31
2	19.396	19.413	19.429	19.446	19.454	19.462	19.471	19.479	19.487	19.496
3	8.7855	8.7446	8.7029	8.6602	8.6385	8.6166	8.5944	8.5720	8.5494	8.5264
4	5.9644	5.9117	5.8578	5.8025	5.7744	5.7459	5.7170	5.6877	5.6581	5.6281
5	4.7351	4.6777	4.6188	4.5581	4.5272	4.4957	4.4638	4.4314	4.3985	4.3650
6	4.0600	3.9999	3.9381	3.8742	3.8415	3.8082	3.7743	3.7398	3.7047	3.6689
7	3.6365	3.5747	3.5107	3.4445	3.4105	3.3758	3.3404	3.3043	3.2674	3.2298
8	3.3472	3.2839	3.2184	3.1503	3.1152	3.0794	3.0428	3.0053	2.9669	2.9276
9	3.1373	3.0729	3.0061	2.9365	2.9005	2.8637	2.8259	2.7872	2.7475	2.7067
10	2.9782	2.9130	2.8450	2.7740	2.7372	2.6996	2.6609	2.6211	2.5801	2.5379
11	2.8536	2.7876	2.7186	2.6464	2.6090	2.5705	2.5309	2.4901	2.4480	2.4045
12	2.7534	2.6866	2.6169	2.5436	2.5055	2.4663	2.4259	2.3842	2.3410	2.2962
13	2.6710	2.6037	2.5331	2.4589	2.4202	2.3803	2.3392	2.2966	2.2524	2.2064
14	2.6022	2.5342	2.4630	2.3879	2.3487	2.3082	2.2664	2.2229	2.1778	2.1307
15	2.5437	2.4753	2.4034	2.3275	2.2878	2.2468	2.2043	2.1601	2.1141	2.0658
16	2.4935	2.4247	2.3522	2.2756	2.2354	2.1938	2.1507	2.1058	2.0589	2.0096
17	2.4499	2.3807	2.3077	2.2304	2.1898	2.1477	2.1040	2.0584	2.0107	1.9604
18	2.4117	2.3421	2.2686	2.1906	2.1497	2.1071	2.0629	2.0166	1.9681	1.9168
19	2.3779	2.3080	2.2341	2.1555	2.1141	2.0712	2.0264	1.9795	1.9302	1.8780
20	2.3479	2.2776	2.2033	2.1242	2.0825	2.0391	1.9938	1.9464	1.8963	1.8432
21	2.3210	2.2504	2.1757	2.0960	2.0540	2.0102	1.9645	1.9165	1.8657	1.8117
22	2.2967	2.2258	2.1508	2.0707	2.0283	1.9842	1.9380	1.8894	1.8380	1.7831
23	2.2747	2.2036	2.1282	2.0476	2.0050	1.9605	1.9139	1.8648	1.8128	1.7570
24	2.2547	2.1834	2.1077	2.0267	1.9838	1.9390	1.8920	1.8424	1.7896	1.7330
25	2.2365	2.1649	2.0889	2.0075	1.9643	1.9192	1.8718	1.8217	1.7684	1.7110
26	2.2197	2.1479	2.0716	1.9898	1.9464	1.9010	1.8533	1.8027	1.7488	1.6906
27	2.2043	2.1323	2.0558	1.9736	1.9299	1.8842	1.8361	1.7851	1.7306	1.6717
28	2.1900	2.1179	2.0411	1.9586	1.9147	1.8687	1.8203	1.7689	1.7138	1.6541
29	2.1768	2.1045	2.0275	1.9446	1.9005	1.8543	1.8055	1.7537	1.6981	1.6376
30	2.1646	2.0921	2.0148	1.9317	1.8874	1.8409	1.7918	1.7396	1.6835	1.6223
40	2.0772	2.0035	1.9245	1.8389	1.7929	1.7444	1.6928	1.6373	1.5766	1.5089
60	1.9926	1.9174	1.8364	1.7480	1.7001	1.6491	1.5943	1.5343	1.4673	1.3893
120	1.9105	1.8337	1.7505	1.6587	1.6084	1.5543	1.4952	1.4290	1.3519	1.2539
∞	1.8307	1.7522	1.6664	1.5705	1.5173	1.4591	1.3940	1.3180	1.2214	1.0000

Denominator degrees of freedom

Table A-7. Critical Values of the Pearson Correlation Coefficient r.

n	$\alpha = .05$	$\alpha = .01$	n	$\alpha = .05$	$\alpha = .01$
4	.950	.999	20	.444	.561
5	.878	.959	25	.396	.505
6	.811	.917	30	.361	.463
7	.754	.875	35	.335	.430
8	.707	.834	40	.312	.402
9	.666	.798	45	.294	.378
10	.632	.765	50	.279	.361
11	.602	.735	60	.254	.330
12	.576	.708	70	.236	.305
13	.553	.684	80	.220	.286
14	.532	.661	90	.207	.269
15	.514	.641	100	.196	.256
16	.497	.623			
17	.482	.606			
18	.468	.590			
19	.456	.575			

absolute value $\pm$

x	the value of a single score.	$\bar{q}$	the proportion or probability equal to $1 - \bar{p}$.
f	frequency with which a value occurs.		
Σ	(capital sigma) summation.	$P(A)$	the probability of event A.
n	the number of scores in a sample.	$P(A \mid B)$	the probability of event A assuming event B has occurred.
$n!$	factorial.		
N	the number of scores in a finite population.	$\bar{A}$	the complement of event A.
$\bar{x}$	the mean of the scores in a sample.	H_0	null hypothesis.
μ	(mu) the mean of all scores in a population.	H_1	alternative hypothesis.
		α	the probability of a type I error or the area of the critical region.
s	the standard deviation of a set of sample values.	β	the probability of a type II error.
σ	(lower case sigma) the standard deviation of all values in a population.	r	linear correlation coefficient.
		m	slope of the straight line with equation $y = mx + b$.
s^2	the variance of a set of sample values.	b	the y intercept of the straight line with equation $y = mx + b$.
σ^2	the variance of all values in a population.	d	the difference between two paired scores.
z	the standard score.	$\bar{d}$	the mean of the differences d found from paired sample data.
$z(\alpha/2)$	the critical value of z.	s_d	the standard deviation of the differences d found from paired sample data.
t	the t distribution.		
$t(\alpha/2)$	the critical value of t.		
df	the number of degrees of freedom.	R	sum of the ranks for a sample; used in the rank-sum test.
F	the F distribution.	μ_R	expected mean rank; used in the rank-sum test.
χ^2	the chi-square distribution.		
χ_R^2	the right-tailed critical value of chi-square.	σ_R	expected standard deviation of ranks; used in the rank-sum test.
χ_L^2	the left-tailed critical value of chi-square.	$\mu_{\bar{x}}$	the mean of the population of all possible sample means $\bar{x}$.
p	the probability of an event or the population proportion.	$\sigma_{\bar{x}}$	the standard deviation of the population of all possible sample means $\bar{x}$.
q	the probability or proportion equal to $1 - p$.	E	maximum error of the estimate of a population parameter or expected frequency.
p_s	sample proportion.		
q_s	the sample proportion equal to $1 - p_s$.	Q_1, Q_2, Q_3	quartiles.
$\bar{p}$	proportion or probability obtained by pooling two samples.	$D_1, D_2, \ldots, D_9$	deciles.
		$P_1, P_2, \ldots, P_{99}$	percentiles.

APPENDIX B
GLOSSARY

Alternative hypothesis. Denoted H_1, the statement that is equivalent to the negation of the null hypothesis.

Analysis of variance. A method analyzing population variance in order to make inferences about the population.

Attribute data. Data that are qualitative and cannot be measured or counted.

Binomial experiment. An experiment with a fixed number of independent trials. Each outcome falls into exactly one of two categories.

Complementary events. Two events that are opposites.

Compound event. A combination of simple events.

Conditional probability. The probability of an event assuming that another event has already occurred.

Confidence interval. A range of values used to estimate some population parameter with a specific level of confidence.

Contingency table. A table of observed frequencies where the rows correspond to one variable of classification and the columns correspond to another variable of classification.

Continuity correction. An adjustment made when a discrete random variable is being approximated by a continuous random variable (see Section 5-5).

Continuous random variable. A random variable with infinitely many values that can be associated with points on a continuous line interval.

Correlation coefficient. A measurement of the strength of the relationship between two variables.

Countable set. A set with either a finite number of values or values that can be made to correspond to the positive integers.

Critical region. The area under a curve which contains those values that lead to rejection of the null hypothesis.

Data. The numbers or information collected in an experiment.

Decile. The nine deciles divide the ranked data into ten groups with 10% of the scores in each group.

Degrees of freedom. The number of values that are free to vary after certain restrictions have been imposed on all values.

Descriptive statistics. The methods used to summarize the key characteristics of known population data.

Event. A result or outcome of some experiment.

Frequency table. A list of categories of scores along with their corresponding frequencies.

Histogram. A graph of vertical bars representing the frequency distribution of a set of data.

Hypothesis. A statement or claim that some population characteristic is true.

Hypothesis test. A method for testing claims made about populations. Also called test of significance.

Independent events. The case when the occurrence of any one of the events does not affect the probabilities of the occurrences of the other events.

Inferential statistics. The methods of using sample data to make generalizations or inferences about a population.

Interval estimate. See confidence interval.

Left-tailed test. A hypothesis test in which the critical region is located only in the left portion of the curve.

Mean. The sum of a set of scores divided by the number of scores.

Mean deviation. The measure of dispersion equal to the sum of the deviations of each score from the mean, divided by the number of scores.

Median. The middle value of a set of scores arranged in order of magnitude.

Midrange. One-half the sum of the highest and lowest scores.

Mode. The score that occurs most frequently.

Multinomial experiment. An experiment with a fixed number of independent trials and each outcome falls into exactly one of several categories.

Mutually exclusive events. Events which cannot occur simultaneously.

Nonparametric test. A hypothesis test that is not based on population parameters and does not have many of the restrictions of parametric tests.

Normal distribution. A bell-shaped probability distribution which is described algebraically by the equation in Section 5-2.

Null hypothesis. Denoted H_0, it is the claim made about some population characteristic. It usually involves the case of no difference.

Parameter. A measured characteristic of a population.

Percentile. The ninety-nine percentiles divide the ranked data into one hundred groups with 1% of the scores in each group.

Point estimate. A single value that serves as an estimate of a population parameter.

Population. The complete and entire collection of elements to be studied.

Probability distribution. Collection of values of a random variable along with their corresponding probabilities.

Quartile. The three quartiles divide the ranked data into four groups with 25% of the scores in each group.

Random sample. A sample selected in a way that allows every member of the population to have the same chance of being chosen.

Random variable. The values that correspond to the numbers associated with events in a sample space.

Range. The measure of dispersion that is the difference between the highest and lowest scores.

Regression line. A straight line that summarizes the relationship between two variables.

Right-tailed test. A hypothesis test in which the critical region is located only in the right portion of the curve.

Sample. A subset of a population.

Sample space. In an experiment, the set of all possible outcomes or events that cannot be further broken down.

Significance level. The probability that serves as a cutoff between results attributed to chance and results attributed to significant differences.

Simple event. An experimental outcome that cannot be further broken down.

Standard deviation. The measure of dispersion equal to the square root of the variance.

Standard error of the mean. The standard deviation of all possible sample means $\bar{x}$.

Standard normal distribution. A normal distribution with a mean of zero and a standard deviation equal to one.

Standard score. Also called z score, it is the number of standard deviations that a given value is above or below the mean.

Statistic. A measured characteristic of a sample.

Statistics. The collection, organization, description, and analysis of data.

Student t distribution. See t distribution.

t distribution. A bell-shaped distribution usually associated with small sample experiments. Also called the student t distribution.

Test statistic. Used in hypothesis testing, it is the sample statistic based on the sample data.

Two-tailed test. A hypothesis test in which the critical region is found in both the left and right portions of the curve.

Type I error. The mistake of rejecting the null hypothesis when it is true.

Type II error. The mistake of failing to reject the null hypothesis when it is false.

Uniform distribution. A distribution of values evenly distributed over the range of possibilities.

Variable data. Quantitative data that can be counted or measured.

Variance. The measure of dispersion found by using Formula 2-4 in Section 2-4.

z score. See standard score.

APPENDIX C
BIBLIOGRAPHY

Adler, I. 1966. *Probability and Statistics for Everyman*. New York: New American Library.

Anderson, R. and T. Bancroft. 1952. *Statistical Theory in Research*. New York: McGraw-Hill.

Armore, S. 1975. *Statistics: A Conceptual Approach*. Columbus, Ohio: Charles E. Merrill.

Bacheller, M. 1978. *The Hammond Almanac*. Maplewood, New Jersey: Hammond Almanac, Inc.

Balsley, H. Editor. 1978. *Basic Statistics for Business and Economics*. Columbus, Ohio: Grid.

Barnett, V. 1973. *Comparative Statistical Inference*. New York: John Wiley.

Bashaw, W. 1969. *Mathematics for Statistics*. New York: John Wiley.

Braverman, J. 1978. *Fundamentals of Business Statistics*. New York: Academic Press.

Chou, Y. 1975. *Statistical Analysis*. 2nd ed. New York: Holt, Rinehart and Winston.

Christensen, H. 1977. *Statistics Step by Step*. Boston: Houghton Mifflin.

Congelosi, V., P. Taylor, and P. Rice. 1979. *Basic Statistics*. 2nd ed. St. Paul, Minnesota: West.

Conover, W. 1971. *Practical Nonparametric Statistics*. New York: John Wiley.

Dixon, W. and F. Massey. 1969. *Introduction to Statistical Analysis*. 2nd ed. New York: McGraw-Hill.

Draper, N. and H. Smith. 1966. *Applied Regression Analysis*. New York: John Wiley.

Dyckman, T. and L. Thomas. 1977. *Fundamental Statistics for Business and Economics*. Englewood Cliffs, New Jersey: Prentice-Hall.

Elzey, F. 1966. *A Programmed Introduction to Statistics*. Belmont, California: Brooks/Cole.

Fairley, W. and F. Mosteller. 1977. *Statistics and Public Policy*. Reading, Massachusetts: Addison-Wesley.

Fisher, R. 1966. *The Design of Experiments*. 8th ed. New York: Hafner.

Freedman, D., R. Pisani, and R. Purves. 1978. *Statistics*. New York: W. W. Norton.

Freund, J. 1973. *Modern Elementary Statistics*. 4th ed. Englewood Cliffs, New Jersey: Prentice-Hall.

Freund, J. 1976. *Statistics, A First Course*. 2nd ed. Englewood Cliffs, New Jersey: Prentice-Hall.

Grant, E. 1964. *Statistical Quality Control*. 3rd ed. New York: McGraw-Hill.

Guenther, W. 1973. *Concepts of Statistical Inference*. 2nd ed. New York: McGraw-Hill.

Haber, A. and R. Runyon. 1973. *General Statistics*. 2nd ed. Reading, Massachusetts: Addison-Wesley.

Hamburg, M. 1977. *Statistical Analysis for Decision Making*. 2nd ed. New York: Harcourt Brace Jovanovich.

Hauser, P. 1975. *Social Statistics in Use*. New York: Russell Sage Foundation.

Heerman, E. and L. Braskam. 1970. *Readings in Statistics for the Behavioral Sciences*. Englewood Cliffs, New Jersey: Prentice-Hall.

Hoel, P. 1976. *Elementary Statistics*. 4th ed. New York: John Wiley.

Hollander, M. and D. Wolfe. 1973. *Nonparametric Statistical Methods*. New York: John Wiley.

Huff, D. 1954. *How to Lie with Statistics*. New York: W. W. Norton.

Johnson, R. 1976. *Elementary Statistics*. North Scituate, Massachusetts: Duxbury.

Kirk, R. Editor. 1972. *Statistical Issues: A Reader for the Behavioral Sciences*. Belmont, California: Brooks/Cole.

Langley, R. 1970. *Practical Statistics Simply Explained*. New York: Dover.

Lapin, L. 1975. *Statistics: Meaning and Method*. New York: Harcourt Brace Jovanovich.

Lindley, D. 1971. *Making Decisions*. New York: John Wiley

McClave, J. and P. Benson. 1978. *Statistics for Business and Economics*. San Francisco: Dellen.

Mendenhall, W. 1975. *Introduction to Probability and Statistics*. 4th ed. North Scituate, Massachusetts: Duxbury.

Mood, A. et al. 1974. *Introduction to the Theory of Statistics*. 3rd ed. New York: McGraw-Hill.

Moore, D. 1979. *Statistics: Concepts and Controversies*. San Francisco: W. H. Freeman.

Mosteller, F., R. Rourke, and G. Thomas. 1970. *Probability with Statistical Applications*. 2nd ed. Reading, Massachusetts: Addison-Wesley.

Neter, J., W. Wasserman, and G. Whitmore. 1973. *Fundamental Statistics for Business and Economics*. 4th ed. Boston: Allyn and Bacon.

Nobile, P. and J. Deedy. Editors. 1972. *The Complete Ecology Fact Book*. Garden City, New York: Doubleday.

Noether, G. 1967. *Elements of Nonparametric Statistics*. New York: John Wiley.

Owen, D. 1962. *Handbook of Statistical Tables*. Reading, Massachusetts: Addison-Wesley.

Raiffa, H. 1968. *Decision Analysis: Introductory Lectures on Choices Under Uncertainty*. Reading, Massachusetts: Addison-Wesley.

Reichard, R. 1974. *The Figure Finaglers*. New York: McGraw-Hill.

Reichmann, W. 1962. *Use and Abuse of Statistics*. New York: Oxford University Press.

Roscoe, J. 1975. *Fundamental Research Statistics for the Behavioral Sciences*. 2nd ed. New York: Holt, Rinehart and Winston.

Siegal, S. 1956. *Nonparametric Statistics for the Behavioral Sciences*. New York: McGraw-Hill.

Snedecor, G. and W. Cochran. 1967. *Statistical Methods*. 6th ed. Ames, Iowa: Iowa State University Press.

Spear, M. 1969. *Practical Charting Techniques*. New York: McGraw-Hill.

Steger, J. Editor. 1971. *Readings in Statistics for the Behavioral Sciences*. New York: Holt, Rinehart and Winston.

Tanur, J. Editor. 1972. *Statistics: A Guide to the Unknown*. San Francisco: Holden-Day.

Ukena, A. 1978. *Statistics Today*. New York: Harper & Row.

Walker, H. and J. Lev. 1969. *Elementary Statistical Methods*. 3rd ed. New York: Holt, Rinehart and Winston.

Weinberg, G. and J. Schumaker. 1969. *Statistics, An Intuitive Approach*. 2nd ed. Monterey, California: Brooks/Cole.

Winkler, R. and W. Hays. 1975. *Statistics: Probability, Inference and Decision*. 2nd ed. New York: Holt, Rinehart and Winston.

Yamane, T. 1973. *Statistics: An Introductory Analysis*. 3rd ed. New York: Harper & Row.

Zeisel, H. 1968. *Say It with Figures*. 5th ed. New York: Harper & Row.

Zuwaylif, F. 1974. *General Applied Statistics*. 2nd ed. Reading, Massachusetts: Addison-Wesley.

APPENDIX D
ANSWERS TO SELECTED EXERCISES

Chapter 2: Section 2.1

2-1 (a).

Class limits	Class boundaries	Class marks
110–129	109.5–129.5	119.5
130–149	129.5–149.5	139.5
150–169	149.5–169.5	159.5
170–189	169.5–189.5	179.5
190–209	189.5–209.5	199.5
210–229	209.5–229.5	219.5
230–249	229.5–249.5	239.5
250–269	249.5–269.5	259.5
270–289	269.5–289.5	279.5
290–309	289.5–309.5	299.5

The class width is 20.

(b)

Class limits	Class boundaries	Class marks
110–134	109.5–134.5	122
135–159	134.5–159.5	147
160–184	159.5–184.5	172
185–209	184.5–209.5	197
210–234	209.5–234.5	222
235–259	234.5–259.5	247
260–284	259.5–284.5	272
285–309	284.5–309.5	297

The class width is 25.

(c)

Class limits	Class boundaries	Class marks
106–118	105.5-118.5	112
119–131	118.5-131.5	125
132–144	131.5–144.5	138
145–157	144.5–157.5	151
158–170	157.5–170.5	164
171–183	170.5–183.5	177
184–196	183.5–196.5	190
197–209	196.5–209.5	203
210–222	209.5–222.5	216
223–235	222.5–235.5	229

Class limits	Class boundaries	Class marks
236–248	235.5–248.5	242
249–261	248.5–261.5	255
262–274	261.5–274.5	268
275–287	274.5–287.5	281
288–300	287.5–300.5	294
301–313	300.5–313.5	307

The class width is 13.

2-3 (a).

Class limits	Class boundaries	Class marks
17.0–19.2	16.95–19.25	18.1
19.3–21.5	19.25–21.55	20.4
21.6–23.8	21.55–23.85	22.7
23.9–26.1	23.85–26.15	25.0
26.2–28.4	26.15–28.45	27.3
28.5–30.7	28.45–30.75	29.6
30.8–33.0	30.75–33.05	31.9
33.1–35.3	33.05–35.35	34.2
35.4–37.6	35.35–37.65	36.5
37.7–39.9	37.65–39.95	38.8
40.0–42.2	39.95–42.25	41.1
42.3–44.5	42.25–44.55	43.4

The class width is 2.3.

(b).

Class limits	Class boundaries	Class marks
17.0–18.7	16.95–18.75	17.85
18.8–20.5	18.75–20.55	19.65
20.6–22.3	20.55–22.35	21.45
.	.	.
.	.	.
40.4–42.1	40.35–42.15	41.25
42.2–43.9	42.15–43.95	43.05

The class width is 1.8.

2-5 (a).

x	f
34–54	6
55–75	8
76–96	16

	97–117	17
	118–138	25
	139–159	15
	160–180	8
	181–201	2
	202–222	2
	223–243	1

(b).

x	f
34–47	2
48–61	5
62–75	7
76–89	15
90–103	7
104–117	11
118–131	17
132–145	13
146–159	10
160–173	5
174–187	3
188–201	2
202–215	2
216–229	0
230–243	1

(c).

x	f
30–47	2
48–65	6
66–83	12
84–101	14
102–119	13
120–137	25
138–155	13
156–173	7
174–191	4
192–209	1
210–227	2
228–245	1

(d).

x	f
Less than 55	5
Less than 76	14
Less than 97	30
Less than 118	47
Less than 139	72
Less than 160	87
Less than 181	95
Less than 202	97
Less than 223	99
Less than 244	100

2-7.

x	f
4.92	17
4.93	8
4.94	12
4.95	11
4.96	8
4.97	8
4.98	15
4.99	9
5.00	10
5.01	13
5.02	8
5.03	14
5.04	17

2-9. The data of Exercise 2-7 appear to be uniformly distributed over the range of values while the data of Exercise 2-8 tend to center about a value near the middle.

2-13.

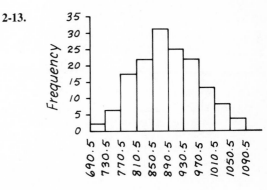

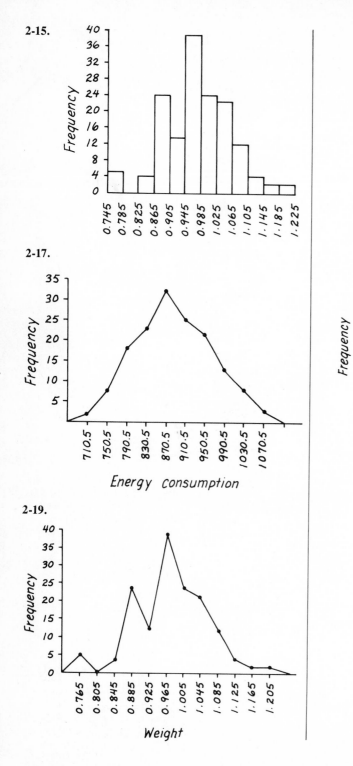

2-15.

2-17.

Energy consumption

2-19.

Weight

2-21.

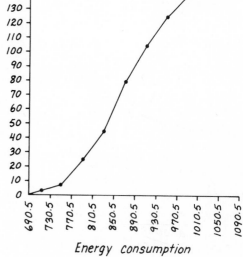

Consumption	Cumulative frequency
Less than 731	1
Less than 771	8
Less than 811	26
Less than 851	48
Less than 891	80
Less than 931	105
Less than 971	127
Less than 1011	140
Less than 1051	148
Less than 1091	150

Energy consumption

2-23.

Weight	Cumulative frequency
Less than 0.785	5
Less than 0.825	5
Less than 0.865	9
Less than 0.905	33
Less than 0.945	46
Less than 0.985	84
Less than 1.025	108
Less than 1.065	130
Less than 1.105	142
Less than 1.145	146
Less than 1.185	148
Less than 1.225	150

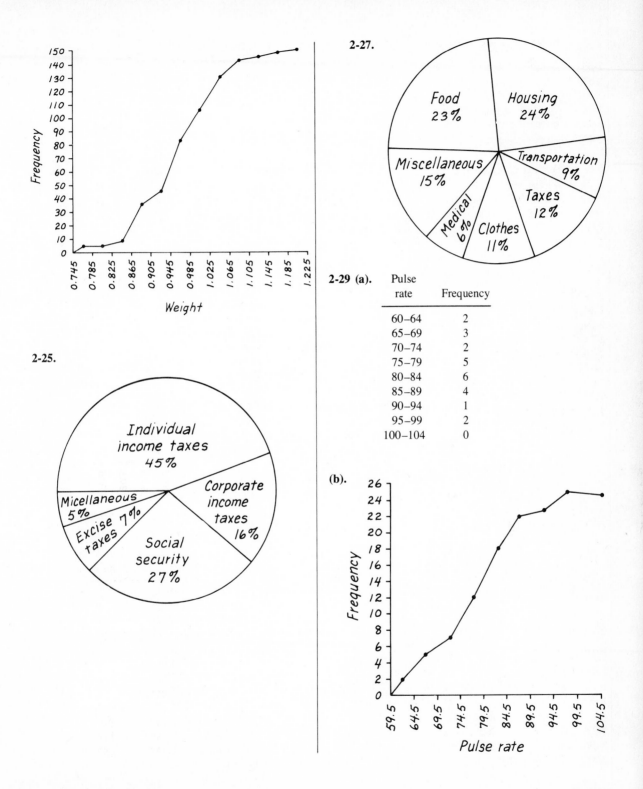

2-27.

2-25.

2-29 (a).

Pulse rate	Frequency
60–64	2
65–69	3
70–74	2
75–79	5
80–84	6
85–89	4
90–94	1
95–99	2
100–104	0

(b).

	Mean	Median	Mode	Midrange
2-35.	7.2	6.9	6.4	8.2
2-37.	48.1	46.0	—	50.0
2-39.	71.5	72.0	77	71.0
2-41.	389.9	403.0	—	421.0
2-43.	199.6	202.5	213	199.5
2-45.	40.5	8.0	1	250.5

2-47. The class marks are 3, 8, 13, 18, 23. The mean is 15.5.   **2-49.** The class marks are 45.5, 55.5, 65.5, 75.5, 85.5, 95.5. The mean is 71.5.   **2-51 (a).** 828, 883, 948.   **(b).** 786, 819.5, 846, 861, 883, 902.5, 931, 956.5, 994.5.   **2-53 (a).** 13982, 18630, 24100.   **(b).** 10000, 12555, 15047.50, 17214, 18630, 20330, 22111, 25615, 28880.   **2-55.** The median is the average which best represents the given scores.   **2-57.** 887.5. **2-61 (a).** 5.4.   **(b).** 48 miles per hour.   **2-63.** 2.9.

	Range	Variance	Standard Deviation
2-65.	9.1	5.62	2.37
2-67.	22	39.1	6.3
2-69.	58	331.8	18.2
2-71.	422	11094.4	105.3
2-73.	81	507.4	22.5
2-75.	449	16193.6	127.3
2-76.	6	5.1	2.3
2-77.	6	5.1	2.3
2-79.	60	511.1	22.6

Comparing 2-76 and 2-77 we see that adding a constant to all scores has no effect on the range, variance, or standard deviation. Comparing 2-76 and 2-79 we see that multiplying all scores by a constant causes the range and standard deviation to be multiplied by that same constant but the variance is multiplied by the square of the constant. **2-81.** The statistics students are a more homogeneous group and should therefore have a smaller variance. **2-83.** If all scores are the same, then the variance is zero. The variance can never be negative.   **2-85 (a).** 1.33.   **(b).** 0. **(c).** −2. **(d).** 1.8. **(e).** −1.2.   **2-87.** Range: 396; variance: 5964.1; standard deviation: 77.2. **2-89.** Mean deviation: 1.5; semi-interquartile range: 2; 10–90 percentile range: 4.   **2-91.** 93.8.   **2-93 (a).** At least 3/4 of the scores are between 70 and 130.   **(b).** At least 8/9 of the scores are between 55 and 145.   **(c).** At least 3/4 of the scores are between 300 and 700.   **(d).** At least 8/9 of the scores are between 200 and 800.

2-95.

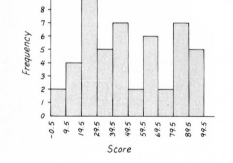

2-97 (a). 27.   **(b).** 73   **(c).** 28.   **(d).** 44.   **2-99.** Mean: 50.1; standard deviation: 27.8.

Chapter 3

3-1. 0.482.   **3-3 (a).** Boy-boy,boy-girl,girl-boy,girl-girl   **(b).** 1/4.   **(c).** 1/2.   **3-5.** 2, −1/2, 5, 1.11, 1.0001.   **3-7.** 0. 772.   **3-9.** 0.549.   **3-11.** 31/366.   **3-13.** 1/19.   **3-15.** Equally likely: (b), (f), (g), (h).   **3-19.** Mutually exclusive: (a), (h), (i).   **3-21.** 7/12.   **3-23.** 11/15.   **3-25.** 17/27. **3-27.** 3410/5673.   **3-29.** 0.845.   **3-31.** 0.8.   **3-33.** 0.8.   **3-35.** 0.63.   **3-37.** 7/13. **3-39.** 1/12.   **3-41 (a).** $P(A \text{ or } B) = 0.9$.   **(b).** $P(A \text{ or } B) < 0.9$.   **3-43.** 0.   **3-45.** 1/144. **3-47 (a).** 1/16.   **(b).** 1/17.   **3-49 (a).** 1/169.   **(b).** 1/221. **3-51.**1/216.   **3-53.** 1/133225.

3-55. 0.0289. **3-57.** 1/36. **3-59.** 1/28. **3-61.** 690/5673. **3-63 (a).** 0.779. **(b).** 0.777.
(c). Sample without replacement. **3-65.** 0.0356 **3-67.** The probability of no two people sharing the same birthday is 0.030. **3-69.** $P(A) = 1/6; P(\overline{A}) = 5/6$. **3-71.** $P(C) = 0.3; P(\overline{C}) = 0.7$. **3-73.** $P(E) = 0.5; P(\overline{E}) = 0.5$. **3-75.** $P(G) = 2/7; P(\overline{G}) = 5/7$. **3-77.** $P(I) = 3/5; P(\overline{I}) = 2/5$. **3-79.** 7/8. **3-81 (a).** 3/4. **(b).** 9/16. **(c).** 7/16. **3-83.** 0.605. **3-85.** 0.151. **3-87.** 0.226. **3-89 (a).** $P(B) = 0$. **(b).** $P(B) = 1$. **(c).** $P(B) < 0.3$. **3-91.** $P(A \text{ or } B) = 1 - P(A) - P(B) + P(A \text{ and } B)$. **3-93.** 0.970. **3-95 (a).** 9/400. **(b).** 51/2380. **3-97 (a).** 7/8. **(b).** 7/16. **(c).** 1. **3-99 (a).** 1/13. **(b).** 57/65. **3-101.** 1/16.

Chapter 4

4-1 (a). 0, 1, 2, 3. **(b).** 1/8, 3/8, 3/8, 1/8.

(c).

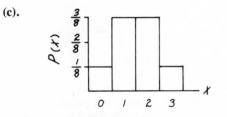

(d). 1/8, 3/8, 3/8, 1/8. **4-3 (a).** 2, 3, 4, 5, 6, 7, 8, 9, 10, 11, 12. **(b).** 1/36, 2/36, 3/36, 4/36, 5/36, 6/36, 5/36, 4/36, 3/36, 2/36, 1/36.

(c).

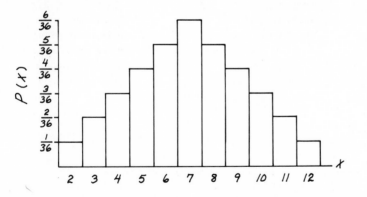

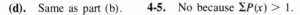

(d). Same as part (b). **4-5.** No because $\Sigma P(x) > 1$.

4-7.

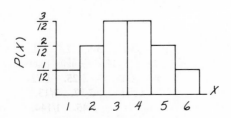

4-9. No because $\Sigma P(x) > 1$. **4-13.** Mean: 1.5; variance: 0.75; standard deviation: 0.87. **4-15.** Mean: 12.9; variance: 87.49; standard deviation: 9.35. **4-17.** Mean: 7/2; variance: 1.92; standard deviation: 1.38. **4-19.** Mean: 5.9; variance: 1.33; standard deviation: 1.15. **4-21.** Mean: 2.3; variance: 1.21; standard deviation: 1.1. **4-23.** Mean: 2.75; variance: 1.11; standard deviation: 1.05. **4-25.** Mean: 5.16; variance: 5.79; standard deviation: 2.41.

4-27.
$$\begin{aligned}
\Sigma(x - \mu)^2 \cdot P(x) &= \Sigma(x^2 - 2\mu x + \mu^2) \cdot P(x) \\
&= \Sigma x^2 \cdot P(x) - 2\mu \Sigma x \cdot P(x) + \mu^2 \Sigma P(x) \\
&= \Sigma x^2 \cdot P(x) - 2\mu \cdot \mu + \mu^2 \cdot 1 \\
&= \Sigma x^2 \cdot P(x) - \mu^2
\end{aligned}$$

4-29 (a). 31.25 **(b).** 0.05 **(c).** 1.25 **(d).** 1.25 **4-31 (a).** 0.117 **(b).** 0.215 **(c).** 0.932 **(d).** 0+ **(e).** 0+ **(f).** 0.003 **(g).** 0.017 **(h).** 0+ **(i).** 0.005 **(j).** 0.229 **4-33.** $n = 10$, $x = 4$, $p = 0.5$, $q = 0.5$, and $P(4) = 0.205$. **4-35.** $n = 4$, $x = 4$, $p = 0.5$, $q = 0.5$, and $P(4) = 0.062$ so that we expect the number of families with 4 girls to be $0.062 \times 5000 = 310$. **4-37.** $n = 10$, $x = 6$, $p = 0.6$, $q = 0.4$, and $P(6) = 0.251$.
4-39 (a). 0.951 **(b).** 0.048 **(c).** 0.001 **4-41.** 0.324 **4-43.** 0.311 **4-45.** 0.375 **4-47.** 0.078 **4-49 (a).** 781.25 **(b).** 3125 **(c).** 4687.5 **(d).** 3125 **(e).** 781.25 **4-51.** 0.010. **4-53.** $\mu = 18$, $\sigma^2 = 9$, $\sigma = 3$. **4-55.** $\mu = 3$, $\sigma^2 = 2.4$, $\sigma = 1.5$. **4-57.** $\mu = 1.1$, $\sigma^2 = 0.99$, $\sigma = 0.99$. **4-59.** $\mu = 6.25$, $\sigma^2 = 4.69$, $\sigma = 2.17$. **4-61.** $\mu = 425$, $\sigma^2 = 63.75$, $\sigma = 7.98$. **4-63.** $\mu = 0.64$, $\sigma = 0.80$. **4-65.** $\mu = 15$, $\sigma = 1.94$. **4-67.** $\mu = 2.4$, $\sigma = 1.53$. **4-69.** $\mu = 0.35$, $\sigma = 0.58$. **4-71.** $\mu = 4$, $\sigma = 1.55$, $\bar{x} = 3.954$, $s = 2.15$. **4-73.** Yes. **4-75.** No. **4-77 (a).** 0. **(b).** 2/5. **(c).** 1/2. **(d).** 1509/5000. **(e).** 999/5000. **4-79 (a).** 26.5 inches. **(b).** 23.5 inches. **(c).** 2/3. **4-81 (a).** 52.1 cubic centimeters. **(b).** 50.9 cubic centimeters. **(c).** 5/6. **4-83.** Skewed to the left. **4-85.** Normal. **4-87.** Uniform. **4-89.** Uniform. **4-91.** Bimodal. **4-93 (a).** 9/25. **(b).** 12/25. **4-97.** No since $\Sigma P(x) > 1$. **4-99 (a).** 0.010. **(b).** 0.230. **(c).** 2. **(d).** 1.2. **(e).** 1.10. **4-101 (a).** 60. **(b).** 50, 70. **(c).** 0.7.

Chapter 5

5-1. 0.3133. **5-3.** 0.5. **5-5.** 0.4474. **5-7.** 0.9146. **5-9.** 0.9802. **5-11.** 0.0466. **5-13.** 0.2483. **5-15.** 0.0668. **5-17.** 0.0017. **5-19.** 0.1401. **5-21.** 0.9573. **5-23.** 0.8413. **5-25.** 0.5351.

5-27.

x	y
-4	0.0001
-3	0.0046
-2	0.0549
-1	0.2434
0	0.4000
1	0.2434
2	0.0549
3	0.0046
4	0.0001

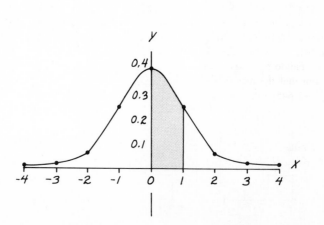

5-29 (a). 0.4452. **(b).** 0.0793. **5-31.** 0.4743. **5-33 (a).** 77.45%. **(b).** 30.85%.
5-35 (a). 86.4%. **(b).** 12.1%. **(c).** 87.9%. **5-37.** 8.38%. **5-39.** 92.7%. **5-43.** 2.33.
5-45. −2.33. **5-47.** −1.645, 1.645. **5-49.** −0.84, 1.04. **5-51.** −0.92. **5-53.** 79,875,000.
5-55. 119.2. **5-57.** 15.29 meters. **5-59.** 44.038, 44.662. **5-61.** About 86 inches. **5-63.** 24.874
millimeters, 25.396 millimeters. **5-65.** 62.8, 55.2, 44.8, 37.2. **5-67.** The approximation is suitable in
parts (b), (g), (h), (i). **5-69 (a).** 0.22722. **(b).** 0.2266. **5-71 (a).** 0.1222. **(b).** 0.1210.
5-73. 0.0287. **5-75.** 0.0119. **5-77.** 0.9949. **5-79.** 0.8461. **5-81.** 0.0329. **5-83.** 0.0485;
Yes. **5-85.** 0.8365. **5-87.** 0.0222. **5-89 (a).** 0.001. **(b).** 0.0012. **(c).** 0.0011. **5-91.** 262.
5-93 (a). 1.06. (b). 0.454. **(c).** 0.628. **(d).** 0.217. **(e).** 2.12. **(f).** 2.05. **(g).** 2.65. **(h).** 1.681.
(i). 0.134. **(j).** 0.462. **5-95.** It doubles. **5-97.** 0.4979. **5-99.** 0.2119. **5-101.** 0.9452.
5-103. 0.1056. **5-105.** 0.4878. **5-107.** 0.8294. **5-109.** 0.5636. **5-111.** 0.9177.
5-113. 0.0456. **5-115.** Quadrupled. **5-117 (a).** 0.2704. **(b).** 0.0392. **(c).** 0.8599. **(d).** 0.9544.
(e). 1.645. **5-119.** 0.0146. **5-121.** 0.0344.

Chapter 6

6-1 (a). $H_0: \mu \le 30$ **(c).** $H_0: \mu \ge 100$ **(e).** $H_0: \mu \le 16000$ **(g).** $H_0: \mu \le 16000$ **(i).** $H_0: \mu = 3.2$
$H_1: \mu > 30$ $H_1: \mu < 100$ $H_1: \mu > 16000$ $H_1: \mu > 16000$ $H_1: \mu \ne 3.2$

(k). $H_0: \mu = 0.73$ **(m).** $H_0: \mu = 3271$ **(o).** $H_0: \mu \ge 75$
$H_1: \mu \ne 0.73$ $H_1: \mu \ne 3271$ $H_1: \mu < 75$

6-2 (a). Reject the claim that the mean age of professors is 30 years or less when their mean age is actually 30 years
or less. **(c).** Reject the claim that the mean I.Q. of college students is at least 100 when that mean is actually 100 or
more. **(e).** Reject the claim that the mean salary is $16,000 or less when it really is. **(g).** Reject the claim that
the mean salary is $16,000 or less when it really is. **(i.)** Reject the claim that the mean weight equals 3.2 kilograms
when it does equal that amount. **(k).** Reject the claim that the mean reaction time equals 0.73 second when it does equal
that value. **(m).** Reject the claim that the mean cost equals $3271 when it does equal that amount. **(o).** Reject the claim
that the mean I.Q. is 75 or more when it actually is 75 or more. **6-3 (a).** Fail to reject the claim that the mean age of
professors is 30 years or less when that mean is actually more than 30 years. **(c).** Fail to reject the claim that the mean I.Q.
of college students is at least 100 when it is actually less than 100. **(e).** Fail to reject the claim that the mean salary is
$16,000 or less when it is really greater than $16,000. **(g).** Fail to reject the claim that the mean salary is $16,000 or less
when it is more than $16,000. **(i).** Fail to reject the claim that the mean weight equals 3.2 kilograms when it does not equal
that amount. **(k).** Fail to reject the claim that the mean reaction time equals 0.73 second when it does not equal that
value. **(m).** Fail to reject the claim that the mean cost equals $3271 when it does not equal that amount. **(o).** Fail to
reject the claim that the mean I.Q. is 75 or more when it is actually below 75. **6-5 (a).** 1.645. **(c).** −1.96,
1.96. **(e).** −1.645. **(g).** −1.645, 1.645. **(i).** 1.96.

6-7. Test statistic: $z = 1.44$
Critical value: $z = 2.33$
Fail to reject $H_0: \mu \le 100$. We fail to reject the claim that $\mu \le 100$.

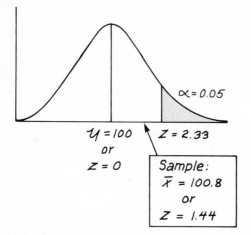

$\alpha = 0.05$

$\mathcal{Y} = 100$
or
$Z = 0$

$Z = 2.33$

Sample:
$\bar{X} = 100.8$
or
$Z = 1.44$

6-9. Test statistic: $z = 1.77$. Critical value: $z = 1.645$. Reject H_0: $\mu \leq 40$. We reject the claim that $\mu \leq 40$.
6-11. Test statistic: $z = 1.73$. Critical value: $z = -1.645$ and $z = 1.645$. Reject H_0: $\mu = 500$. We reject the claim that $\mu = 500$.   **6-13.** Test statistic: $z = -3.33$. Critical value: $z = -2.33$. Reject H_0: $\mu \geq 5.00$. The sample data support the claim that the mean weight is less than 5 pounds.   **6-15.** Test statistic: $z = -8.63$. Critical value: $z = 1.645$. Fail to reject H_0: $\mu \leq 420$. The sample data do not indicate an improved reliability.   **6-17.** Test statistic: $z = 4.74$. Critical value: $z = 2.33$. Reject H_0: $\mu \leq 50$. The new model does appear to increase productivity.
6-19. Test statistic: $z = 2.26$. Critical value: $z = -2.33$. Fail to reject H_0: $\mu \geq 125.3$. The data do not support the claim that the pianists are faster.   **6-21.** Test statistic: $z = -0.47$. Critical value: $z = -1.645$. Fail to reject H_0: $\mu \geq 18.24$. The sample data do not support the claim that there is less fat.   **6-23.** $\bar{x} = 104.3$ and $s = 13.5$. Test statistic: $z = 1.86$. Critical value: $z = 1.88$. Fail to reject H_0: $\mu \leq 100$. The sample data do not support the claim that the mean is above 100.   **6-25.** The test becomes insensitive since we avoid a type I error by rarely rejecting H_0.
6-26 (a). ± 2.056. **(c).** -3.365. **(e).** -2.467. **(g).** ± 2.132. **(i).** 1.943.   **6-27.** Test statistic: $t = 1.5$. Critical value: $t = 1.860$. Fail to reject H_0: $\mu \leq 10$.   **6-29.** Test statistic: $t = 2.01$. Critical value: $t = 1.761$. Reject H_0: $\mu \leq 75$.   **6-31.** Test statistic: $t = -1.30$. Critical value: $t = -1.721$. Fail to reject H_0: $\mu \geq 100$.
6-33. Test statistic: $t = -0.71$. Critical value: $t = -3.250$. Fail to reject H_0: $\mu \geq 3000$. The sample data do not warrant rejection of the manufacturer's claim.   **6-35.** Test statistic: $t = 1.83$. Critical values: $t = -2.462$ and $t = 2.462$. Fail to reject H_0: $\mu = 20.0$ The pills appear to be acceptable.   **6-37.** Test statistic: $t = -2.40$. Critical value: $t = -2.681$. Fail to reject H_0: $\mu \geq 0.75$. The data do not support the instructor's claim.   **6-39.** Test statistic: $t = 3.97$. Critical values: $t = -2.052$ and $t = 2.052$. Reject H_0: $\mu = 60$. The mean appears to be different from 60.   **6-41.** Test statistic: $t = -4.84$. Critical value: $t = -2.861$. Reject H_0: $\mu \geq 154$. The new tire is purchased.   **6-43.** Test statistic: $t = 2.00$. Critical values: $t = -2.132$ and $t = 2.132$. Fail to reject H_0: $\mu = 5.4$. The data support the claim that the mean equals 5.4 microvolts.   **6-45.** Test statistic: $t = 3.61$. Critical value: $t = 1.796$. Reject H_0: $\mu \leq 500$. The sample data do not support the claim that the mean is more than 500 meters.   **6-47.** $\bar{x} = 15.89$ and $s = 0.26$. The test statistic is $t = -1.27$. The critical value depends on the level of significance, but the test statistic is not in the critical region for any reasonable choice of α. Fail to reject H_0: $\mu \geq 16.0$. There is insufficient evidence to conclude that the consumer is being cheated.   **6-51.** Parts (b), (c), (e), (f), (g), and (j) involve attribute data.   **6-53.** Test statistic: $z = 1.37$. Critical values: $z = -1.96, 1.96$. Fail to reject H_0: $p = 0.6$. The sample data do not warrant rejection of the claim that the proportion equals 0.6.   **6-55.** Test statistic: $z = 1.02$. Critical value: $z = 2.33$. Fail to reject H_0: $p \leq 0.2$. The sample data do not support the claim that the percentage of Catholics is more than 20%.   **6-57.** Test statistic: $z = -2.11$. Critical value: $z = -1.645$. Reject H_0: $p \geq 0.8$. The sample data cause us to reject the claim that at least 80% pass.
6-59. Test statistic: $z = -1.26$. Critical value: $z = -1.645$. Fail to reject H_0: $p \geq 0.22$. There are insufficient sample data to reject the sociologist's claim.   **6-61.** Test statistic: $z = -1.16$. Critical value: $z = -2.33$. Fail to reject H_0: $p \geq 0.012$. There is insufficient evidence to support the claim of a lowered failure rate.   **6-63.** Test statistic:

$z = -1.63$. Critical value: $z = -1.645$. Fail to reject $H_0: p \geq 0.6$. There is insufficient sample data to reject the Senator's claim.   **6-65.**   Test statistic: $z = -5.40$. Critical value: $z = -2.575, 2.575$. Reject $H_0: p = 0.95$. The sample data warrant rejection of the claim that there is a 95% recognition rate.   **6-67.**   Test statistic: $z = -2.61$. Critical value: $z = -1.645$. Reject $H_0: p \geq 0.12$. The failure rate does appear to be lower.   **6-69.**   Test statistic: $z = 2.77$. Critical values: $z = -2.33, 2.33$. Reject $H_0: p = 1/8$. The Mendelian law does not appear to be working.   **6-71.**   Test statistic: $z = -1.56$. Critical values: $z = -2.33, 2.33$. Fail to reject $H_0: p = 0.56$. There is insufficient evidence to reject the candidate's claim.   **6-75.**   From the table of binomial probabilities with $n = 15$ and $p = 0.1$ we get $P(0) = 0.206$. If p is really 0.1, then there is a good chance (0.206) that none of the 15 residents will feel that the mayor is doing a good job. Since that result could easily occur by chance, there is not enough evidence to reject the reporter's claim. We would reject the claim only if the probability was found to be less than 0.05.   **6-76 (a).**   8.907, 32.852.   **(c).**   8.643, 42.796.   **(e).**   4.075.   **(g).**   55.758.   **(i)**   27.991, 79.490.   **6-77.**   Test statistic: $\chi^2 = 50.44$. Critical value: $\chi^2 = 38.885$. Reject $H_0: \sigma^2 \leq 100$.   **6-79.**   Test statistic: $\chi^2 = 108.889$. Critical value: $\chi^2 = 101.879$. Reject $H_0: \sigma^2 \leq 9.00$.   **6-81.**   Test statistic: $\chi^2 = 11.778$. Critical value $\chi^2 = 12.443$. Reject $H_0: \sigma^2 \geq 90$.   **6-83.**   Test statistic: $\chi^2 = 14.82$. Critical values: $\chi^2 = 13.844, 41.923$. Fail to reject $H_0: \sigma^2 = 100$.   **6-85.**   Test statistic: $\chi^2 = 31.871$. Critical values: $\chi^2 = 7.564, 30.191$. Reject $H_0: \sigma = 52$.   **6-87.**   Test statistic: $\chi^2 = 10.222$. Critical values: $\chi^2 = 9.260, 44.181$. Fail to reject $H_0: \sigma = 15$.   **6-89.**   Test statistic: $\chi^2 = 38.609$. Critical values: $\chi^2 = 16.047, 45.722$. Fail to reject $H_0: \sigma = 52$. There is insufficient evidence to reject the claim that the standard deviation is 52 hours.   **6-91.**   Test statistic: $\chi^2 = 43.862$. Critical value: $\chi^2 = 35.172$. Reject $H_0: \sigma \leq 0.21$. The sample data support the claim that the standard deviation is greater than 0.21 second.   **6-93.**   Test statistic: $\chi^2 = 44.8$. Critical value: $\chi^2 = 51.739$. Reject $H_0: \sigma^2 \geq 0.0225$. The new machine appears to produce less variance.   **6-95.**   Test Statistic: $\chi^2 = 32.813$. Critical values: $\chi^2 = 10.283, 35.479$. Fail to reject $H_0; \gamma = 8$. There is insufficient evidence to reject the claim that the standard deviation is 8.   **6-97.**   Test statistic: $z = -3.00$. Critical values: $z = -1.96, 1.96$. Reject $H_0: \mu = 10.0$.   **6-99.**   Test statistic: $\chi^2 = 9.31$. Critical values: $\chi^2 = 6.844, 38.582$. Fail to reject $H_0: \sigma = 4.0$.
6-101.   Test statistic: $t = -0.922$. Critical value: $t = -2.625$. Fail to reject $H_0: \mu \geq 100$. $\bar{x} = 98.0$ and $s = 8.4$.
6-103.   Test statistic: $t = -1.36$. Critical value: $t = -1.740$. Fail to reject $H_0: \mu \geq 0.700$. The data do not support the claim that the mean is less than 0.700 second.   **6-105 (a).** $t = -1.761, 1.761$.   **(b).** $\chi^2 = 6.571, 23.685$.   **(c).** $z = -1.88, 1.88$.   **(d).** $\chi^2 = 2.700, 19.023$.   **(e).** $\chi^2 = 14.257$.

Chapter 7

7-1 (a).   1.96.   **(c).**   2.575.   **(e).**   1.15.   **(g).**   1.96.   **(i).**   2.093.   **(k).**   1.833.   **7-3.** $98.09 < \mu < 99.12$.   **7-5.** $72.70 < \mu < 78.90$.   **7-7.** $0.593 < \mu < 0.627$.   **7-9.** $299.76 < \mu < 345.04$.   **7-11.** $34.81 < \mu < 37.59$.   **7-13.**   683.   **7-15.**   1492.   **7-17.**   185.   **7-19.** $119.78 < \mu < 127.62$.   **7-21.** $12.85 < \mu < 16.55$. (use $z(\alpha/2)$ since σ is known.)   **7-23.**   61.   **7-25.**   119.   **7-27.**   As long as the sample size exceeds 30, we can use the normal distribution.

7-29.

	p_s	q_s	Point estimate
(a).	0.450	0.550	0.450
(b).	0.608	0.392	0.608
(c).	0.050	0.950	0.050
(d).	0.880	0.120	0.880
(e).	0.466	0.534	0.466

7-31. $0.412 < p < 0.477$.   **7-33.** $0.267 < p < 0.287$.   **7-35.**   4145.   **7-37.**   355.   **7-39.** $0.545 < p < 0.595$.   **7-41.** $0.728 < p < 0.858$.   **7-43.** $0.828 < p < 0.852$.   **7-45.** $0.647 < p < 0.733$.   **7-47.**   16,517.   **7-49.**   88%.   **7-51.** $p = 0.5, q = 0.5$.   **7-53.**   63%.   **7-55 (a).**   12.5.   **(b).**   234.09.   **(c).**   1.44.   **(d).**   3.61.   **7-57.**   5.629, 26.119.   **7-59.**   15.379, 38.885.   **7-61.** $106.45 < \sigma^2 < 750.00$.

7-63. $4.801 < \sigma^2 < 17.306.$ **7-65.** $3.346 < \sigma < 6.426.$ **7-67.** $0.4845 < \sigma^2 < 1.3805.$ **7-69 (a).** 100. **(b).** 4200. **(c).** 4200. **(d).** 5150. **(e).** 300. **(f).** 125. **(g).** 750. **7-71.** $2.367 < \sigma^2 < 48.533.$ **773.** $6.307 < \sigma < 15.117.$ **7-75 (a).** 1.645. **(b).** 2.700, 19.023. **(c).** 2.262. **(d).** 0.25. **7-77 (a).** 4.1. **(b).** $3.450 < \sigma < 4.950.$ **7-79 (a).** 4.1. **(b).** $3.029 < \sigma < 6.346.$ **7-81.** 601. **7-83.** $0.033 < p < 0.087.$

Chapter 8

8-1. Test statistic: $F = 2.0000.$ Critical value: $F = 4.0260.$ Fail to reject H_0: $\sigma_1^2 = \sigma_2^2.$ There is not sufficient evidence to reject the claim that the variances are equal. **8-3.** Test statistic: $F = 2.4091.$ Critical value: $F = 2.4523.$ Fail to reject H_0: $\sigma_1^2 = \sigma_2^2.$ There is not sufficient evidence to reject the claim that the variances are equal. **8-5.** Test statistic: $F = 3.446.$ Critical value: $F = 3.8919.$ Fail to reject H_0: $\sigma_1^2 = \sigma_2^2.$ There is not sufficient evidence to reject the claim that the variances are equal. **8-7.** Test statistic: $F = 4.0000.$ Critical value: $F = 3.1789.$ Reject H_0: $\sigma_1^2 \leq \sigma_2^2.$ The variance of population A does appear to exceed that of population B. **8-9.** Test statistic: $F = 1.4063.$ Critical value: $F = 1.7505.$ Fail to reject H_0: $\sigma_1^2 \leq \sigma_2^2.$ There is not sufficient evidence to support the claim that the variance of population A exceeds that of population B. **8-11.** Test statistic: $F = 1.7333.$ Critical value: $F = 2.5848.$ Fail to reject H_0: $\sigma_1^2 = \sigma_2^2.$ There is not sufficient evidence to reject the claim of equal variances. **8-13.** Test statistic: $F = 1.8526.$ Critical value: $F = 1.8055.$ Reject H_0: $\sigma_1^2 \leq \sigma_2^2.$ The second scale does appear to produce greater variance. **8-15.** 60. **8-17.** 24.5. **8-19.** Test statistic: $F = 1.0596.$ Critical value: $F = 3.0527.$ Fail to reject H_0: $\sigma_1^2 = \sigma_2^2.$ The standard deviations appear to be equal. **8-21.** $\bar{d} = 11.9$ and $s_d = 11.2.$ Test statistic: $t = 3.360.$ Critical values: $t = -2.262, 2.262.$ Reject H_0: $\mu_1 = \mu_2.$ We reject the claim that the pill was ineffective. **8-23.** Test statistic: $z = -6.66.$ Critical values: $z = -1.96, 1.96.$ Reject H_0: $\mu_1 = \mu_2.$ We reject the claim that the means are equal. **8-25.** Test statistic: $z = -1.79.$ Critical values: $z = -1.96, 1.96.$ Fail to reject H_0: $\mu_1 = \mu_2.$ There is not sufficient evidence to reject the claim that both tests produce the same mean. **8-27.** F test results: Test statistic $F = 3.8118.$ Critical value $F = 3.5879.$ We conclude that $\sigma_1 \neq \sigma_2.$ Test of means: Test statistic $t = -3.07.$ Critical value $t = -2.262,$ 2.262. Reject H_0: $\mu_1 = \mu_2.$ The two models appear to have different means. **8-29.** F test results: Test statistic $F = 1.6198.$ Critical value $F = 2.0677.$ We conclude that $\sigma_1 = \sigma_2.$ Test of means: Test statistic $t = -1.412.$ Critical value $t = -1.960, 1.960.$ Fail to reject H_0: $\mu_1 = \mu_2.$ There is not sufficient evidence to reject the claim that the means are equal. **8-31.** $\bar{d} = -1.67$ and $s_d = 3.14.$ Test statistic: $t = -1.840.$ Critical values: $t = -2.201, 2.201.$ Fail to reject H_0: $\mu_1 = \mu_2.$ There is not sufficient evidence to reject the claim that the diet has no affect. **8-33.** F test results: Test statistic $F = 3.7539.$ Critical value $F = 3.2497.$ We conclude that $\sigma_1 \neq \sigma_2.$ Test of means: Test statistic $t = -1.916.$ Critical values $t = -2.201, 2.201.$ Fail to reject H_0: $\mu_1 = \mu_2.$ There is not sufficient evidence to reject the claim that the mean down times are equal. **8-35.** F test results: Test statistic $F = 1.3176.$ Critical value $F = 2.9493.$ We conclude that $\sigma_1 = \sigma_2.$ Test of means: Test statistic $t = -1.200.$ Critical value $t = -1.701.$ Fail to reject H_0: $\mu_1 \geq \mu_2.$ There is not sufficient evidence to support the claim that girls have a higher mean than boys. **8-37.** Test statistic: $z = 3.10.$ Critical values: $z = -1.96, 1.96.$ Reject H_0: $\mu_1 = \mu_2.$ We reject the claim that the means for the two schools are equal. **8-39.** Test statistic: $z = 5.58.$ Critical values: $z = -1.96, 1.96.$ Reject H_0: $\mu_1 = \mu_2.$ We reject the claim that both models have the same mean. **8-41.** String A: $n = 9, \bar{x} = 26.9, s = 4.7.$ String B: $n = 12, \bar{x} = 18.8, s = 3.4.$ F test results: Test statistic $F = 1.911.$ Critical value $F = 3.6638.$ We conclude that $\sigma_1 = \sigma_2.$ Test of means: Test statistic $t = 4.593.$ Critical value $t = -2.093, 2.093.$ Reject H_0: $\mu_1 = \mu_2.$ The two types of string do appear to have different mean breaking points. **8-43.** F test results: Test statistic $F = 1.0330.$ Critical value $F = 2.3072.$ We conclude that $\sigma_1 = \sigma_2.$ Test of means: Test statistic $t = -2.104.$ Critical values $t = -1.960, 1.960.$ Reject H_0: $\mu_1 = \mu_2.$ We reject the claim that there is no difference between the two brands. **8-45.** $\bar{d} = 0.429$ and $s_d = 1.089.$ Test statistic: $t = 1.474.$ Critical values: $t = -2.160, 2.160.$ Fail to reject H_0: $\mu_1 = \mu_2.$ There is not sufficient evidence to reject the claim that the means are equal. **8-47.** $\bar{d} = 2.9$ and $s_d = 3.3.$ Test statistic: $t = 2.779.$ Critical value: $t = 1.833.$ Reject H_0: $\mu_1 \leq \mu_2.$ The diet appears to be effective. **8-49.** 15.11; s_{x+y}^2 is roughly equal to $s_x^2 + x_y^2.$

8-51.

	n_1	n_2	x_1	x_2	$\bar{p}$	$\bar{q}$
(a).	200	400	67	148	0.358	0.642
(b).	250	300	95	138	0.424	0.576
(c).	250	250	50	60	0.220	0.780
(d).	300	400	159	212	0.530	0.470
(e).	50	100	21	57	0.520	0.480

8-52 **(a).** -0.84. **(c).** -1.08. **(e).** -1.73. **8-53 (a).** -2.05. **(b).** $-1.96, 1.96$. **(c).** Reject H_0: $p_1 = p_2$. **8-55.** Test statistic: $z = -0.87$. Critical value: $z = -1.96, 1.96$. Fail to reject H_0: $p_1 = p_2$. There is not sufficient evidence to reject the claim that the two proportions are equal. **8-57.** Test statistic: $z = 0.83$. Critical value: $z = 2.33$. Fail to reject H_0: $p_1 \leq p_2$. There is not sufficient evidence to support the claim that the younger age group has a greater proportion of fatal accidents. **8-59.** Test statistic: $z = 1.43$. Critical values: $z = -1.96, 1.96$. Fail to reject H_0: $p_1 = p_2$. There is not sufficient evidence to reject the claim that both groups have the same proportions. **8-61.** Test statistic: $z = 2.97$. Critical value: $z = 1.645$. Reject H_0: $p_1 \leq p_2$. The first county appears to have a higher percentage of voters in favor. **8-63.** Test statistic: $z = -2.02$. Critical values: $z = -1.96, 1.96$. Reject H_0: $p_1 = p_2$. There is sufficient evidence to reject the claim that the proportions are equal. **8-65.** Test statistic: $z = -5.35$. Critical value: $z = -2.33$. Reject H_0: $p_1 \geq p_2$. It appears that a greater proportion of viewers did not have a college education. **8-67.** Test statistic: $z = 0.847$. Critical value: $z = 1.645$. Fail to reject H_0: $p_1 \leq p_2$. There is not sufficient evidence to reject the equality of the two proportions. **8-69.** Test statistic: $z = 2.03$. Critical values: $z = -1.96, 1.96$. Reject H_0: $p_1 = p_2$. We reject the claim that the proportions are equal. **8-71.** F test results: Test statistic $F = 1.8225$. Critical value $F = 3.9639$. We conclude that $\sigma_1 = \sigma_2$. Test of means: Test statistic $t = 1.421$. Critical values $t = -2.086, 2.086$. Fail to reject H_0: $\mu_1 = \mu_2$. There is not sufficient evidence to reject the claim of equal means. **8-73.** Test statistic: $z = 3.67$. Critical value: $z = 1.645$. Reject H_0: $p_1 \leq p_2$. The proportion of correct answers by good students appears to be greater. **8-75.** $\bar{d} = -234.0$ and $s_d = 199.5$. Test statistic: $t = -3.709$. Critical value: $t = -1.833$. Reject H_0: $\mu_1 \geq \mu_2$. The program appears to be effective. **8-77.** F test results: Test statistic $F = 27.5625$. Critical value $F = 2.1952$. We conclude that $\sigma_1 \neq \sigma_2$. Test of means: Test statistic $t = -4.843$. Critical values $t = -2.093, 2.093$. Reject H_0: $\mu_1 = \mu_2$. We reject the claim that the means are equal. **8-79.** Test statistic: $z = -2.45$. Critical values: $z = -1.645, 1.645$. Reject H_0: $p_1 = p_2$. The sample proportions appear to be different.

Chapter 9

9-1. Positive correlation: (a), (b), (c), (f), (g), (i). Negative correlation: (d), (h), (j). No correlation: (e). **9-3.** **(b).** 4. **(c).** 8. **(d).** 18. **(e).** 64. **(f).** 22 **(g).** -0.956. **9-5 (b).** 6. **(c).** 21. **(d).** 85. **(e).** 441. **(f).** 33. **(g).** -0.957. **9-6 (a).** Significant positive linear correlation. **(c).** No significant linear correlation. **(e).** No significant linear correlation. **(g).** No significant linear correlation. **(i).** Significant negative linear correlation. **9-7 (b).** -0.726. **(c).** 0.444. **(d).** Significant negative linear correlation. **9-9 (b).** 0.600. **(c).** 0.707. **(d).** No significant linear correlation. **9-11 (b).** 0.999. **(c).** 0.632. **(d).** Significant positive linear correlation. **9-13 (b).** 0.310. **(c).** 0.514. **(d).** No significant linear correlation. **9-15 (b).** 0.450. **(c).** 0.811. **(d).** No significant linear correlation. **9-17.** In attempting to calculate r we get a denominator of zero, so a real value of r does not exist. However, it should be obvious that the value of x is not at all related to the value of y. **9-19.** Same result as in Exercise 9-23. **9-23.** $y = -2x + 7.25$. **9-25.** $y = -1.09x + 5.97$. **9-27.** $y = -1.87x + 78.49$. **9-29.** $y = 0.000124x - 0.210196$. **9-31 (c).** $y = 0.26x + 37.66$. **(d).** 55.86 degrees. **9-33.** $y = 0.034x + 3.880$. **9-35.** $y = 0.45x + 64.64$. **9-36 (a).** 6. **(c).** -13. **(e).** 6. **9-37.** We want to show that $\bar{y} = m\bar{x} + b$. Replace $\bar{x}$ by $\Sigma x/n$, replace $\bar{y}$ by $\Sigma y/n$, and replace m and b by their expressions as given in Section 9-3 and show that an identity is created. **9-39.** Note that s_x and s_y are never negative. **9-43 (a).** Correlation. **(b).** Regression. **(c).** Regression. **(d).** Correlation. **(e).** Correlation. **9-45 (b).** 0.887. **(c).** 0.707. **(d).** There is a significant positive linear correlation. **(e).** $y = 0.95x +$

7.01.　**(g).**　92.51.　　**9-47 (b).**　0.285.　**(c).**　0.707.　**(d).**　No significant linear correlation.　　**(e).**　$y = 0.015x$ + 1.380.　**(g).**　2.594.

Chapter 10

10-1.　The expected frequencies are 20, 20, 20, 20, 20.　**(a).**　5.900.　**(b).**　13.277.　**(c).**　We fail to reject the claim that absences occur on the five days with equal frequency.　　**10-3.**　The expected frequencies are 84, 180, 84, 84, 84, 84. The test statistic is $\chi^2 = 14.470$ and the critical value is $\chi^2 = 11.071$. We reject the manufacturer's claim. **10-5.**　The test statistic is $\chi^2 = 4.800$ and the critical value is $\chi^2 = 14.067$. We fail to reject the claim that the aspirins are equally effective.　　**10-7.**　The test statistic is $\chi^2 = 18.600$ and the critical value is $\chi^2 = 9.488$. We reject the claim that the photographs are equally pleasant.　　**10-9.**　The test statistic is $\chi^2 = 5.957$ and the critical value is $\chi^2 = 9.488$. We fail to reject the manager's claim.　　**10-11.**　The test statistic is $\chi^2 = 9.561$ and the critical value is $\chi^2 = 14.067$. We fail to reject the claim that the figures agree.　　**10-13.**　180 rolls; reject the claim of equal frequencies. **10-15.**　The test statistic is $\chi^2 = 4.556$ and the critical value is $\chi^2 = 5.991$. Fail to reject the claim that teachers have the same accident rate.　　**10-17.**　The test statistic is $\chi^2 = 2.911$ and the critical value is $\chi^2 = 7.815$. Fail to reject the claim that the actual number of classes agreed with the teacher's expectation.　　**10-19.**　Combining A with B, E with F, G with H, and I with J, we get a test statistic of $\chi^2 = 4.118$ and a critical value of $\chi^2 = 11.071$ with $\alpha = 0.05$. We fail to reject the claim that the observed and expected frequencies are compatible.　　**10-21. (a).**　27.778, 22.222, 22.222, 17.778.　**(b).**　11.025.　**(c).**　3.841.　**(d).**　Reject the claim that voting is independent of sex. **10-23.**　The test statistic is $\chi^2 = 18.726$ and the critical value is $\chi^2 = 6.635$. Reject the claim that adjustment is independent of sex.　　**10-25.**　The test statistic is $\chi^2 = 4.776$ and the critical value is $\chi^2 = 9.210$. Fail to reject the claim that grade category is independent of sex.　　**10-27.**　The test statistic is $\chi^2 = 11.825$ and the critical value is $\chi^2 = 9.488$. Reject the claim that day of the week is independent of the number of defects.　　**10-29.**　The test statistic is $\chi^2 = 8.373$ and the critical value is $\chi^2 = 21.026$. Fail to reject the claim that grade distribution is independent of the subject.　　**10-31.**　Use categories of East, Central, and West. The test statistic is $\chi^2 = 10.362$ and the critical value is $\chi^2 = 9.488$. Reject the claim that region and opinion are independent.　　**10-32 (a).**　Reject equality. The critical F is 3.8853.　**(c).**　Fail to reject equality. The critical F is 3.8853.　**(e).**　Reject equality. The critical F is 3.8853. **(g).**　Fail to reject equality. The critical F is 6.9266.　　**10-33.**　The test statistic is $F = 1.5297$ and the critical value is $F = 3.0556$. Fail to reject the claim of equal means.　　**10-35.**　The test statistic is $F = 8.2086$ and the critical value is $F = 3.8853$. Reject the claim that the population means are equal.　　**10-37.**　The test statistic is $F = 1.6373$ and the critical value is $F = 3.2317$. Fail to reject the claim of equal means.　　**10-39.**　The test statistic is $F = 31.1111$ and the critical value is $F = 3.2389$. Reject the claim of equal means.　　**10-41.**　The test statistic is $F = 49.3187$ and the critical value is $F = 3.2592$. Reject the claim of equal means.　　**10-43.**　The test statistic is $F = 1.5430$ and the critical value is $F = 2.6060$. Fail to reject the claim of equal means.　　**10-45 (a).**　10.　**(b).**　0.774. **(c).**　0.05.　**(d).**　We get better results with analysis of variance.　　**10-47.**　The test statistic is $\chi^2 = 27.325$ and the critical value is $\chi^2 = 12.592$. Reject the claim that customers arrive on the different days with equal frequencies. **10-49.**　The test statistic is $\chi^2 = 3.620$ and the critical value is $\chi^2 = 13.277$. Fail to reject the claim that the responses are independent of the student group.　　**10-51.**　The test statistic is $\chi^2 = 6.050$ and the critical value is $\chi^2 = 9.488$. Fail to reject the claim that I.Q. score and category of forgetfulness are independent.

Chapter 11

11-1.　We assume in the null hypothesis that the pill has no affect, so the probability of a plus sign is 0.5. The probability of 8 or more plus signs out of 9 is 0.020, which is less than 0.025. Thus we conclude that these results are not likely to occur by chance. We reject the null hypothesis and conclude that the pill does have an effect.　　**11-3.**　We assume in the null hypothesis that the diet has no effect, so the probability of a plus sign is 0.5. The probability of 8 or more plus signs out of 10 is 0.055, which is greater than 0.025. Thus we conclude that these results could easily occur by chance. We fail to reject the null hypothesis that the diet has no effect.　　**11-5.**　We assume in the null hypothesis that there is no difference between the median ages, so the probability of a plus sign is 0.05. The probability of 6 or more plus signs out of 10 is 0.377, which is greater than 0.025. Thus we conclude that these results could easily occur by chance. We fail to reject the null hypothesis that there is no difference between the median ages of the two groups. **11-7.** The probability of 9 or more plus signs out of 13 is 0.134. This probability is not less than 0.05 and we therefore conclude that the results could easily occur by chance. We fail to reject the null hypothesis that half (or fewer) of all department heads feel that they understand the purpose of their council. That is, there is not sufficient evidence to support the claim that more than half feel that they understand. **11-9.** The alternate hypothesis is the claim that girls score higher on the verbal portion, so $M - V$ should produce a significantly large number of negative signs. The null hypothesis is the claim that girls score equally well or lower on the verbal portion. The probability of 10 or more negative signs out of 15 is 0.152, which is not less than 0.05. Thus we fail to reject the null hypothesis. **11-11.** The probability of 7 or more plus signs out of 10 is 0.172. This probability is not less than 0.05 so the results can easily occur by chance. We fail to reject the null hypothesis that at most half of all dentists favor Covariant toothpaste. **11-13.** The probability of 32 or more plus signs out of 50 is 0.0336. This probability is less than 0.05 so there is a significantly large number of plus signs. We therefore reject the null hypothesis that production was unchanged or lowered by the new fertilizer. **11-15.** The probability of 22 or more plus signs out of 28 is 0.0023. This probability is less than 0.005 so the number of plus signs does represent a significant difference. We therefore reject the null hypothesis of no effect. **11-17.** The probability of 28 or more plus signs out of 50 is 0.2389. This probability is greater than 0.10 so the results could easily occur by chance. We fail to reject the null hypothesis that at most half of the voters favor the bill. **11-19.** The probability of 32 or fewer plus signs out of 75 is 0.1251. This probability is greater than 0.05 so the results could easily occur by chance. We fail to reject the null hypothesis that the median life is at least 40 hours. **11-21.** 79. **11-23.** $\mu_R = 210$, $\sigma_R = 20.49$, $R = 160.5$, $z = -2.42$. The test statistic $z = -2.42$ is in the critical region bounded by $z = 1.96$ or $z = -1.96$, so we reject the null hypothesis of no difference. The groups appear to have different reaction times. **11-25.** $\mu_R = 138$, $\sigma_R = 15.17$, $R = 111$, $z = -1.78$. The test statistic $z = -1.78$ is not in the critical region bounded by $z = 1.96$ or $z = -1.96$, so we fail to reject the null hypothesis of no difference. **11-27.** $\mu_R = 232.5$, $\sigma_R = 24.11$, $R = 307$, $z = 3.09$. The test statistic $z = 3.09$ is in the critical region bounded by $z = 1.96$ or $z = -1.96$ so we reject the null hypothesis of no difference. **11-29.** $\mu_R = 188.5$, $\sigma_R = 71.71$, $R = 135.5$, $z = -2.44$. The test statistic $z = -2.44$ is in the critical region bounded by $z = 1.96$ or $z = -1.96$ so we reject the null hypothesis of no difference. **11-31.(a).** $\mu_R = 105$, $\sigma_R = 13.23$, $R = 55$, $z = -3.78$. The test statistic $z = -3.78$ is in the critical region bounded by $z = 1.96$ or $z = -1.96$ so we reject the null hypothesis of no difference. **(b).** $\mu_R = 105$, $\sigma_R = 13.23$, $R = 100$, $z = -0.38$. The test statistic of $z = -0.38$ is not in the critical region bounded by $z = 1.96$ or $z = -1.96$ so we fail to reject the null hypothesis of no difference. **11-33.** $\mu_R = 137.50$, $\sigma_R = 17.26$, $R = 177.5$, $z = 2.32$. The test statistic $z = 2.32$ is in the critical region bounded by $z = 1.96$ or $z = -1.96$ so we reject the null hypothesis of no difference. **11-35.** The probability of 30 or more plus signs out of 40 is 0.0013, which is less than 0.025. Thus we reject the null hypothesis of equal abilities.

INDEX